Art and Technology in Early Modern Europe

***Art History* Book Series**

Series Editors
Genevieve Warwick (University of Edinburgh) and
Natalie Adamson (University of St Andrews)

This distinctive series, developed from special issues of *Art History*, publishes new work by leading scholars on a full range of key themes and breaking issues for the discipline. Generously illustrated, and in full colour, each volume is designed to engage critical debate and to reframe the boundaries of art-historical thought.

Titles in the series include

Art and Technology in Early Modern Europe
Edited by Richard Taws and Genevieve Warwick

Objects in Motion in the Early Modern World
Edited by Daniela Bleichmar and Meredith Martin

To Scale
Edited by Joan Kee and Emanuele Lugli

Theorizing Imitation in the Visual Arts: Global Contexts
Edited by Paul Duro

Andrea Mantegna: Making Art (History)
Edited by Stephen J. Campbell and Jérémie Koering

The Clever Object
Edited by Matthew C. Hunter and Francesco Lucchini

The Erotics of Looking: Early Modern Netherlandish Art
Edited by Angela Vanhaelen and Bronwen Wilson

British Art in the Cultural Field, 1939-69
Edited by Lisa Tickner and David Peters Corbett

Anglo-American: Artistic Exchange between Britain and the USA
Edited by David Peters Corbett and Sarah Monks

Creative Writing and Art History
Edited by Catherine Grant and Patricia Rubin

Theatricality in Early Modern Art and Architecture
Edited by Caroline van Eck and Stijn Bussels

Photography After Conceptual Art
Edited by Diarmuid Costello and Margaret Iversen

Art History: Contemporary Perspectives on Method
Edited by Dana Arnold

Art and Architecture in Naples, 1266–1713: New Approaches
Edited by Cordelia Warr and Janis Elliott

About Mieke Bal
Edited by Deborah Cherry

Spectacle and Display
Edited by Deborah Cherry and Fintan Cullen

Location
Edited by Deborah Cherry and Fintan Cullen

About Stephen Bann
Edited by Deborah Cherry

Between Luxury and the Everyday: Decorative Arts in Eighteenth-Century France
Edited by Katie Scott and Deborah Cherry

Art: History: Visual: Culture
Edited by Deborah Cherry

Difference and Excess in Contemporary Art: The Visibility of Women's Practice
Edited by Gill Perry

Tracing Architecture: The Aesthetics of Antiquarianism
Edited by Dana Arnold and Stephen Bending

Other Objects of Desire: Collectors and Collecting Queerly
Edited by Michael Camille and Adrian Rifkin

Fingering Ingres
Edited by Susan Siegfried and Adrian Rifkin

The Metropolis and its Image: Constructing Identities for London, c. 1750–1950
Edited by Dana Arnold

About Michael Baxandall
Edited by Adrian Rifkin

Image: Music: Text
Edited by Marcia Pointon, Simon Miller and Paul Binski

Art and Technology in Early Modern Europe

Edited by Richard Taws and Genevieve Warwick

AAH
ASSOCIATION OF ART HISTORIANS

WILEY Blackwell

This edition first published 2016
Originally published as Volume 39, Issue 2 of *Art History*

Registered Office
John Wiley & Sons Ltd, The Atrium, Southern Gate, Chichester, West Sussex, PO19 8SQ, UK

Editorial Offices
350 Main Street, Malden, MA 02148-5020, USA
9600 Garsington Road, Oxford, OX4 2DQ, UK
The Atrium, Southern Gate, Chichester, West Sussex, PO19 8SQ, UK

For details of our global editorial offices, for customer services, and for information about how to apply for permission to reuse the copyright material in this book please see our website at www.wiley.com/wiley-blackwell.

Library of Congress Cataloging-in-Publication Data applied for.

9781119291688 (paperback)

A catalogue record for this book is available from the British Library.

Cover image: John Russell, *The Face of the Moon*, 1793-1797. Pastel on paper strained over a wooden stretcher, 60.7 x 50.4 cm. Birmingham: Birmingham Art Gallery. Photo: © Birmingham Museums Trust.

Set in 10.8/13.5 Joanna by Laserwords Private Limited, Chennai, India
Printed and bound in Singapore by Markono Print Media Pte Ltd

1 2016

Contents

Notes on Contributors

Ann Bermingham is a specialist in British art of the eighteenth and nineteenth centuries and a research professor in art history at the University of California, Santa Barbara. She is the author of Landscape and Ideology: The English Rustic Tradition, 1740–1860 (University of California Press, 1986), and Learning to Draw: Studies in the Cultural History of a Polite and Useful Art (Yale University Press, 2000); the editor of Sensation and Sensibility: Viewing Gainsborough's Cottage Door (Yale University Press, 2005), and, with John Brewer, The Consumption of Culture: Image, Object, Text, 1600–1800 (Routledge, 1995).

Jan Blanc is Professor of Art of the Early Modern Period at the University of Geneva. His research focuses on Flemish, Dutch and British art and on artistic theory and practice in the early modern period. His publications include: Peindre et penser la peinture au XVIIe siècle: la théorie de l'art de Samuel van Hoogstraten (Peter Lang, 2008); 'Rembrandt and the Historical Reconstruction of his "Conspiracy of Claudius Civilis"', in Myth in History/History in Myth (Brill, 2009); and Vermeer: la fabrique de la gloire (Citadelles & Mazenod, 2014).

Hanneke Grootenboer is a Professor of the History of Art at the University of Oxford. She is author of: The Rhetoric of Perspective: Realism and Illusionism in Seventeenth-Century Dutch Still Life Painting (Chicago University Press, 2005); and Treasuring the Gaze: Intimate Vision in Eighteenth-Century British Eye Miniatures (Chicago University Press, 2013, winner of the Kenshur Prize for 18th Century Studies). Her articles have been published in The Art Bulletin, Oxford Art Journal and Art History among other venues. She is currently completing a book entitled The Pensive Image, which argues that painting is a mode of thinking.

Etienne Jollet is a professor in early modern art history at the University Paris I Panthéon-Sorbonne. He works mainly on the relationship between art and knowledge on gravity in painting: Figures de la pesanteur: Fragonard, Newton et les hasards de l'escarpolette (Jacqueline Chambon, 1998); on the notion of place in still-life: La Nature morte et la place des choses (Hazan, 2007)), and on the link between art and power as in Jean et François Clouet (Lagune, 1997) or in the book on the French royal monuments of the Ancien Régime that he is currently completing.

Amy Knight Powell is Associate Professor of Art History at the University of California, Irvine and author of Depositions: Scenes from the Late Medieval Church and the Modern Museum (Zone Books, 2012). She is currently working on two projects tentatively titled The Whitewashed Image: Iconoclasm and Seventeenth-Century Dutch Landscapes and A History of the Picture as Box. Along with the essay that appears here, she has written another on the morphology of the circle for postmedieval: a journal of medieval cultural studies, 7: 1, 2016.

Pamela H. Smith is Seth Low Professor of History and founding director of the Center for Science and Society at Columbia University. Her books include: The Business of Alchemy (Princeton University Press, 1994); Merchants and Marvels: Commerce, Science and Art in Early Modern Europe, (Routledge, 2002); The Body of the Artisan (University of Chicago Press, 2004); Making Knowledge in Early Modern Europe (University of Chicago Press, 2008); Ways of Making and Knowing: The Material Culture of Empirical Knowledge (University of Michigan Press, 2014); and The Matter of Art: Materials, Practices and Cultural Logics c. 1250–1750 (Manchester University Press, 2015).

Barbara Maria Stafford is William B. Ogden Distinguished Service Professor Emerita at the University of Chicago and until recently Distinguished University Visiting Professor at Georgia Institute of Technology. Her work has consistently explored the intersections between the visual arts and the physical and biological sciences from the early modern to the contemporary era. Her current research charts the revolutionary ways that the neurosciences are changing our views of the human and animal sensorium, shaping our fundamental assumptions about perception, sensation, emotion, mental imagery, and subjectivity.

Richard Taws is Reader in the History of Art Department at University College London. The recipient of fellowships from the Getty, the Institute for Advanced Study, Princeton, and the Bard Graduate Center, New York, he was recently awarded a Philip Leverhulme Prize. He is the author of The Politics of the Provisional: Art and Ephemera in Revolutionary France (Penn State University Press, 2013), and is currently writing a book about art and technology in post-revolutionary France.

Genevieve Warwick is Editor of Art History and author of The Arts of Collecting (University of Cambridge Press, 2000 and 2012), Bernini: Art as Theatre (Yale University Press, 2012), and Picturing Venus in the Renaissance Print (The Hunterian, 2014), as well as collections of essays on Poussin, prints and drawings and Caravaggio, and numerous articles on art collections, art and science, and architecture and urbanism. Her research has received awards from the Kress Foundation, the Leverhulme Trust, the Getty Foundation, the Arts and Humanities Research Council UK, and the Renaissance Society of America.

Giulia Martina Weston completed her PhD, entitled 'Niccolò Tornioli: The Life and Times of a Sienese Painter in Seventeenth-Century Rome', at the Courtauld Institute of Art in 2015. She is an independent art consultant and contributing editor of several online journals. She co-organized the Courtauld Fifth Early Modern Symposium and co-edited the volume I Pittori del Dissenso (Artemide, 2014). Her forthcoming publication concerns the practice of painting on stone in the age of the Baroque.

Bryan Wolf is the Jones Professor Emeritus in American Art and Culture at Stanford University. His books include: Vermeer and the Invention of Seeing (2001) Romantic Re-Vision: Culture and Consciousness in Nineteenth-Century American Painting and Literature (1982); and American Encounters: Art, History, and Culture (2007), a co-authored textbook. His current project, Hidden Histories, explores three artists who grapple, in unexpected ways, with historical events that have shaped them: Philip Guston and the Holocaust; Martin Puryear and the Civil Rights movement; and South African artist William Kentridge and apartheid.

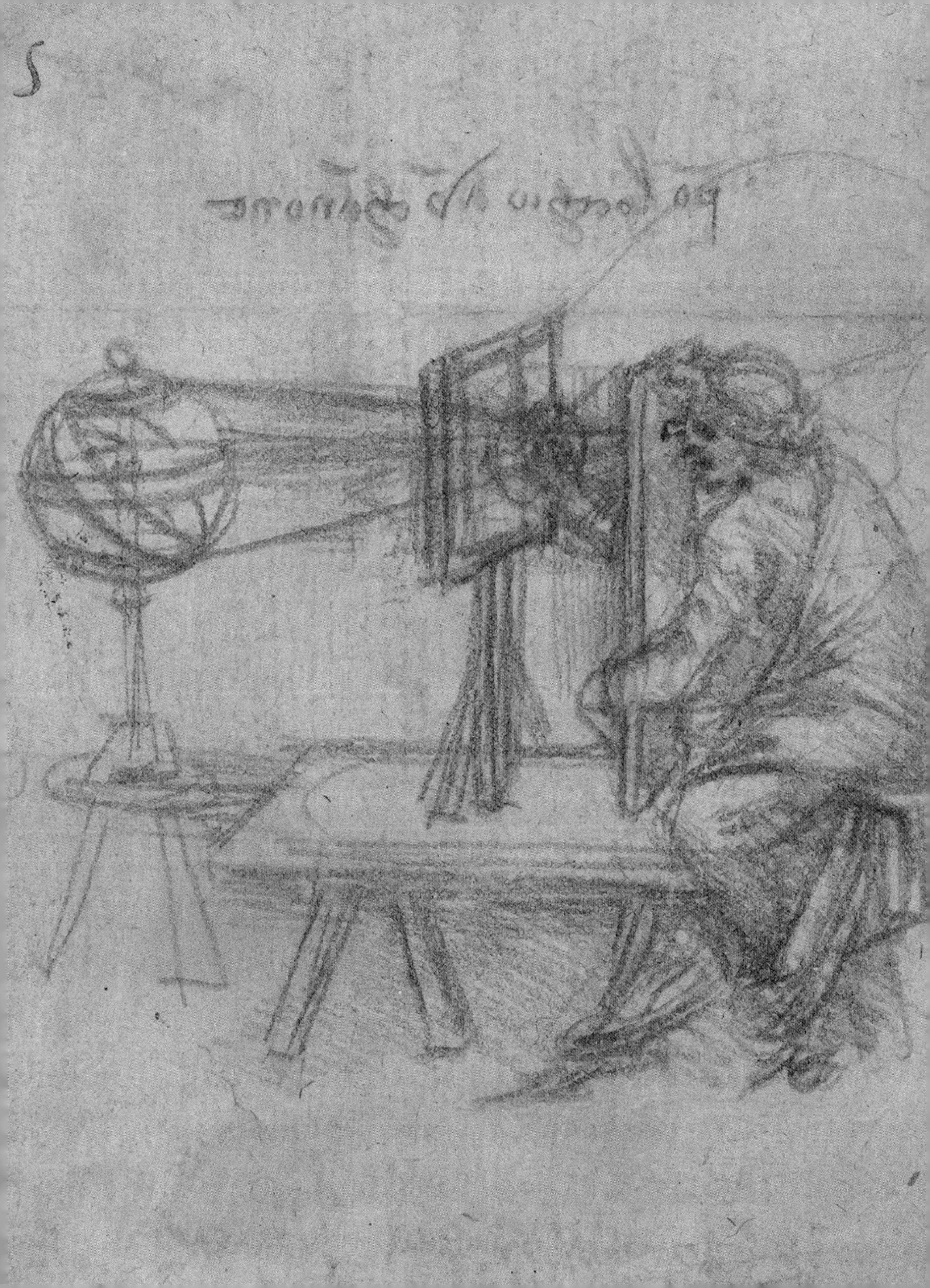

Chapter 1
After Prometheus: Art and Technology in Early Modern Europe

Genevieve Warwick and Richard Taws

In a celebrated letter to Ludovico Sforza, Duke of Milan, Leonardo offered his complement of services as both an artist and an engineer. His stated abilities comprised a wide array of skills, from the construction of portable bridges, tunnels under rivers, methods of draining water for agricultural as well as military purposes; fortifications, weapons and armoured vehicles; architecture for both public and private uses; sculpture and casting in metal; and finally, painting, which he offered to do 'as well as any other'.[1]

The thousands of pages of Leonardo's notebooks provide evidence of the full range of interests and abilities that his letter claims, from hydraulic engineering to anatomy, astronomy, biology, optics, and art. Throughout the notebooks, his studies in art and technology sit side by side, often on the same sheet. Early sketches for the figural composition of a painting appear beside mechanical designs for water wheels, or machinery for the transport of heavy goods, regularly interspersed with notes of his further thoughts on their function and design. Other pages bring together the different facets of his varied interests in technology, for example, designs for industrial machinery and for astronomical investigation appear alongside visual technology deployed by an artist at work (*plate 1*). In the centre left of this sheet is the deep cylinder of a reciprocal displacement pump for moving water, driven by mechanized bellows, with a detail of the valve mechanism above. It configures one of the leading threads of Leonardo's notes, the study of water as a force of nature to be harnessed, as in Giorgio Vasari's account of Leonardo designing pumps 'to draw up water from great depths'.[2] To the right, a seated draughtsman sketches after a globe-shaped astrolabe or armillary sphere, an astronomical model of the circuits of the planets used to derive mathematical measures of space and time. Together the pump and the astrolabe demonstrate Leonardo's lifelong interest in the development of technologies as extensions of the human ability to master and study nature. In order to render the visual appearance of the astrolabe's spherical volume on the two-dimensional surface of a diagram, the draughtsman works by means of a further instrument apparently of Leonardo's devising, a perspectograph.[3] The artist looks through a sighting device to steady the eye onto a framed pane of glass placed before the object to be depicted. By this means the perspectograph allows the draughtsman to establish the outline and geometrical relation of the parts directly on the glass before him.[4] As an ensemble, the instrument guided and thereby extended the capacities of the human hand and eye in the accurate observation of volume, and facilitated the visual demonstration of this knowledge in a graphic model.

Detail from Leonardo, page of sketches including bellows for raising water and a draughtsman drawing after an astrolabe through a perspectograph, *Codex Atlanticus*, f. 5r, c. 1480 (*plate 1*).

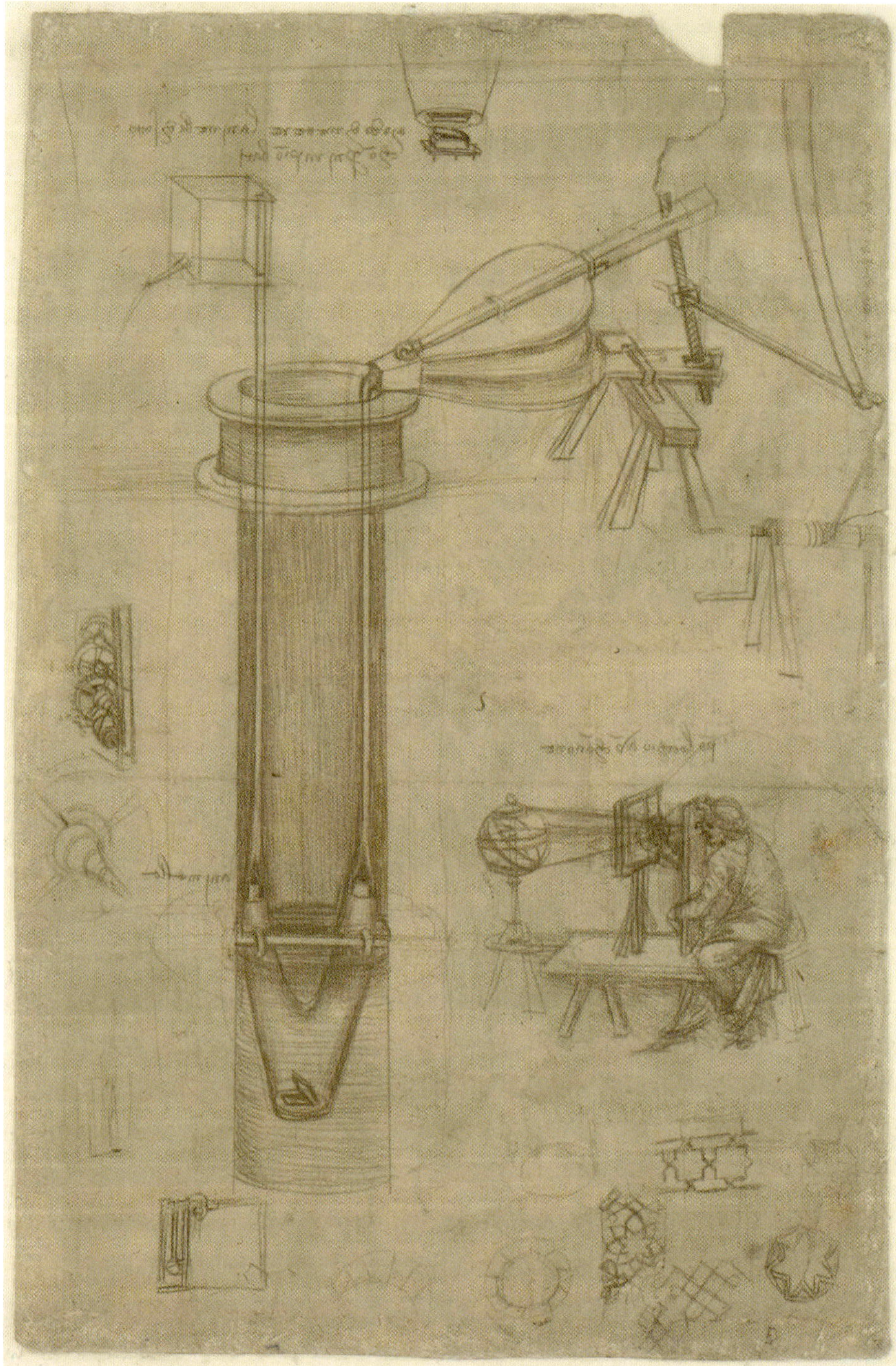

1 Leonardo, page of sketches including bellows for raising water and a draughtsman drawing after an astrolabe through a perspectograph, *Codex Atlanticus*, f. 5r, c. 1480. Silverpoint on prepared paper, 297 × 198 mm. Milan: Ambrosiana. Photo: Veneranda Biblioteca Ambrosiana.

Collectively, the sketches on the page testify to the growing role of early modern technology across all realms of human endeavour, from the pump and bellows of the hydraulic engineer to the artist's drawing frame. They demonstrate how skill as a draughtsman served the development of new technological designs, just as new technologies shaped the development of art.

As Leonardo's drawings and notebooks amply disclose, the writing of art's histories rests substantially, if for the most part tacitly, on an underlying account of technological change and development. This volume embarks on a history of that technological substrate as it pertains to the making and viewing of art in early modern Europe, *c.* 1420–1820. That is to say, it examines artists' instruments, tools, machines, technologies, crafts, materials, skills, and techniques in their historic applications, to consider how they shaped the course of early modern art. The analytical endeavour is to knit together the history of early modern visual technology with the history of its art, and to make manifest the far-reaching connections between the two. The concern is a history of practice-based skills and devices intrinsic to early modern art's production and use, and in technology's relationship with a theoretical conceptualization of 'art' in the broader visual field.[5]

The temporal scope of the volume is framed by two metanarratives in the history of visual technology: the development of the printed image in the mid-fifteenth century; and of the photographically produced image in the early nineteenth century. The essays that follow do not treat the history of the print or the photograph as such, however, but rather the period that lies between, in order to map a distinctively early modern history of art and technology. Throughout, we use the term 'invention' advisedly, acknowledging Marc Bloch's longstanding critique of its artificial relief cut against the grain of a more gradual history of technological development.[6] We remain equally wary of a teleological or determinist history of technological change, and of the assumptions regarding a triumphal narrative of human achievement implicit in such an account.[7] Yet the model of history prompted by chronicles of 'invention', as marked by moments of fundamental rupture with the past that align with Thomas Kuhn's conception of the paradigm shifts distinguishing scientific 'revolutions', also helps delineate our field of study.[8] Both the print and the photograph facilitated a vastly greater diffusion of visual knowledge than ever before. Such historic augmentation of the circuits of knowledge exchange, then as now, fuelled intellectual developments and discovery in all aspects of human enquiry, including the artistic.

If our own information age is defined by the digital structures of electronic communication, early modern culture was inextricably bound to the medium of print. Printed text and image arose within a few years of each other in the mid-fifteenth century, credited to the German goldsmith, Johannes Gutenberg, who seemingly drew together a series of extant yet disparate technologies into a new machine that could print several thousand sheets a day. The ancient oil or wine press, coupled with the goldsmith's craft in fine metal carving, the late-medieval development of plentiful rag paper, and the recent formulation of more stable oil-based inks enabled Gutenberg's 'revolution'. Similarly, early photography developed from a coming together of two otherwise disparate technologies; on the one hand the pinhole camera through which to capture a reflected view of the world as an image, and on the other the chemical means to fix the effects of light exposure on paper. In both cases these technologies shared aesthetic resources with other media available at the time, while also producing forms of representation that were uniquely theirs, and which offered access to new ways of seeing, and enabled new forms of subjectivity. The greatly expanded flow of visual information facilitated by these technological breakthroughs worked to quicken the circulation of knowledge, and so the foundations of thought itself.[9]

To understand the far-reaching consequences of Gutenberg's legacy is pressing today, as we inhabit a new heuristic landscape of technologically enabled knowledge flow structured by the internet. From Gutenberg to Google, these technological 'revolutions' have brought about a fundamental reordering of the structures of knowledge in images and texts alike. Today, the parallel interfaces of Google Books and Google Images represent coterminous curations of textual and visual knowledge in a new *Encyclopédie* of the screen.[10] By contrast, the story of the print as a bearer of

2 Etienne-Claude Voysard after Claude-Louis Desrais, *L'homme de nature est un faible animal. Mais la Philosophie, aux Dieux le rend égal*, 1799. Etching and engraving, 15.7 × 9.7 cm. Paris: Bibliothèque nationale de France. Photo: BnF.

visual information, now increasingly superseded by the digital image, has become the historian's domain. If this is a relatively new prospect for the printed book, in many respects the printed image began the process of historical contingency some 150 years earlier, with the emergence of photography in all its varied early experimental forms.[11] Photography did not, of course, replace other media, rather functioning alongside and in many cases in collaboration with older technologies of visual representation, from painting to print.[12] Yet as Nicéphore Niépce, Louis Daguerre, and Henry Fox Talbot simultaneously developed the first photographs, the cultural position of the printed image changed irrevocably under the weight of this new, fully mechanized process of image reproduction.

The focus of this study, however, lies between these great shifts in the technological and material, but also ontological, status of the image. It is an interwoven history, marked by narratives of technological development, to be sure, but also vexed by instances of rupture, reversal, and obsolescence.[13] Our own historical position at the threshold of 'new media' in the arts rests on an awareness of this historical process as technological shifts restructure art itself in the digital medium.[14] Meanwhile, recent scholarship in the emerging field of media archaeology, and arguments for 'remediation' as an essential aspect of media change, as each new medium quotes from and refashions its predecessors, have demonstrated not only the persistence of past technologies for image-making in the present, but also the contemporaneity of historical images and processes that might otherwise have appeared obsolete.[15]

With contemporary visual practices and debates fully in mind, we are here committed to an examination of these issues within early modernity, and in their historical specificity. Rather than isolate the great gateposts of the print and the photograph, our intention is to open up for study the place of changing artistic technologies in the centuries that separated them, and thus the incursion of technology on artistic decision-making during early modernity. We are, then, concerned with the history of art's technologies from the Renaissance to the Industrial Revolution. The volume runs from the largely craft-based methods of the early Renaissance, to the wide plethora of automated viewing instruments and mechanical devices that characterized the years immediately preceding the advent of photography. The growing spread of technological marvels in all areas of human endeavour in the wake of early industrialization is, for example, given pictorial representation in a print by Etienne-Claude Voysard's *c.* 1800 print after Claude-Louis Desrais (*plate* 2). Balloons, parachutes, telegraphs, astrolabes, cannon, warships, telescopes, chemical experiments, and other 'philosophical' objects converge in this condensed allegorical reflection on both the wonder and the darker implications of new technologies for narratives of European imperialism, as laid bare in the background image of a plantation, and the focus on the Atlantic trade enabled by these devices.[16]

The array of technological artefacts under review in this volume likewise brings to light the scale of historical transformation at stake. For the early Renaissance, artistic production drew on a range of ancillary domestic crafts from which it borrowed materials and methods. In the eighteenth century, by contrast, definitions of visual technology resemble our own, signifying a spectrum of automated devices and viewing machines that mechanized the image in various ways. The encroaching instrumentalization of artistic practice across the period may be tied to broader historical processes of industrialization, but also to related developments in the history of science. For botany, astronomy and anatomy alike, an early modern

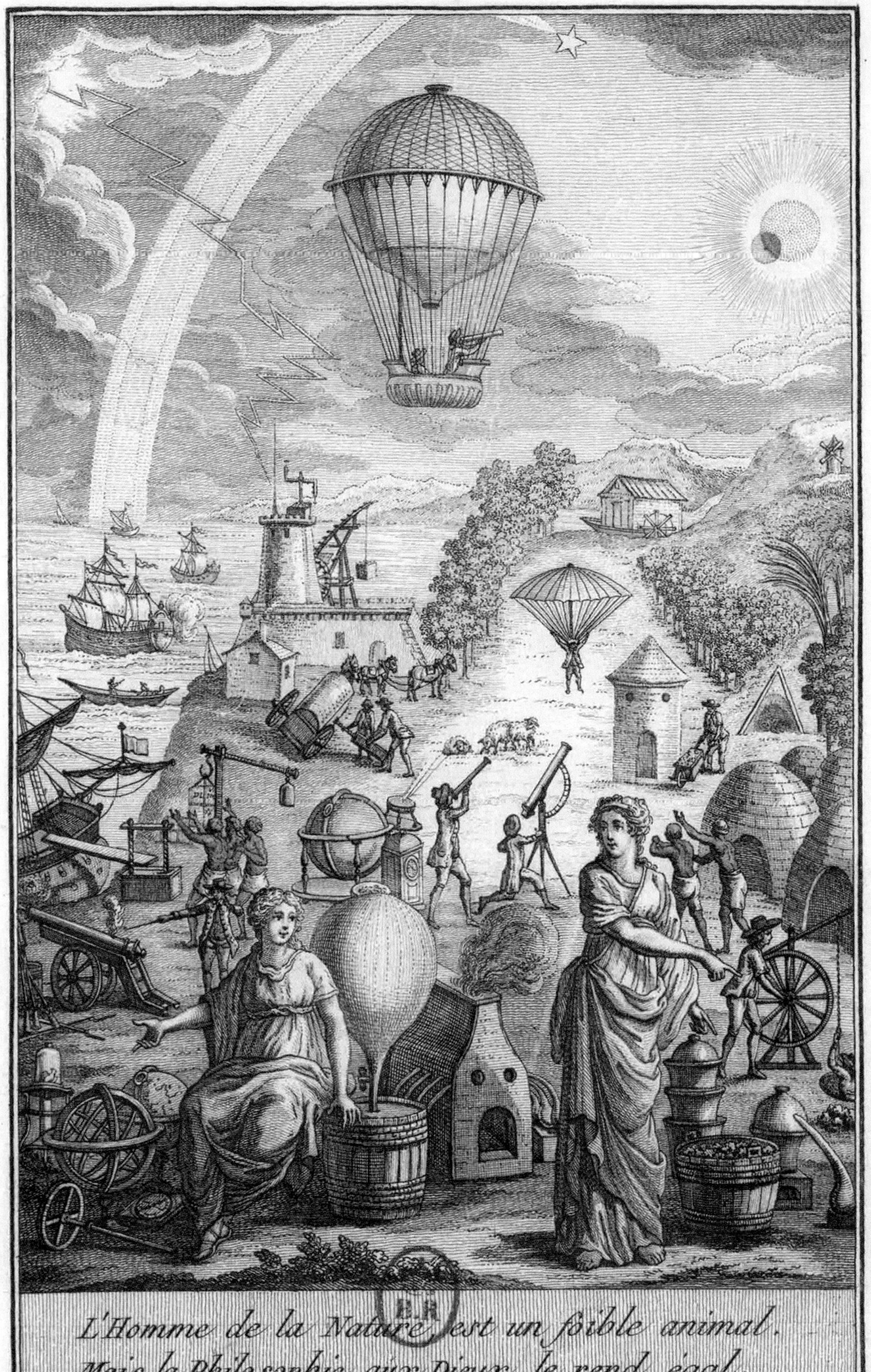
L'Homme de la Nature est un foible animal.
Mais la Philosophie, aux Dieux le rend égal.
C.L. Desrais. del.
C. Voysard Sculpsit.

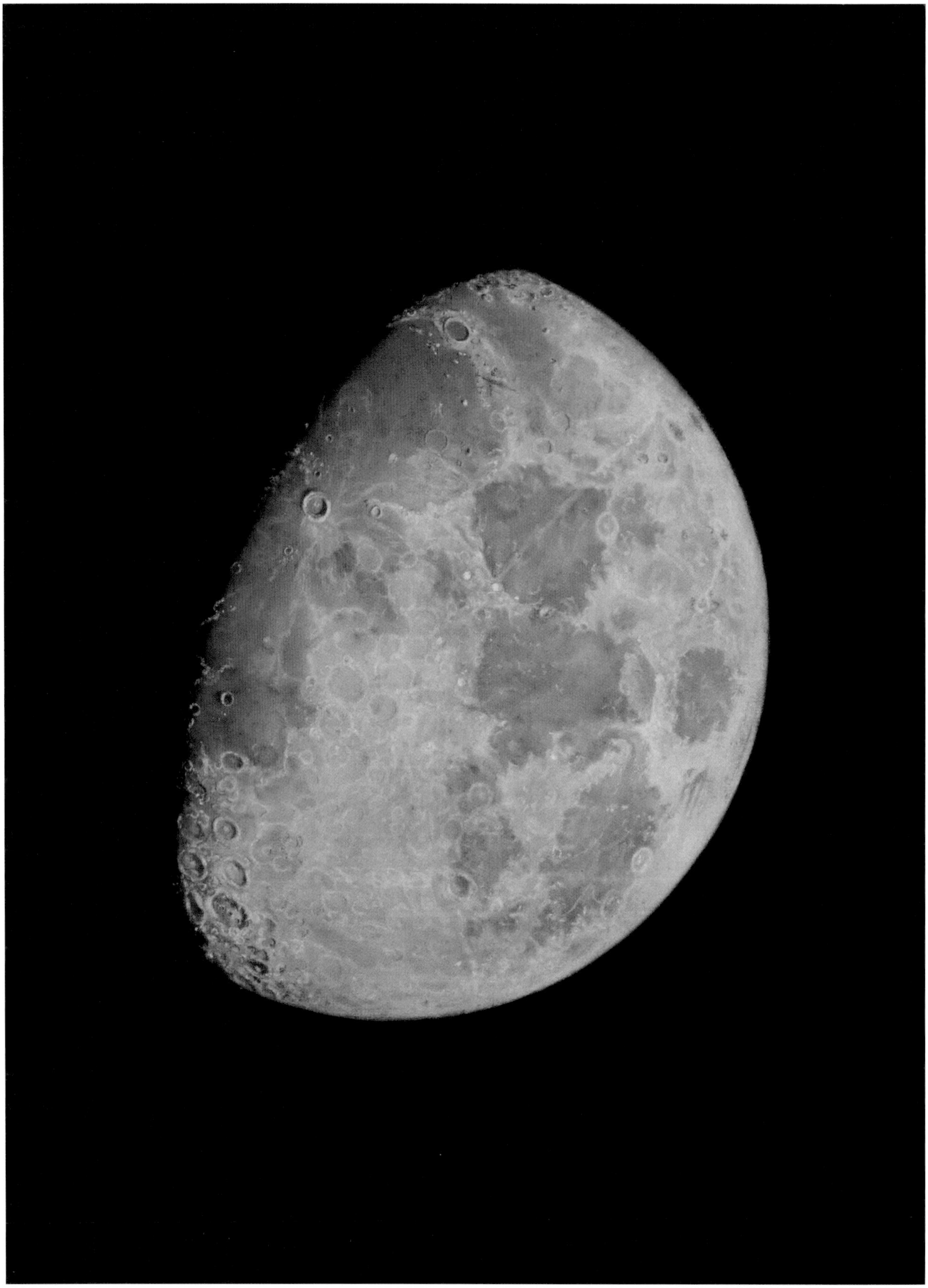

epistemic shift from a largely text-based to a predominantly observation-based mode of study engendered through new lens-based visual technologies. Exemplified in the linked invention of the microscope and the telescope, these viewing devices in turn produced new visual paradigms in all aspects of learning, including the artistic.[17] Towards the end of our period, for instance, portraitist John Russell's pastel drawing of the gibbous moon, made between 1793 and 1797, seems to anticipate, in that most fragile of mediums, the permanence and fidelity of a 'photographic effect' by way of observation through a telescope obtained from one of his sitters, the royal astronomer Frederick William Herschel (*plate* 3).

To use the terms of Marcel Mauss's now-classic anthropological analysis, our study charts this broad temporal shift from manual techniques to mechanized technologies.[18] Yet the complexity of the historical material inevitably complicates any such trajectory, requiring us to attend both to change over time and to the contextualized analysis of specific moments within the longer spectrum, to survivals of techniques across broad stretches of time that coexisted alongside new instruments and technologies. Michel Foucault's far-reaching history of thought, which he termed an archaeology of knowledge, structured by a matrix of instruments, architectures, machines and procedures, draws together the terms 'technique' and 'technology' in ways that defy any easy distinction.[19] Similarly, the sociologist Jacques Ellul defined both technique and technology as constituted from an ensemble of methods designed for the greatest possible efficiency in their practical application, in any given historical period, thus undermining any straightforward developmental relationship between the two.[20] Furthermore, the historical range of early modern visual technologies expanded the possible meanings of technology 'proper', complicating our understanding of the relationship between technique and technology, and specifically in their visual forms.

Yet an overarching view of the historic development of tools, instruments and machines also signals key differences in the relationship between these terms over time. Turning back for a moment to our earliest human technological endeavours, archaeologists constate these in the material remains of archaic tools shaped by fire.[21] Such prehistoric technology is mythologized in the story of Prometheus, the Titan god who first lit fire for human application. The originary technology of the flame is manifest in the god's use of fire to fashion the first human figure from clay as the exordium of art. Just as fundamentally, the story testifies to the central role of art as the locus of material and technological experimentation.[22]

Thus as a term, technology encompasses an array of diverse, even conflicting definitions, from different disciplines, languages, and historical periodizations. This lexical complexity is, for example, central to Martin Heidegger's celebrated if dystopian 1954 essay on the ethical-philosophical 'question of technology', which draws on the dual meaning of '*Technik*' in German to encompass both technology and technique. At the core of his essay is an understanding of thought itself as a craft, thus erasing any neat separation between concept and instrumentation, or its more familiar art-historical designations of theory and practice.[23]

As a Promethean endeavour, this volume too is concerned with *techné*, those embodied forms of knowledge central to the manual/mechanical work of artistic production and use. Yet the history of visual technology is, following Heidegger, equally concerned with *logos*, or a theory of the image, and indeed with its changing conceptualization over time. The term 'techn[o]logy' seems to have first appeared in print in English in the early years of the seventeenth century, precisely to connote a form of knowledge comprising both *techné* and *logos*.[24]

3 John Russell, *The Face of the Moon*, 1793–97. Pastel on paper strained over a wooden stretcher, 60.7 × 50.4 cm. Birmingham: Birmingham Art Gallery. Photo: © Birmingham Museums Trust.

Thus the study of technology is not only concerned with practice-based forms of knowledge, but also with the changing conceptual paradigms that drive technological change. Analysis of technology brings together *homo faber* with *homo sapiens* – the human manufacture of tools and instruments as the material extensions of productive bodily labour, but also of patterns of thought and intention. It is the points of intersection between craft and concept that drive the development of new technological means. We may understand technological development as practical problem-solving, to be sure, but always within the folds of larger patterns of cultural enquiry.

Such expansive definitions of technology retain the meaning of techniques so resonant for the arena of artistic production in any period, as well as the instrumentation of its display. For the Renaissance workshop, at the outset of our study, art-making emerged through a process of thought predicated on the possibilities of the media at hand, a practice-based way of thinking through materials. From the mixing of paint solutions to the grinding of minerals and the weaving of canvas, these craft technologies were as fundamental to the making of early modern art as the grids, squaring and sighting devices that constituted the instruments of a mathematically theorized perspectival vision, or the mechanics that would later enable magic lanterns, panoramas, and early moving images. As technological prowess itself became a sign of wonder, machinic visibility as part of the presentation of an art object was often enhanced. Yet there is also evidence of the suppression of art's technical and technological means, its very invisibility understood as a token of art's skill.

By the end of the eighteenth century, as technology became more indelibly associated with industrialization, it came to be viewed as counter to the 'true' aims of art, a critique that became ever more pronounced as the nineteenth century progressed. Yet such commentaries only served to highlight technology's interface with a developing social critique of the 'machinic', in the arts as for society more broadly. As the writings of Jean-Jacques Rousseau embody, and later the works of the Arts and Crafts movement, the heralding of new technologies was met with cultural anxiety as much as industrial acclaim.[25]

How then may this historic interplay of social critique, analysis and definition be brought to bear on the study of early modern technologies pertaining to art? Like Leonardo's perspectograph, the development of visual technology in the realm of art was linked to a technical history of the image as the means to instrumentalize the replication of nature. The role of artistic technology was understood as the method by which to render an ever-more exacting imitation of the visible world in the form of art. At the same time, there was a marked shift in the means to achieve this, from a predominantly but certainly not exclusively technical paradigm of manual execution in the early fifteenth century, to an increasingly mechanical or industrialized instrumentation of art over the course of the period covered. Moreover, there was a discernible shift towards technologies concerned with the display of art – phantasmagoria, stereoscopes, and other apparatuses – in addition to its production. Thus our study moves spatially as well as temporally, between art-making and art-viewing, from the craftsman's workshop to the collector's gallery, the artist's studio to civic space, the scientist's laboratory and the fairground's popular entertainments. It also takes into account those technologies not directly implicated in the making or display of art, but which had a bearing on it nonetheless, often in unanticipated or unexpected ways. Above all, it traces a history of early modern art in its complex relationship with technology, as forms of instrumentation to be sure, but also as systems of knowledge. In so doing, it lends new visibility to the often occluded history of art's means, and situates the study of early modern art and technology within a complex network of historical relations – between practice and concept, mechanics and epistemes, *ars* and *scienza*.

The following essays in this volume all, in different ways, contribute to a new analysis of the role of visual technologies in the development of early modern art. Together they map a history of visual instrumentation devised for the making and viewing of art interwoven with a broader historical shift towards industrial modernity. Individually, they signpost new avenues for enquiry, singling out those objects and images possessed of a particularly dense, heightened ability to break fresh ground in our understanding of early modern developments in art and technology. Four opening essays study a range of early modern artists' instruments. The Renaissance artist's 'recipe book' is Pamela H. Smith's example, a collection of 'how-to' instructions in materials and techniques. Smith's mode of analysis in itself instantiates *techné*, conducted through practice-based reconstructions of the recipes undertaken as an aspect of graduate training, much like the Renaissance workshop. The example of artists' mannequins and maquettes, used by Tintoretto, Poussin, and Vermeer among others, allows Jan Blanc to argue for a finely nuanced understanding of artistic process as a coupling together of working 'from life' and from art-historical memory. Genevieve Warwick takes up the case of the mirror, both as a workshop instrument used to translate the painter's view of the world into a two-dimensional surface, and as a motif within painting to signify the painter's art. Meanwhile, Amy Knight Powell takes up the telescopic view of landscape manifest in roundel paintings to argue for a new attention to the representation of distance in painting at the moment in which prosthetic devices for the extension of human vision first appeared.

The ensuing five essays take an individual painting or monument as their focus, collectively demonstrating a growing emphasis on technologies of artistic display. Giulia Martina Weston's study of Niccolò Tornioli's *c.* 1645 depiction of astronomers binds the representation of astronomical viewing devices within the painting to early modern collections of such instruments of visual wonder, to argue for a renewed defence of the role of epistemic images in the immediate wake of Galileo's death. Etienne Jollet considers pictorial representations of the casting and elevation of the colossal bronze equestrian monument to Louis XIV by François Girardon at the Place Vendôme, completed in 1699. These images formed part of a widespread interest in technical knowledge surrounding artistic production, but also of royal power, through the patronage of mechanical technologies understood as manifestations of the marvellous within a broader early modern culture of *theatrum mechanicum*. Hanneke Grootenboer also takes up the example of the mechanical, through a small automated landscape painting framed by a display of clock faces produced in 1739 for a *Cabinet de Mécanique*, as a pictorial commentary on Descartes' disquisition on the nature of the universe as clockwork. Joseph Wright of Derby's celebrated pictorial representation of a scientific experiment, *The Bird in an Air Pump* of 1768, is the subject of Bryan Wolf's essay, newly analysed as a primer in the ideological and pedagogical underpinnings of visual observation and the cultural power of science. Ann Bermingham interrogates Philippe de Loutherbourg's Eidophusikon, a public 'moving picture' attraction staged in 1781 in a London square, and fêted for its combination of technical accomplishment and mimetic visual magic at the threshold of a new, technologically driven, display culture of images in motion. By contrast, Richard Taws studies an historical instance of a new information technology in post-revolutionary France, that of optical telegraphy, which for some fifty years was a highly visible if ostensibly 'secret' means of disseminating political news and messages in the charged years during and after the French Revolution. Finally, Barbara Stafford's coda takes us from early modern technological wonder to contemporary instances of technological fascination in the visual realm. It dwells on the changing quality of human mental attention in relation to a history of new technological devices in the digital age. Tacitly recalling Ernst Kapp's foundational treatment of technology as prosthetic, it acknowledges this Aristotelian view of technologies as 'the extensions of man', in Marshall McLuhan's phrase. Thus the hammer is an extension of the fist, be it of a sculptor or a carpenter; the pencil and paintbrush of the artist's hand; while eyeglasses, microscopes, telescopes and lenses of all kinds extend the power and functions of the eye, both artistic and scientific. Technology as such, Leo Marx reminds us, makes nothing happen – its agency lies in its human applications.[26] Stafford concludes with current considerations concerning digital memory and electronic data, leading her to reflect on how the computer has now seemingly become an extension of the brain itself. As Leonardo well understood, technologies could augment the capacities of hand and eye in the quest to know, and thus master, the elemental forces of nature. In this regard, he likewise recognized in technology a means to extend knowledge, and thus the mind.[27]

Notes

Warm thanks are extended to all our contributors for their essays and interest in this project, and especially to each other for sharing the editorial collaboration, to Sam Bibby for his manifold patience, and to the attentive contributions of our anonymous peer reviewer.

1 *Codex Atlanticus*, f. 1082r, Milan: Biblioteca Ambrosiana, c. 1480s; cited in Martin Kemp, *The Science of Art: Optical Themes in Western Art from Brunelleschi to Seurat*, New Haven, CT and London, 1990, 170–1.

2 Giorgio Vasari, *Le vite* ..., ed. Luciano Bellosi and Aldo Rossi, Turin, 1986, 550.

3 On the diagram as a means of producing technical and artistic knowledge in early modernity, see John Bender and Michael Marrinan, *The Culture of Diagram*, Stanford, CA, 2010.

4 B. N. Ms. Ital. 2038, f. 24r, Paris: Bibliothèque de l'Institut de France, 1480; cited in *The Literary Works of Leonardo da Vinci*, ed. Jean Paul Richter, Oxford, 1977, vol. 1, 317.

5 See Horst Bredekamp, *The Lure of Antiquity and the Cult of the Machine: The Kunstkammer and the Evolution of Nature, Art and Technology*, trans. Allison Brown, Princeton, NJ, 1995; Horst Bredekamp, Vera Dunkel and Birgit Schneider, eds, *The Technical Image: A History of Styles in Scientific Imagery*, Chicago, IL, 2015; Martin Kemp, *The Science of Art: Optical Themes in Art from Brunelleschi to Seurat*, New Haven, CT and London, 1990; Jonathan Crary, *Techniques of the Observer: On Vision and Modernity in the Nineteenth Century*, Cambridge, MA and London, 1992; Barbara Maria Stafford, *Artful Science: Enlightenment Entertainment and the Eclipse of Visual Education*, Cambridge, MA and London, 1994; Barbara Maria Stafford and Frances Terpak, *Devices of Wonder: From the World in a Box to Images on a Screen*, Los Angeles, CA, 2001; Caroline A. Jones and Peter Galison, eds, with Amy Slaton, *Picturing Science, Producing Art*, London and New York, 1998; Pamela H. Smith, *The Body of the Artisan: Art and Experience in the Scientific Revolution*, Chicago, IL and London, 2004; Alina Payne, ed., *Vision and its Instruments: Art, Science, and Technology in Early Modern Europe*, University Park, PA, 2015. On pre-modern technology and culture more broadly, see Lewis Mumford, *Technics and Civilization*, New York, 1934; Sigfried Giedion, *Mechanization Takes Command: A Contribution to Anonymous History*, New York, 1948; Lynn T. White Jr, *Medieval Technology and Social Change*, Oxford, 1962; David S. Landes, *The Unbound Prometheus: Technological Change and Industrial Development in Western Europe from 1750 to the Present*, Cambridge, 1969; Lorraine Daston, *Biographies of Scientific Objects*, Chicago, IL and London, 2000; Lorraine Daston, *Histories of Scientific Observation*, Chicago, IL and London, 2011; and Lissa Roberts, Simon Schaffer and Peter Dear, eds, *The Mindful Hand: Inquiry and Invention from the Late Renaissance to Early Industrialisation*, Amsterdam, 2007.

6 Marc Bloch, 'Les inventions médiévales', *Annales d'histoire économique et sociale*, 7, 1935, 634–43.

7 See Lisa Gitelman, *Always Already New: Media, History, and the Data of Culture*, Cambridge, MA, 2006; and the essays in Lisa Gitelman and Geoffrey B. Pingree, eds, *New Media, 1740–1915*, Cambridge, MA, 2003. Gitelman argues that claims to novelty are neither fixed nor inevitable, for all media were once themselves 'new', thus complicating claims for technological progress and agency.

8 Thomas S. Kuhn, *The Structure of Scientific Revolutions*, Chicago, IL, 1962.

9 See the classic study by William M. Ivins Jr, *Prints and Visual Communication*, Cambridge, MA, 1953; and Roger Chartier, *The Culture of Print: Power and the Uses of Print in Early Modern Europe*, Princeton, NJ, 1989.

10 Jay David Bolter, *Turing's Man: Western Culture in the Computer Age*, Chapel Hill, NC, 1984.

11 The historical study of the invention of photography has been instrumental to the analysis of the technological conditions of image production more broadly, and to the revival of Walter Benjamin's now-classic 1936 essay, *The Work of Art in the Age of Mechanical Reproduction*, London, 2008. From a vast bibliography, see in particular: Paul Valéry, 'And then came Daguerre...', in *The Collected Works of Paul Valéry*, ed. Roger Shattuck and Frederick Brown, Princeton, NJ, 1970, vol. 2, especially 159; and Victor Burgin, ed., *Thinking Photography*, London, 1982. More recently, the multiple 'inventions' of photography have been the subject of revived scholarly interest. See, for example, the essays in Tanya Sheehan and Andrés Zervigón, eds, *Photography and Its Origins*, New York, 2015. For a critique of 'photographic exceptionalism', see Stephen Bann's essay, 'Against photographic exceptionalism', in that volume, 94–103.

12 See Stephen Bann, *Parallel Lines: Printmakers, Painters, and Photographers in Nineteenth-Century France*, New Haven, CT and London, 2001.

13 Michael Werner and Bénédicte Zimmermann, 'Vergleich, Transfer, Verflechtung. Der Ansatz der Histoire croisée und der Herausforderung des Transnationalen', *Geschichte und Gesellschaft*, 28, 2002, 607–36.

14 See James Elkins, *The Domain of Images*, Ithaca, NY, 1999; James Elkins, *Six Stories from the End of Representation: Images in Painting, Photography, Astronomy,*

Microscopy, Particle Physics, and Quantum Mechanics, 1980–2000, Stanford, CA, 2008; and Julian Stallabrass, *Internet Art: The Online Clash of Culture and Commerce*, New York, 2003.

15 See Jay David Bolter and Richard Grusin, *Remediation: Understanding New Media*, Cambridge, MA, 1998; Siegfried Zielinski, *Deep Time of the Media: Toward an Archaeology of Hearing and Seeing by Technical Means*, Cambridge, MA, 2008; and for an introduction to these issues, Jussi Parikka, *What is Media Archaeology?*, Cambridge, 2012.

16 On the background to the development of many of these technologies, see Charles Coulston Gillispie, *Science and Polity in France: The Revolutionary and Napoleonic Years*, Princeton, NJ, 2004.

17 See Svetlana Alpers, *The Art of Describing: Dutch Art of the Seventeenth Century*, Chicago, IL, 1984; Catherine Wilson, *Early Modern Philosophy and the Invention of the Microscope*, Princeton, NJ, 1995; Eileen Reeves, *Painting the Heavens: Art and Science in the Age of Galileo*, Princeton, NJ, 1997; and Domenico Laurenza, *Art and Anatomy in Renaissance Italy: Images from a Scientific Revolution*, New Haven, CT and London, 2012.

18 Marcel Mauss, *Techniques, Technology and Civilisation*, ed. Nathan Schlanger, New York and Oxford, 2006

19 Michel Foucault, *Power/Knowledge: Selected Interviews and Other Writings*, New York, 1980

20 Jacques Ellul, *The Technological Society*, New York, 1964, xxv.

21 Earl Swanson, ed., *Lithic Technology: Making and Using Stone Tools*, The Hague, 1975.

22 Hesiod, *Theogony*, ed. M. L. West, Oxford, 2008, 507–616; Aeschylus, *Prometheus Bound: The Complete Greek Tragedies*, ed. David Greene and Richmond Lattimore, Chicago, IL, 1960, vol. 1, 309–51.

23 Martin Heidegger, *The Question Concerning Technology and Other Essays*, New York, 1977. For a sustained engagement with Heidegger's thought on these issues, and on the repressed role of technics in the history of philosophy, see Bernard Stiegler, *Technics and Time*, 3 vols, Stanford, CA, 1998–2010.

24 'Technology', in C. T. Onions, ed., *The Oxford English Dictionary on Historical Principles*, Oxford, 1972, 2140.

25 See, for example, Jean-Jacques Rousseau, *Discourse on the Sciences and Arts*, in Roger D. Masters and Christopher Kelly, eds, *Collected Writings of Jean-Jacques Rousseau*, Hanover, NH and London, 1992, vol. 1; and William Morris, 'Art and Socialism', in G. D. H. Cole, ed., *William Morris: Stories in Prose, Stories in Verse, Shorter Poems, Lectures, and Essays*, New York, 1934, 624–45.

26 Leo Marx, 'Technology: The emergence of a hazardous concept', *Technology and Culture*, 51: 3, July 2010, 577.

27 Ernst Kapp, *Grundlinien einer Philosophie der Technik: Zur Entstehungeschichte der Kultur aus neuen Gesichtspunkten*, Brauenschweig, 1877; Marshall McLuhan, *Understanding Media: The Extensions of Man*, London, 1964. The influential work of Friedrich Kittler offers a more pessimistic, anti-humanist counter-argument to this position. Arguing against McLuhan's 'extensions of man', for Kittler technology is autonomous and directs human activity, rather than the other way around. See, especially *Discourse Networks 1800/1900*, Stanford, CA, 1990; *Gramophone, Film, Typewriter*, Stanford, CA, 1999; and, with particular reference to the visual, *Optical Media: Berlin Lectures 1999*, Cambridge, 2009.

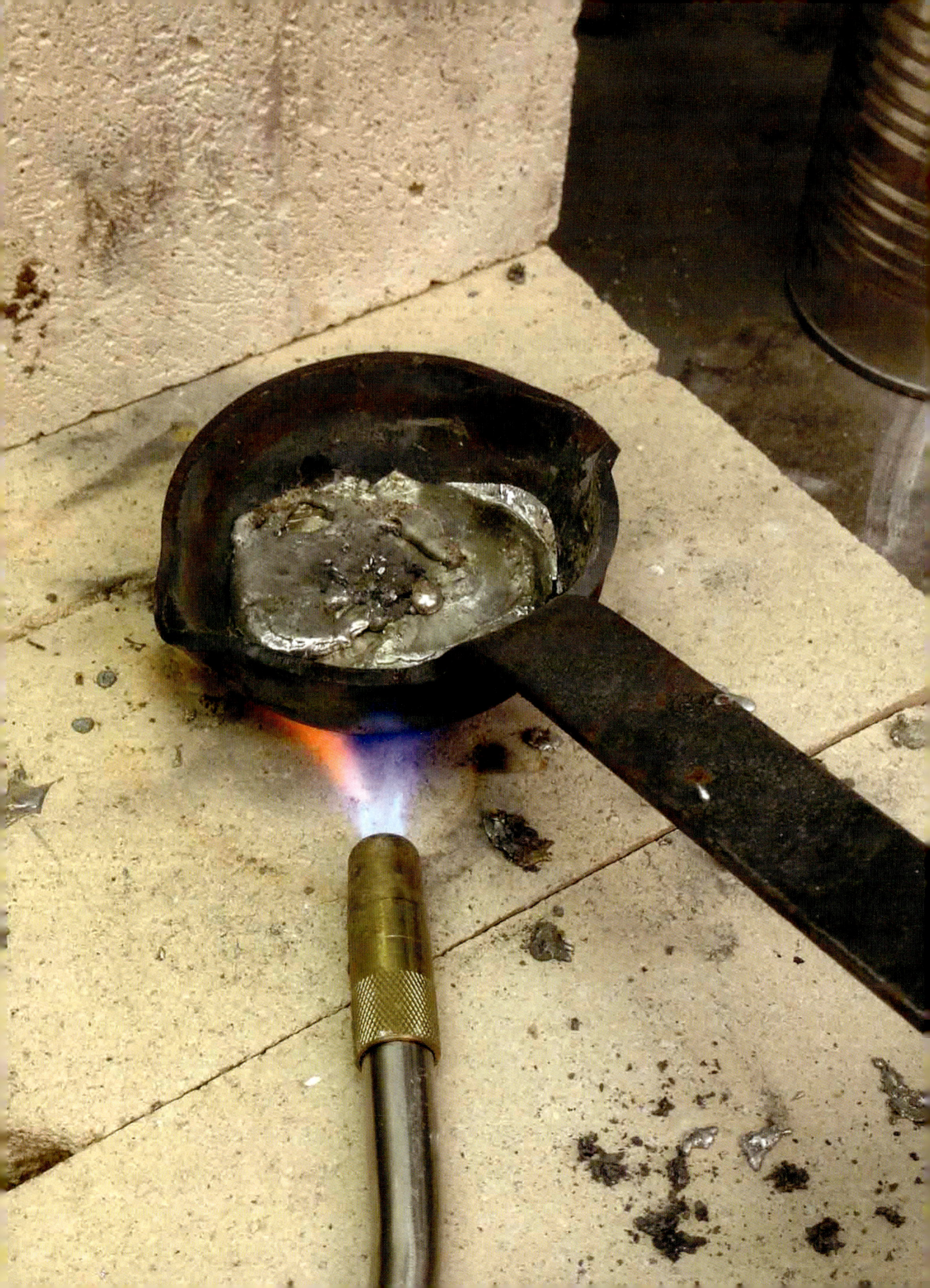

Chapter 2
Historians in the Laboratory: Reconstruction of Renaissance Art and Technology in the Making and Knowing Project

Pamela H. Smith and The Making and Knowing Project

The Inadequacy of Words

In the sixteenth and seventeenth centuries, European scholars began to search for a new kind of knowledge, what Francis Bacon (1561–1626) in 1620 would call a 'New Philosophy; or Active Science' (*The Great Instauration*, 1620), and what we have come to see as the beginnings of the modern natural sciences. These scholars sought to engage with the things of nature, in addition to the words of texts, and, as they looked about for models of this new kind of enquiry, they took up the case history used by their medical colleagues. They also looked to the methods of history, for history involved gathering observations and experiences about the human world, just as the new type of investigation these scholars sought would observe and collect experiences of the natural realm. They began to call what they did 'natural history'.[1] These scholars also looked to the handwork of craftspeople and their ability to manipulate natural materials in order to produce valuable products. Where these newly self-described 'natural historians' and 'experimental philosophers' could read the texts of their medical and historian colleagues, they generally had no such familiar entry point into handwork, for craftspeople produced things, and only rarely recorded their work in words and texts that the scholars could read.[2] As Francis Bacon complained in the *Novum Organum*, 'experience is illiterate'. Of course, craftspeople were not illiterate, but were fluent, rather, in a different kind of language and knowledge, one that posed problems for sixteenth-century scholars, and continues to make life difficult for the historians who study them. In the following essay, we suggest that one means for overcoming this problem is to bring historians and natural scientists back into conversation with each other, as they were in the sixteenth century when exploration of the human world provided a model for the just emerging study of the natural realm.

Between 1400 and 1700, European craftspeople similarly found words to be inadequate: they paradoxically declared *in writing* that writing was inadequate to convey their skills, and that book learning was inferior to bodily experience. Benvenuto Cellini (1500–71) in *Two Treatises on Goldsmithing and Sculpture*, wrote, 'How careful you have to be with this cannot be told in words alone – you'll have to learn that by experience.'[3] Bernard Palissy (c. 1510–90), who created extraordinary ceramics in France declared:

> Even if I used a thousand reams of paper to write down all the accidents that have happened to me in learning this art, you must be assured that, however

Detail from the Making and Knowing Laboratory, the paper test (*plate 8*).

Art and Technology in Early Modern Europe *Edited by Richard Taws and Genevieve Warwick*.

> good a brain you may have, you will still make a thousand mistakes, which cannot be learned from writings, and even if you had them in writing, you would not believe them until practice has given you a thousand afflictions.[4]

In his defence of craftsmen's knowledge, the religious, medical, and intellectual reformer, Theophrastus von Hohenheim, called Paracelsus (1493–1541), declared in writing that writings teach nothing: 'For who could be taught the knowledge of experience from paper? Since paper has the property to produce lazy and sleepy people, who are haughty and learn to persuade themselves and to fly without wings. . . . Therefore the most fundamental thing is to hasten to experience.'[5]

These statements form part of a remarkable development in European history, commencing around 1400, during which craftspeople and practitioners, used to looking, learning, and practising on the shop floor, suddenly transformed their lived experience and embodied – often tacit – knowledge into writing and compiled it into texts. Many well-known artists, such as Cellini and Leonardo da Vinci (1452–1519), wrote about their techniques, but many more lesser-known artisans, including gunpowder makers, fortification experts, and navigators, who previously had been happy to live out their lives without recording their experiences and knowledge, and creating and producing in relative obscurity, suddenly began to write. A flood of technical writing followed with the advent of printing in the 1460s, as recipe collections and technical treatises became some of the biggest bestsellers. Works in the vernacular appeared on distillation, mining, assaying and metalworking, and then there appeared 'books of secrets', which sound esoteric, but actually, in effect, were collections of all kinds of technical recipes, which, while practical and useful, also hinted at the tacit dimension and initiation process of much technical knowledge. The pseudonymous Alessio Piemontese's *Secrets*, a true early modern bestseller, is the exemplar of such texts, published and republished in varying editions from 1555. This boom in what we have come to call 'how-to' books continued into the eighteenth century.[6] Arguably, this how-to literature fostered on the part of its readers new attitudes to the creative process and to technical ability.

Writing Down Art

Why did craftspeople begin to write down their knowledge, especially in texts that declare written accounts to be useless? Writing down techniques is not an easy thing to do, as it is enormously inefficient, and steals time away from actually making a livelihood. Moreover, craftspeople did not (and still do not) learn skills from books, or even sometimes by language at all, but by experience, working alongside skilled practitioners, observing and imitating. The knowledge they gained was known in their bodies, and written descriptions could only point to bodily activity, but could not teach it or even describe it fully. In addition, craft work involved unpredictable qualities of materials, always-changing workshop conditions, and rapidly transforming matter, all of which the craftsperson had to respond to in real time. Little of this could be captured or codified in writing, for a text is just not an optimal means for conveying technique. Interestingly, this problem also plagues the natural sciences today, and has led to innovations such as the *Journal of Visualized Experiments*, which is 'devoted to publishing scientific research in a visual format to help researchers overcome … poor reproducibility, and the time and labor-intensive nature of learning new experimental techniques'.[7]

What should historians make of this how-to literature in early modern Europe? Why did practitioners write it down? Is it really a record of practice? Who is it for? Why did it sell so well? Is it really meant for use, or for teaching 'how to'?[8] How might historians make use of this genre – often simply compilations of technical recipes – as historical evidence, and what sort of evidence does

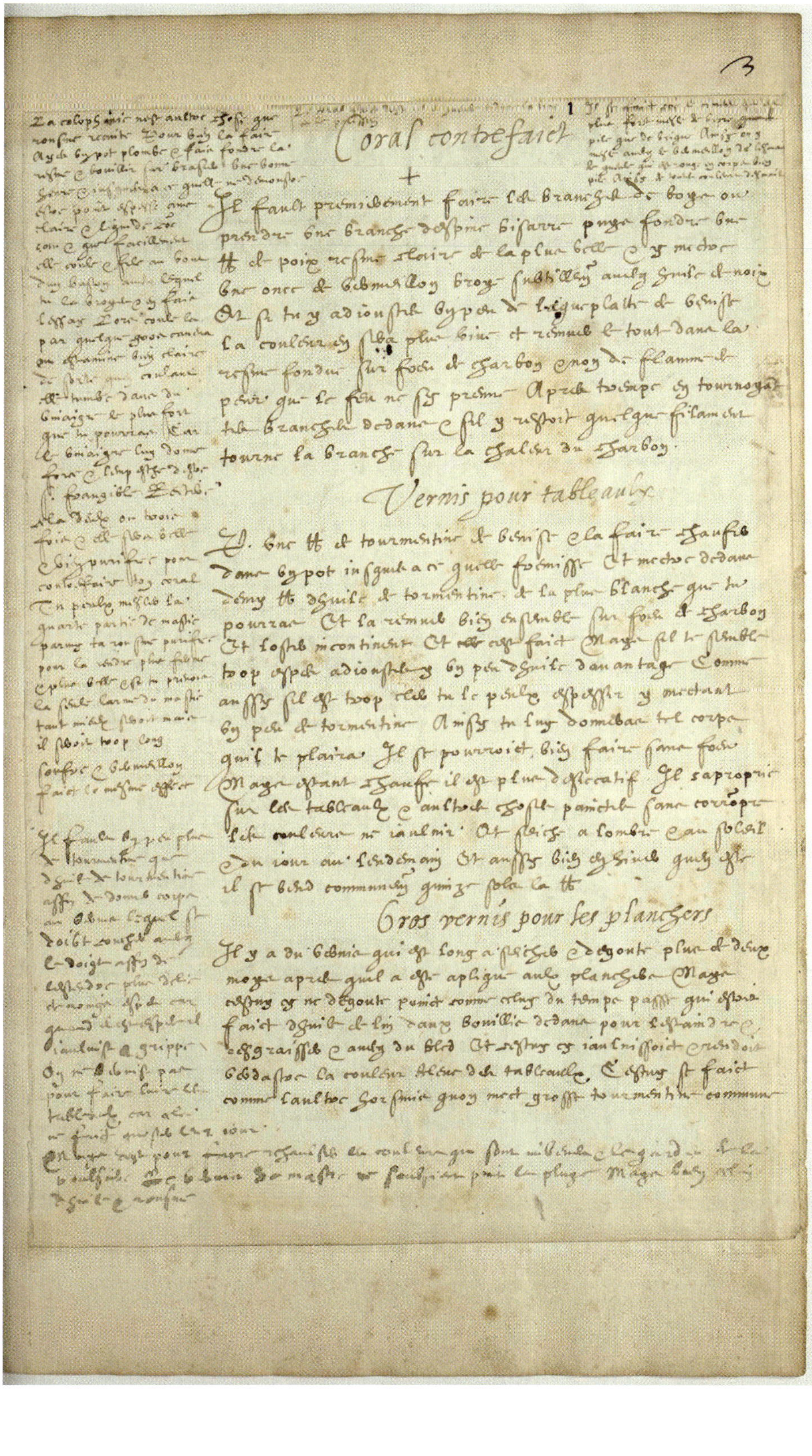

Coral contrefaict

Vernis pour tableaulx

Gros vernis pour les planchers

1 BnF Ms. Fr. 640, f. 3r, probably late sixteenth century. Photo: Bibliothèque nationale de France, Paris. ['Coral contrefaict' (imitation coral) is the first recipe in the text block. Also included on this page are two recipes for varnish.]

it supply? These writings would seem to provide evidence to help understand materials and techniques employed by artists in the past, and they have been used in this way by conservators and some art historians, but historians have also determined that such technical literature does not always reflect practices current at the time of writing.[9] What then is the nature and the status of the evidence such literature provides to a historian? To analyse this complex of issues, and to obtain a better sense of the purpose and ambitions of this literature, I began a close examination of a sixteenth-century technical manuscript mostly in French held by the Bibliothèque nationale in Paris, Ms. Fr. 640 (*plate 1*).[10] This manuscript exists in a fair copy, almost exclusively in a single hand, and has been well preserved since the seventeenth century, when it entered the King's Library as part of the donation of their library by the Dukes de Béthune.[11] The manuscript's 170 folios consist of a collection of mainly technical recipes in no apparent order for objects that might have formed the contents of a *Kunstkammer* (art chamber or cabinet), a type of collection brought together by European elites in the fifteenth through seventeenth centuries.[12] The manuscript's recipes would have pleased such collectors who delighted in objects that demonstrated the interplay of nature's artifice with that of the human hand, for they contain instructions for the draughtsman, for pigment-making, wood- and metal-colouring, imitation gem production, tree-grafting, land-surveying, a practice of taxidermy to manufacture monstrous composite animals (kittens and bats, among others), making papier mâché masks, and much more. In some respects, this manuscript resembles a 'book of secrets' such as that published in Italian under Alessio Piemontese's name in 1555. Like BnF Ms. Fr. 640, such books of secrets contained recipes for materials and objects ranging from perfumes, pigments, medicines, and cosmetics to small sculpture and portrait medals, as well as techniques for house- and landholders.[13] While Ms. Fr. 640 contains recipes for all such things, the largest numbers of recipes concentrate on techniques of mouldmaking, metal casting, and colour-making (pigments, dyes, varnishes, artificial gems, and colouring woods and metals).

The author of this intriguing manuscript may have been based in Languedoc, as he[14] mentions towns of this area, particularly Toulouse. Although not obviously the master of any trade, he clearly possessed firsthand knowledge of techniques for pigment preparation, painting, casting, and jewellery making, and awareness of many other now obscure European working practices, apparently through travel or exchange in the workshop. It would appear that the author-practitioner had plans to transform the manuscript into a book, presumably for publication. He reads Latin, albeit faultily, and appears to plan to incorporate Latin tags and allusions to ancient authors in his manuscript. This anonymous author-practitioner is a vocal and self-reflective witness to his culture, and his text offers insight into how natural materials and art objects were made, collected, appreciated, and circulated in a period of burgeoning production and consumption of material goods. Because he includes so much information about animals and natural materials, the manuscript also provides a rare view into attitudes to nature at the dawn of the 'new experimental philosophy'.

While the manuscript writings of many well-known artists, including those of Cennino Cennini (*c.* 1360–before 1427), Lorenzo Ghiberti (1378–1455), Leonardo da Vinci, and Cellini, began to be published in the nineteenth century, very few anonymous manuscript sources similar to Ms. Fr. 640 have received attention. Two recent exceptions are the manuscript compilation of illumination techniques stretching over three centuries from the Monastery of Tegernsee, the *Liber illuministarum* (c. 1200–1500), and the anonymous Montpellier *Liber diversarum arcium* (c. 1300).[15] These treatises lack the immediacy, self-reflexivity, and processual character of Ms. Fr. 640. In view of the proliferation of books of secrets during the early modern period, the singular nature of Ms. Fr. 640 renders it all the more interesting as a primary source: it differs from most other such collections of techniques in its apparent lack of formulaic recipes, its constant reference to the writer's own experiences, its extensive observations of animal behaviour, and its illustrations. The margins of the instructions are overflowing with the practitioner's comments on his trials of various technical processes, thus the manuscript forms much more than a prescription of procedure; it is also a rich record of workshop processes and attitudes to the material world.

In spite of the obviously fascinating nature and density of information to be gained from this manuscript, we found it difficult to decipher – not by reason of the script, which is fairly regular and legible, but because of its apparently unordered organization, and the obscurity of its contents – what *are* these materials and techniques? While it is possible to determine whether the materials and processes had a relationship to contemporaneous techniques described by other technical writings of the period, including Cellini, Piemontese (probably the scholar Girolamo Ruscelli), and Vannoccio Biringuccio (1480–c. 1539)[16], it was not immediately obvious whether the apparently firsthand accounts of work were actually so. Perhaps they were just the fruit of years of recipe collecting, with concern neither for the actual technical procedures nor for the production of things. The question of the status of a recipe compilation as evidence of historical practice arose once more.

Reconstruction as Method

Reconstruction of the recipes in the manuscript appeared promising as a means to begin to answer some of the questions that recipe compilations present to the historian. It could help to understand the materials and techniques in this manuscript, so difficult to draw out by reading alone, and whether these recipes were intended to be efficacious, that is, was the author-practitioner actually engaged in working through the recipes, and thus could the manuscript itself be read as a record of practice? We hoped that grappling with the recipes might allow a glimpse into the mental world of the practitioner to gain a stronger sense of a Renaissance understanding of materials and of the ability to transform them. Perhaps it might also be possible to gain a clue to the possible motivations for writing down technique. Finally, whatever the insights into the manuscript, this mode of research would help think through the methodological implications of using reconstruction as a source of evidence for historians.

While techniques of reconstruction are familiar to museum scholars, conservators, and archaeologists, they are less well known, and less interesting, to general historians and many art historians. My own apprenticeship began with classes on historical techniques of painting and metal working. After an all too brief apprenticeship in courses on historical techniques, I began working with Tonny Beentjes, a practising silversmith and Programme Leader of Metals Conservation at the University of Amsterdam, in order to reconstruct techniques of casting from life by trying the very detailed instructions for life-casting techniques in Ms. Fr. 640. In conjunction with the conservation staff, including Joosje van Bennekom, of the Rijksmuseum, Amsterdam, we were able to compare the instructions in the

French manuscript with the objects of the master goldsmith of Nuremberg, Wenzel Jamnitzer (1510–85), famed for his remarkable life-casting, including his metre tall table centrepiece, held by the Rijksmuseum (*plate* 2).[17]

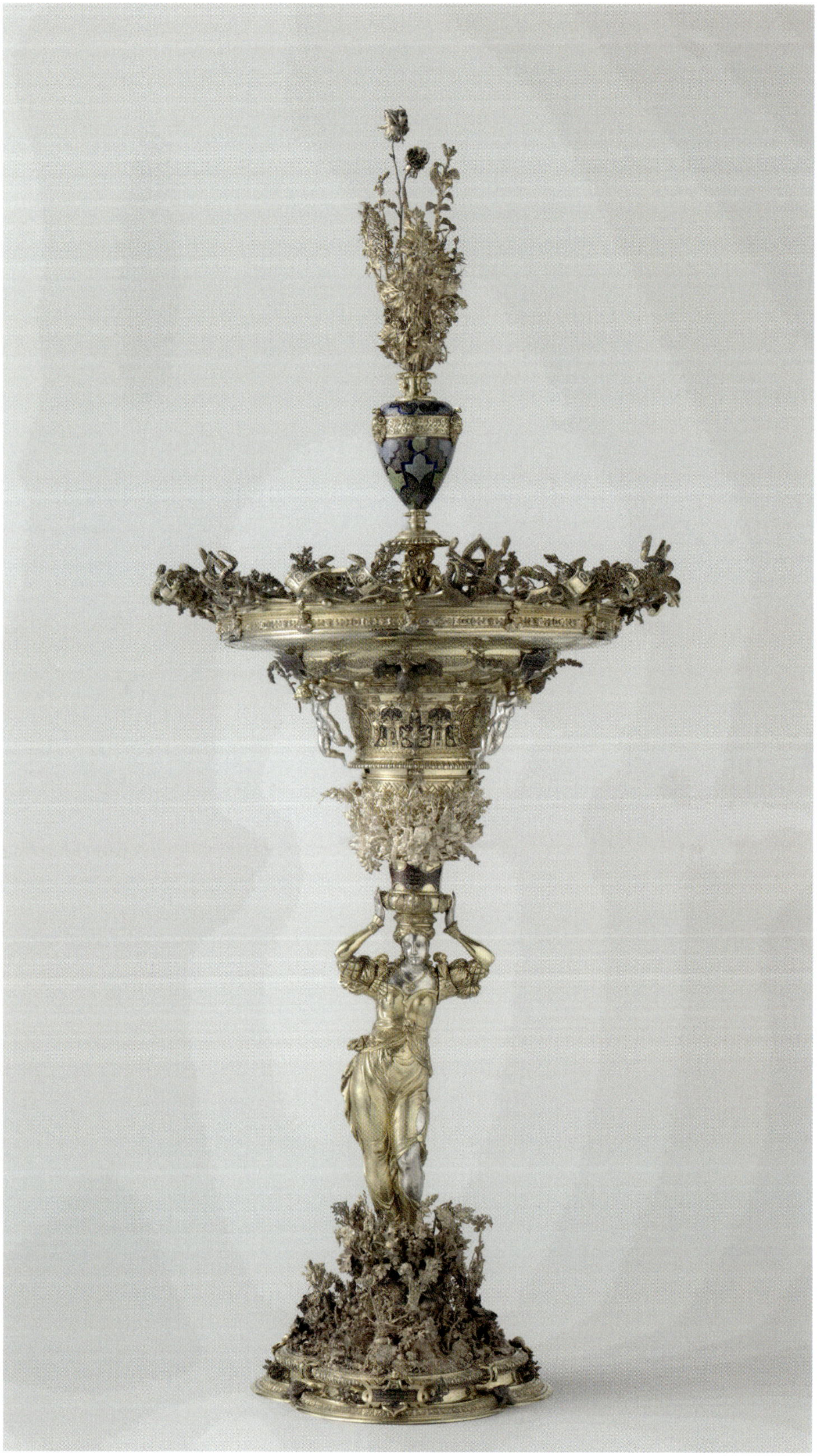

2 Wenzel Jamnitzer, *Table Centerpiece*, 1549. Silver, cast, chased, and stamped, etched, gilded, and painted, height 100 cm. Rijksmuseum: Amsterdam. Photo: Rijksmuseum. [Encrusted with small animals and plants cast from life, this centerpiece is a paean to Terra Mater and her fertility. Plants and small reptiles cluster at the base. Mother Earth supports a basin, with life-cast snakes and lizards around the rim. The floor of the basin is decorated with cornucopias and moresques, while the egg-shaped vase holds plants and flowers, also cast from life.]

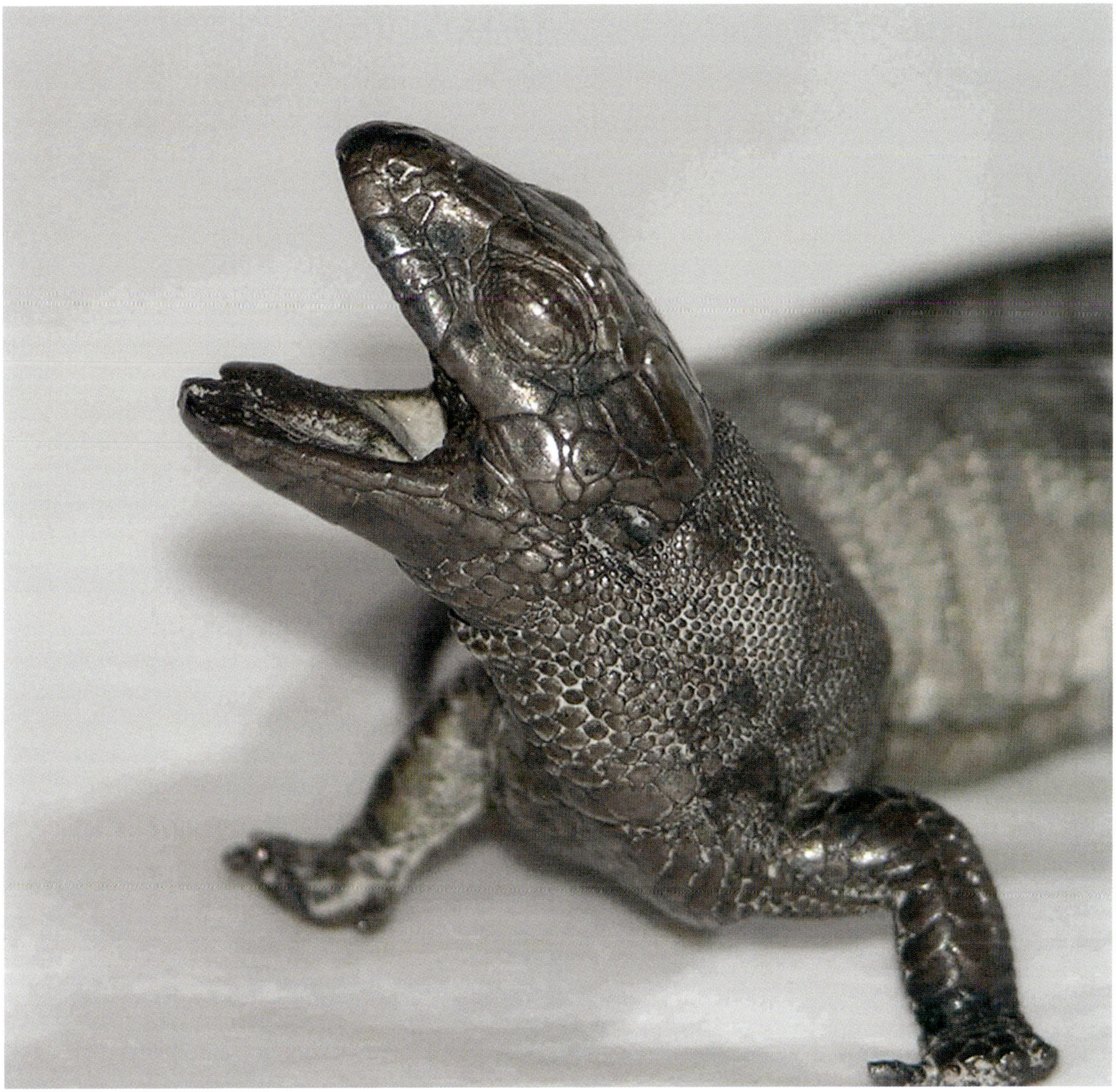

3 Circle of Wenzel Jamnitzer, *Lizard Cast from Life*, c. 1540–50. Silver, 7 × 4.1 cm. Nuremburg: Germanisches Nationalmuseum. Photo: Pamela H. Smith.

The stunningly lifelike objects produced by Renaissance life-casting techniques encouraged conversations in early modern collections about the interplay of human art and nature, and about the multiple valences possessed by the creatures used for life-casting: these animals were often inhabitants of more than one elemental zone, including frogs, toads, salamanders, lizards, crabs, crayfish and some snakes, which inhabited both earth and water; insects and birds were denizens of earth, air, and, in the case of insects, were seen to belong to fire as well, as I have discussed elsewhere.[18] The first stage of our investigation was to examine life-cast objects in museums for the traces of techniques used in Ms. Fr. 640, and then to reconstruct the techniques of the practitioner-author of our manuscript. Where he engaged in a laborious process of translating his making and doing into words and writing, we reverse engineered his words into processes and products. This reverse engineering necessitated conventional textual research, object-based research, and the hands-on research of reconstruction. We started by reconstructing the life-casting of lizards (*plate* 3). We did not, however, re-enact the catching, feeding, and killing of these animals described

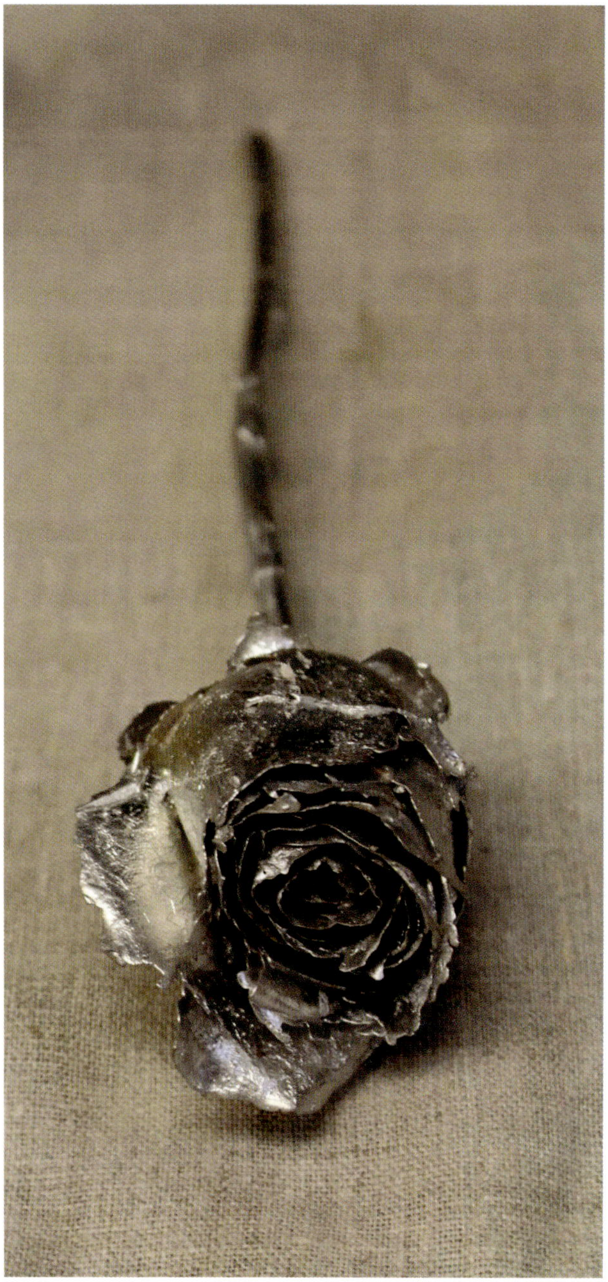

4 The Making and Knowing Laboratory, life-cast rose, the result of experiments by Giulia Chiostrini and Jeffrey Palframan, 2015. Photo: The Making and Knowing Team, Columbia University.

in vivid detail in the manuscript, such as the technique of catching a lizard by suspending a net with a slip knot from a stick, then whistling while dropping it around the lizard's head – a method that herpetologists still use today.[19] Our first lizard reconstruction was quite flawed, but with experience, our castings improved (*plate 4*), and we were able to draw out much detailed technical information from the manuscript about materials and procedures.[20] Our reconstructions gave many kinds of insight into the manuscript: one of the most basic – and most important – for beginning to understand the intentions of the manuscript was that we gained certainty that the manuscript did indeed accurately describe the techniques and materials by which life-casting was carried out in the sixteenth century. It clarified a whole series of techniques, and demonstrated that various (and sometimes implausible) materials could be employed to produce the intended end products of the recipes. The insights I as a text-centric scholar gained into the nature of experience and experiential knowledge, although perhaps familiar to craftspeople and artists even today, were invaluable and would not have been possible simply by reading the manuscript. Straightforward reading of the manuscript was in any case not really possible because it is not a linear text, nor, despite its unusual detail, does it usually provide enough information for a reader to follow the procedures as one does in most modern cookbooks. As I have noted before,[21] understanding the text cannot actually be separated from trying the methods recorded in it, for the manuscript's author-practitioner set down his text at various intervals in a fair copy, followed by more experimentation, which gave rise to failures and new ideas, followed by further observations and trials, all recorded in increasingly cramped script in the margins, and, indeed, in any available space in the manuscript (sometimes, but not always, indicated by insertion marks). It appears that the *composition* of the manuscript itself could not be divorced from actually performing the actions. The text enacts the trying again and again of the author-practitioner, and the essential need to proceed in gaining an understanding of materials and processes by what we would call the method of 'experimentation'. The recipes in the manuscript necessitate imitation and re-enactment in order to be comprehensible. Indeed, it became clear to us that 'reading' the text for understanding in fact meant reconstructing the actions described in it.

Recipes and Emergent Knowledge

These experiences led me to conclude that the modest form of a technical recipe actually forms an efficient vehicle for conveying the nature of practical knowledge in the Renaissance. By their usually anonymous form, recipes assert the collective and cumulative nature of practical knowledge-making.[22] Like other written forms of practical knowledge, recipes can point to bodily action, but obviously cannot accomplish that action, nor provide for every circumstance which might bring about necessary variations in procedure. Especially in the early modern period, recipes were often listed in series, setting out several slightly varying instructions for the same process or object, often simply labelled 'another way'. The recipes thus formed an invitation to action, and, by their serial format, replicated the necessity of repeated hands-on testing and trying in order to work through the variables and resistances of materials to bring about a successful (and replicable) result. Practical knowledge is very different from propositional knowledge, which is easy to capture in a written proof. Practical knowledge always involves emergent phenomena, and its 'proof' is brought into being in real time. A recipe, or better still, a compilation of recipes, thus indicates, in abbreviated form, the particularity, variability, and the emergent quality of material things, and of practical knowledge. Recipes, then, can effectively capture in written form – to the extent that this is possible in *writing* – this characteristic of emergence, as well as setting out a pathway for the acquisition of skill by means of which the emergent phenomena can be channelled and harnessed. Only by engaging in the practice of reconstruction – itself a form of emergent knowledge – would I have recognized this dimension of recipe compilations.

The emergent quality of reconstruction poses a problem for its status as evidence to be employed by the historian – how can a historian bring into being her own evidence and be certain of its validity? This methodological question led me to found 'The Making and Knowing Project' at Columbia University, both to examine this question by collective work, and to produce a digital critical edition and translation of Ms. Fr. 640.[23] For some time, it had seemed to me that this fascinating manuscript merited a critical edition, but I was always stymied by the inadequacy of the printed words of a conventional codex to convey the manuscript's rich contents, not to mention the multiple reconstructions that a critical edition would necessitate. The Making and Knowing Project constitutes an experiment in collaboration, crowd-sourcing, and the integration of pedagogy and original research, and involves teams of students and collective groups of participants Although it has operated for only a year, it has reaffirmed the precept that historians whose object of study is historical materials and techniques, including historians of material culture, of art, and of science in the pre-industrial period, greatly benefit from a type of literacy not generally regarded as a necessary part of their intellectual toolbox, namely, a literacy of materials and techniques. The only efficient – perhaps the sole – means of beginning to acquire this literacy is through hands-on work with materials and techniques. The reconstruction of historical techniques and objects can *begin* to provide such a training.

The Making and Knowing Project

The Making and Knowing project involves teams of graduate student paleographers and researchers who work throughout the course of a semester to reconstruct selected recipes from the manuscript. In 2014–15, reconstructions focused on the mouldmaking and metalworking recipes in Ms. Fr. 640;[24] in 2015–16 they will focus on colour-making recipes (including dyes, pigments, varnishes, techniques of colouring woods, metals, and making imitation stones). In order to reconstruct the recipes, conventional text, document, and object-based historical research is combined in order to begin to understand terminology and process, gain a sense of the object that is the goal of the recipe, and try to tease out a 'genealogy' for the recipe in order to make clear whether the author-practitioner was drawing upon published (or manuscript) texts that might have been available to him. The entire project strives to engender reflection on the methodology and epistemology of reconstruction – how can we responsibly make use of reconstruction as a historical source?[25]

Reconstruction involves both subjective action, self-reporting, and the manufacture of evidence by the historian in the present, a process rife with pitfalls, thus the status and strength of its products as historical evidence must always be at the forefront of the reconstructor's consciousness. It may be useful to think of this procedure as 'historical experiments', drawing in part on the methods of the natural sciences, as Hasok Chang suggests.[26]

Reconstructions in 2014–15 included using the pith of bread ('*mie de pain*') as a mould in which to test a plaster or metal moulding patterns. Various mixtures of beeswax, tallow, and sulfur were then cast into the bread moulds (*plate* 5). In preparing and baking the bread for the moulds, we encountered the changes induced by fermentation – an important component of attitudes to nature and the body in the early modern period – as well as the changes in state that can be brought about by the mixing and tempering of the various casting materials. These exercises also provided insight into the mechanics of moulding in one- and two-piece moulds, training that continued in moulding and casting in cuttlefish bone, and, after that, in producing 'sands' and 'binders' to carry out sand casting for box moulding, and life-casting in plaster (sand and plaster casting are both regarded as varieties of 'sand' casting by the author-practitioner of Ms. Fr. 640). These reconstructions have yielded much information about the manuscript, which includes deciphering marginalia; clues to the author-practitioner's identity, as well as his level of education, knowledge and practice; his familiarity with techniques outside of his French-speaking domain; evidence of his learning new techniques or experiencing failures in his practice.[27] They have also provided insight into specific obscure techniques and materials used in the sixteenth century, such as the so-called 'incuse reverse casting' (or as the author terms it, 'Moulding a hollow on one side, and a relief on the other' ['Mouler cave d'un costé et de relief de laultre'])[28] (*plate* 6) and the material referred to as 'spalt',

5 The Making and Knowing Laboratory, bread moulding for the rapid testing of patterns, following Ms. Fr. 640, ff. 140v and 156r. Photos: The Making and Knowing Team, Columbia University. [Note detail of bottle top cast in wax.]

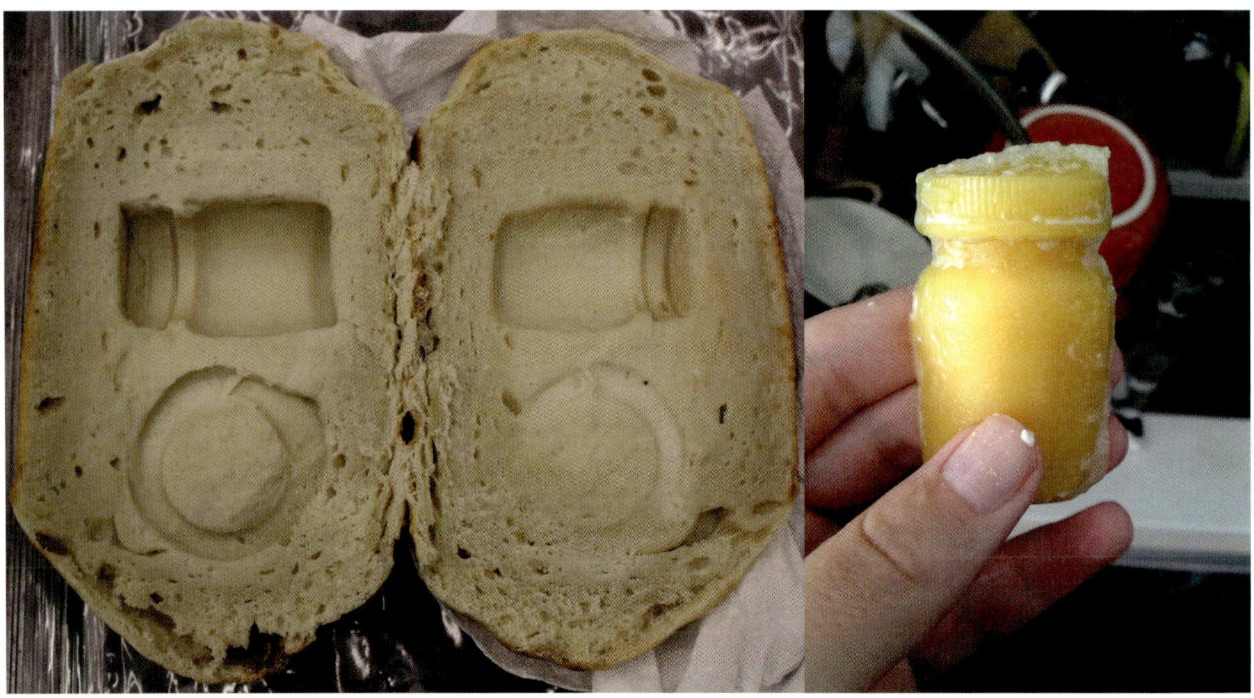

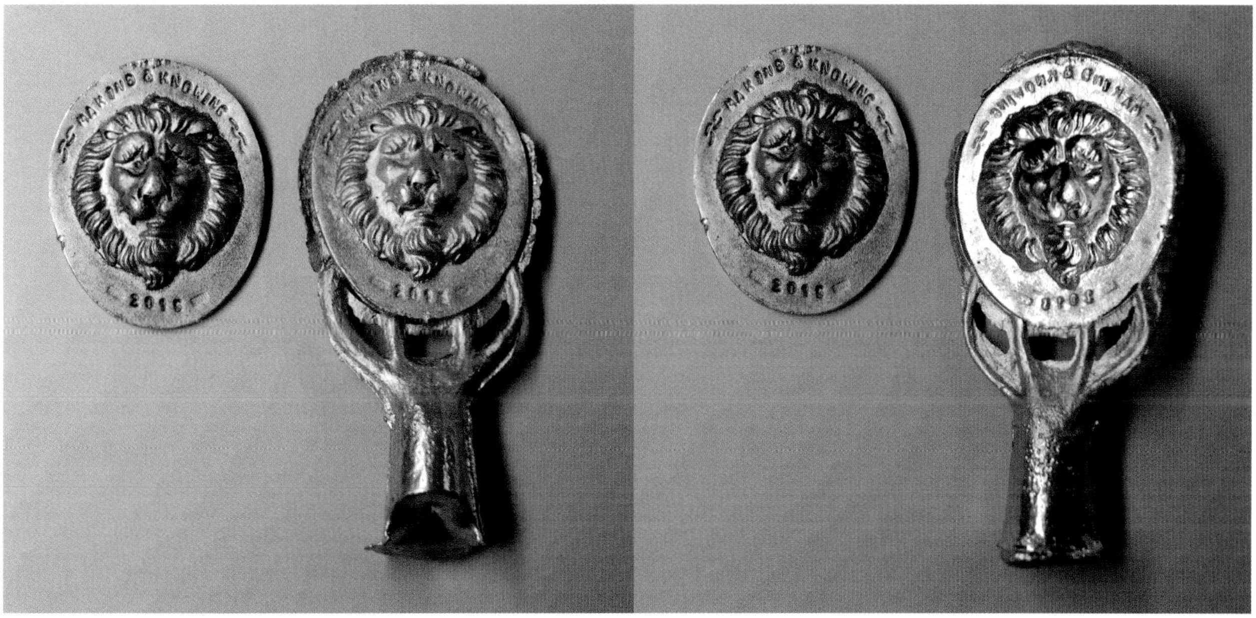

6 Medal pattern and convex side of medal (with casting infrastructure still attached) [left]; and pattern and concave side of medal [right]; following Ms. Fr. 640, f. 92r ('Moulding a hollow on one side and on the other a relief' [known today as 'incuse reverse casting'], the result of experiments by Michaela Groeneveld and Marianne Nuij, University of Amsterdam. Photos: Tamar Davidowitz, The Making and Knowing Project, University of Amsterdam.

'laspalt', and 'spat' by the author[29]; as well as about the nature of experiential or practical knowledge (as revealed, for example, in his use of particular terminology, such as 'impalpable', or 'attack/attach' [*s'attaque*]).[30] The reconstructions have also given insight into daily life, such as those that dealt with mustard; the use of noisy mortars by apothecaries to attract business; a method to train a dog (involving cheese held in the armpit); millet consumption in Languedoc; and the surprising domestic production of sand for hourglasses.[31] Most interesting, perhaps, has been the insight provided by reconstruction into the author-practitioner's system of knowledge about nature and the behaviour of natural materials – his 'material imaginary'.

The digital publication of the critical edition itself – still in preparation – will imitate the reading experience of a codex, but with the ability to compare the four versions of the manuscript, consisting of the high-definition scans of the manuscript, the diplomatic transcription, the (lightly) normalized French transcription, and the English translation formatted as scrollable fluid text.[32] As a reader progresses through the digital edition, they will be able to open 'windows' in the text into the illustrated annotations, containing additional photographs, videos, short essays, glossary entries, and longer essays that set out provenance, context and meaning of the manuscript.[33]

The Material Imaginary of BnF Ms. Fr. 640

Many of our reconstructions of the manuscript's recipes have explicated the author-practitioner's system of natural knowledge, his 'material imaginary', and the remainder of this essay will focus on this dimension of the research. An example of this knowledge system can be found in his instructions to use rusted red iron filings as an ingredient in mould material, in order to make the mould strong. In oxidizing the iron filings, he strives for the deepest red possible and likens the colour to vermilion and cinnabar (fol. 161v). At another point, he mentions that the mould is stronger when it retains the colour of the brick dust and is red (fol. 107v). Of course, in addition to tinging the plaster mould red, the brick dust would also make the plaster mould better able to withstand the burnout and casting process.[34] By this, he might have referred to the 'strength' of blood-red colour. Blood was implicated in metalworking both metaphorically, as in the *Perseus* of Benvenuto Cellini and *Hercules*

Pomarius of Adrian de Vries, and materially in metalworking recipes, where all kinds of red substances appear among the ingredients and operations, including coral, the red pigments vermilion and cinnabar, and even blood. Red and blood are also connected to gold in these processes: red ingredients are specified in making gold pigments, and gold is used to produce red colour in glaze and glass. When taken as potable gold, or even when worn on the body, gold was seen to promote health and healing, just as the most balanced and healthful temperament – the sanguine – was associated with the humour of deep red blood in medical theory. As in other metalworking recipes, lizards, too, crop up in Ms. Fr. 640, as a reported method for turning metals a gold colour (fol. 98r). As I have treated elsewhere, red, blood, gold, and lizards were part of a knowledge system that underpinned metalworking practices and techniques, a kind of 'vernacular science' of matter and transformation, articulated by metalworkers in their works of art, their practices and their texts.[35]

Red is also important on the very first leaf of Ms. Fr. 640, where the author-practitioner has headed the page with a suggestive cross (see *plate* 1), by which he might have been following the practice of commonplace and account book writers who often began their text with a cross, under the sign of Jesus. More significantly, this page contains a recipe for imitation coral, a material that was understood both to imitate natural coral, and to explicate and replicate, in its making, the processes of growth and transformation by which coral was understood to be produced (*plate* 7).[36] The colouring ingredient of the imitation coral of Ms. Fr. 640 is vermilion, a pigment used by the painters and scribes of illuminated manuscripts, who used the mark of a cross to signify where the red pigment vermilion should be applied in the text, thereby connecting the pigment to Jesus's spilling of blood on the cross to redeem humankind.[37] Vermilion was employed to portray the blood of Jesus, such as in a Psalter, now in the British

7 The Making and Knowing Laboratory, imitation coral, following Ms. Fr. 640, f. 3r. Photo: The Making and Knowing Team, Columbia University.

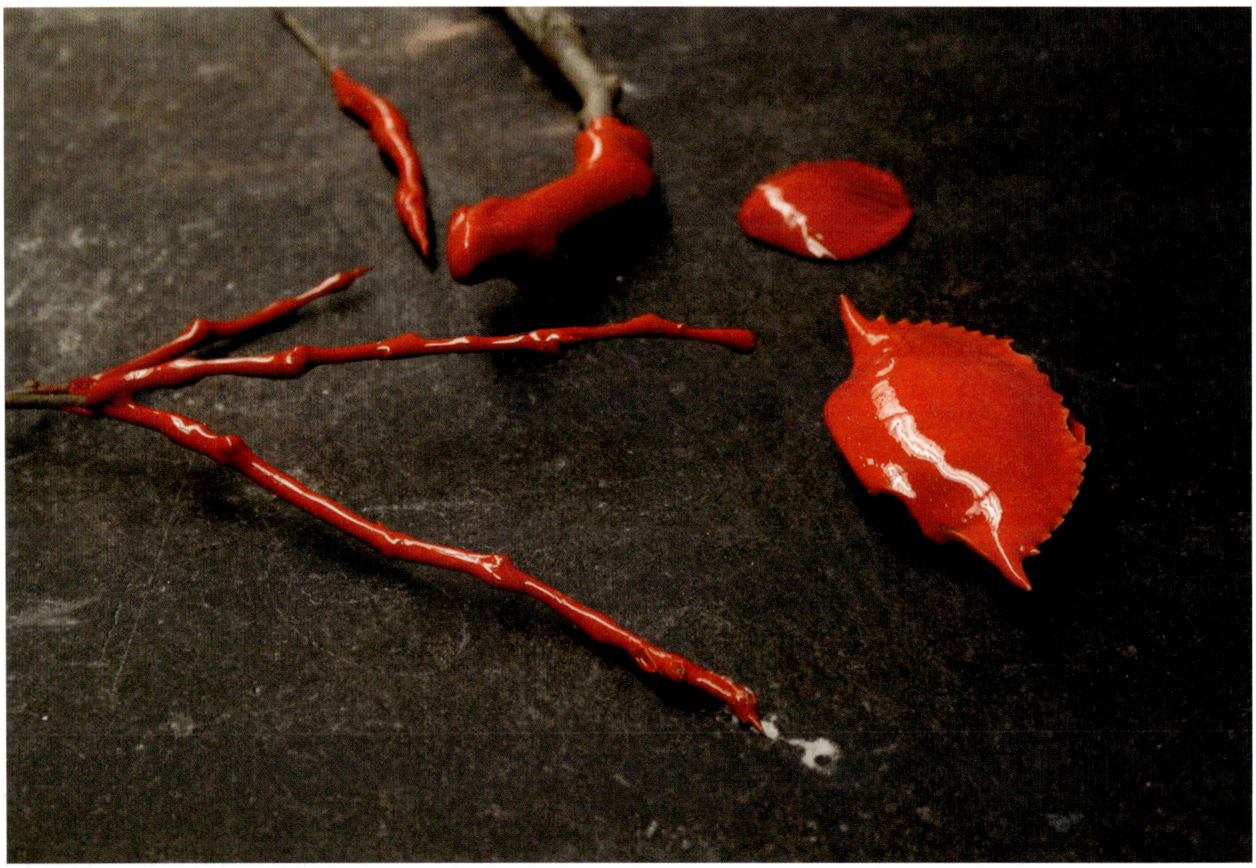

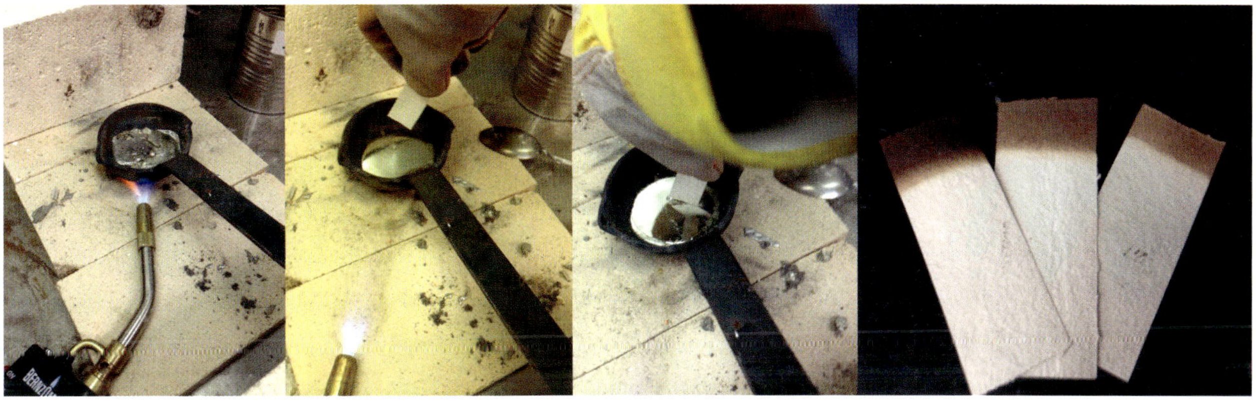

8 The Making and Knowing Laboratory, the paper test. Photos: The Making and Knowing Team, Columbia University. ['To know when it (the 50–50 tin–lead alloy) is the right temperature (for casting into cuttlefish bone), dip a little piece of paper in it. If it turns black without catching fire, it is the right temperature. But if it burns and makes a fire, it is too hot'. The paper employed in this experiment is made from 50–50 hemp and cotton, heavy weight, gelatin sized, and of third quality, used in the repair and conservation of rare books, and supplied by Timothy Barrett, University of Iowa Center for the Book.]

Library, known as Egerton 1821, which opens with drops of Jesus's blood portrayed in vermilion.[38] Such practices of work were oriented to the production of goods, but they also demonstrate that materials, such as iron oxide, vermilion, and even imitation coral, functioned as epistemic things – that is, things that embodied concepts and knowledge systems (but also had real material effects in the world). In other words, craft practices functioned at levels deeper than just the operative.

The Material Imaginary: The Problem of Qualitative Descriptions and Missing Quantities

Ms. Fr. 640 contains very few quantitative measurements. It most often uses ratios (for example, one part lead to two parts tin), as seems to be common more generally in early modern technical writings, as well as other kinds of literature, such as merchant manuals and *abbaco* books that teach practical mathematical techniques for determining variables by working with ratios.[39] In our reconstructions, we soon learned to use other kinds of indicators in the manuscript in place of measured quantities, namely, descriptions of the state of the material in its final form, such as, 'Make your mixture as thin as a potato stock, or as clear as starch water, the one women use to starch' (fol. 113v), or by comparing it to the consistency of mustard (89v, 113v, 121v).[40] Temperature can be measured by feel, for example, a mould only as hot as allows a person to hold it in their hand (115v), or by the combustion point of other materials – straw (72v) – or, 'to know when it is the right temperature [for casting into cuttlefish bone], dip a little piece of paper in it. If it turns black without catching fire, it is the right temperature. But if it burns & makes a fire, it is too hot' (145r) (*plate* 8). In an early modern workshop, materials were not standard, heat was provided by wood that burned with different intensity, and environmental conditions might change with the weather; thus quantitative measures would actually have been *less* useful than qualitative descriptions of the aimed-for consistency, the appearance of the material at certain points of the process, and the testing of heat by feel (among other methods).

Qualities and Properties: Forming a Taxonomy

In general, we have found that the author-practitioner thinks through materials and their properties as he makes decisions about what to 'try' next.[41] He appears to hypothesize about the behaviour of materials on the basis of the properties they exhibit, or the processes through which he puts them, such as calcining (burning to a powdery ash), for example, to produce very fine, white 'impalpable sand' made from oyster shells, bovine bones, wheat flour, and alabaster. He makes generalizations about the properties of materials, based on their appearance, sensory look and feel (as well as smell, taste, and sometimes sound), but most often on their usefulness for his practice. In the process, he forms a taxonomy that cuts across and organizes materials and processes, performing what, arguably, all craftspeople did as they scrounged their local environs for materials they hypothesized might be useful.[42] Early modern things were often organized according to their uses, rather than to their morphology, structure, or appearance. The contrast to today's largely morphological taxonomic classification system was brought home to us as we searched for elm roots in spring 2015 for a binder made with elm roots. Ms. Fr. 640 specifies on fol. 87v that 'Founders harvest the roots of a young elm when it is sappy, and boil it in wine, or better yet vinegar. They prepare a year's worth of it and store it in a cask.'[43] We tried this binder recipe in the fall with the easily available (amazon.com) slippery elm bark, well known for its mucilaginous properties, in spite of slippery elm being native to North America. When the ground thawed and the sap began to flow, we hunted for an authentic sixteenth-century French elm through a cross country email exchange with historical gardens' curators, learning in the process about the nostalgia for the European elm, killed off by elm disease. Finally, we decided to use a hybridized elm species that according to our informants would have been closest to the sixteenth-century elm.[44] When I asked whether this species, considered taxonomically similar today, would necessarily share the properties sought out by a sixteenth-century practitioner, or did taxonomic systems today rather operate by a different set of categories, and got the answer that in hybridization, the property of 'gooey sap from the roots would be one of the last things they were looking for',[45] I came to realize that our taxonomies for classifying species based on morphological properties determined by the visual appearance of leaf, bark, and flower, differed profoundly from the taxonomy of our author-practitioner. His system of classification was based on the properties of the material as they were expressed or manifested in the process of human manipulation, a system also found in early modern herbals, which organize plants according to their 'virtues', or properties useful to humans for healing and other processes (sometimes in an order that proceeds from 'head to toe').

States and Phases of Materials

Another dimension of the author-practitioner's material imaginary came into focus when we investigated his intensive experimentation with different types of sand ('*sable*', by which he meant the dry components of a mould) and binders (the binding medium, sometimes called a *magistra*, a term also used by other craftspeople and alchemical writers). In the metalworking recipes of the manuscript, sands and binders are employed to produce box moulding 'sand', or powder, which will take a fine impression, is cohesive enough to allow the opening of the mould, but releases the pattern easily, and is strong enough to withstand the entry of the metal into the mould. In order to create this ideal moulding matrix, dry and wet materials of various kinds are mixed carefully together to form a sand that gives 'a nice hold', but still comes 'apart easily',[46] what the lab came to refer to as the 'squeeze test' (*plate* 9).[47] In creating both the sand and the binding medium, the author-practitioner is preoccupied with transforming materials from one state to another: hard brittle oyster shells, for example, into fine powdery 'impalpable' sand.[48] Indeed, throughout the text, he is preoccupied with the native properties of materials – liquid, congealed, solid, vaporous, hard, soft, malleable, brittle, among others – and their transformation. This focus is in line with many recipes in early modern metalworking writings that attempt to 'fix' volatile materials such as mercury to make them endure the fire and become malleable or workable, and conversely to increase the plasticity of solid materials and make them capable of being moulded.

Recipes for softening hard substances, such as liquefying stone or horn in order to be able to cast them, and the related process of making brittle materials malleable, as in the transformation of iron into steel, are to be found in large numbers in such recipe collections.[49] This emphasis on changes of state indicates the nature of many of these recipes – they aimed at varying the qualities of metals by putting them through phase processes, sometimes by slow, constant, and long-term heating, such as that of an egg sitting under a broody hen, or by the heat that is given out by the slow fermentation of manure, caused by thermophilic bacteria that maintain a constant temperature. Change of state could also be brought about by dissolution in acid, or transformation through alternating heating and quenching, as in the production of steel.

The author-practitioner of Ms. Fr. 640 also induced changes of state by combining different substances to form a material with new properties, such as mixing melted wax with tallow, which produced a material more capable of being carved,[50] or mixing melted wax with different pigments, or, in a rather complicated process, passing sulfur through melted wax to bring into being a material, which, when cast, produced very smooth, milky, homogeneous surfaces, containing fine detail, with all the fine definition of sulfur casts, but less brittle, and all the malleability of wax, but without its translucency (which makes it difficult to see detail and therefore more challenging to carve fine detail).[51]

This deep preoccupation with transformations of state is developed by the author-practitioner in a series of recipes that might be called implausible, or paradoxical binders – paradoxical because, instead of mixing a dry sand and a wet binder to produce a suitable casting matrix, he used two dry materials, namely ox-bone or alabaster and rock salt, ground together to form a dry powder. This dry

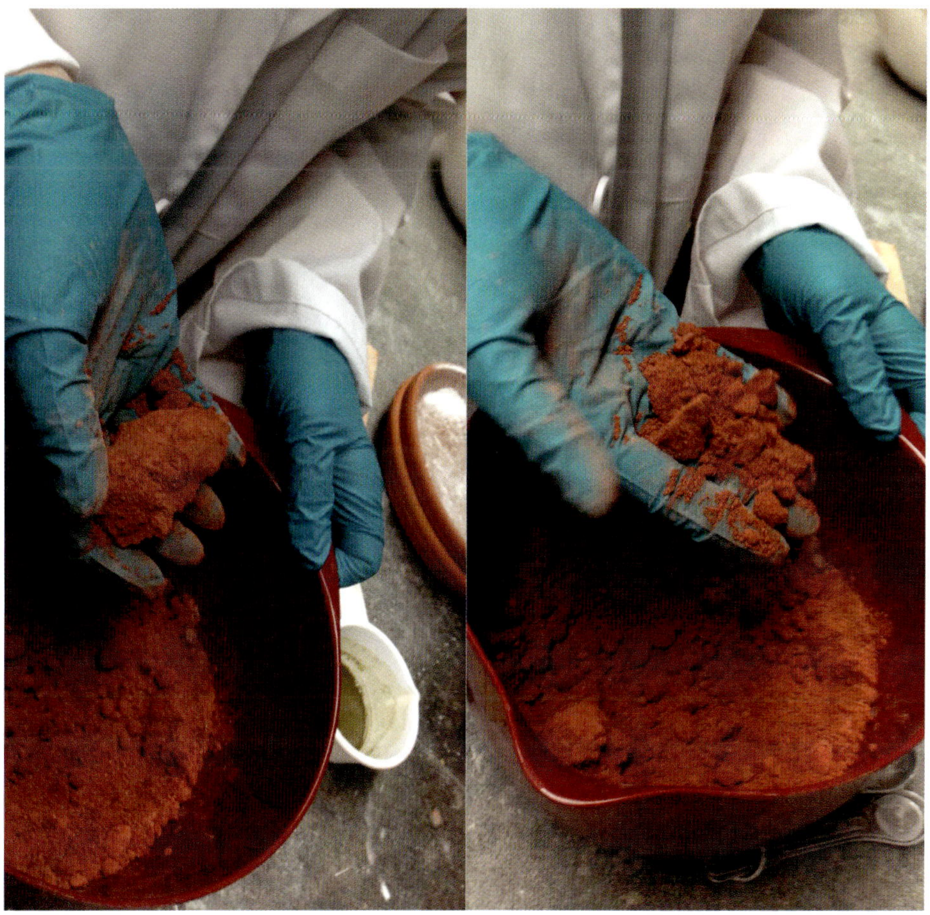

9 The Making and Knowing Laboratory, the 'squeeze test' for moulding with sand in a frame, following Ms. Fr. 640, f. 118v. Photos: Emogene Cataldo and Julianna van Visco. [Moulding sand contains previously used ground moulds (made from plaster and ground brick) held together with egg-white binder (f. 82r).]

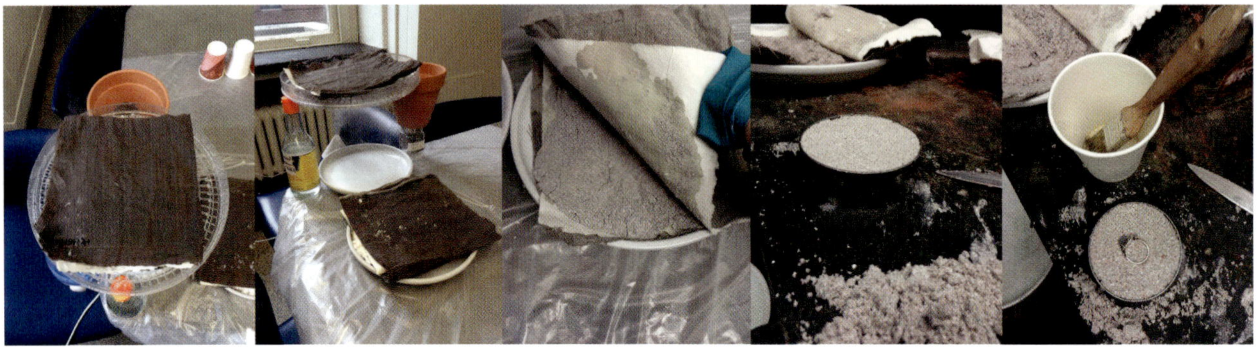

10 The Making and Knowing Laboratory, paradoxical binders: experiments with bovine bone and rock salt, after the damp of one night (in reality, an improvised humidifier), following Ms. Fr. 640, f. 89r. Photos: Michele Goun Lee, Diana Mellon, and Yijun Wang.

crumbly mixture appears entirely implausible as a mould material, until it is left in the 'moisture of the night', or of a cellar, or other damp place, which causes long crystals to form in the salt, by which the unlikely powder is transformed into a very effective mould material that takes detail well and is extremely durable (*plate 10*).[52]

Intimacy with Materials

These paradoxical binder recipes gave us insight into the author-practitioner's bodily intimacy with materials, and the ways in which it shaped the framework of understanding that underpinned and gave meaning to his practices. The body functioned as a tool in the author-practitioner's workshop. Bodily fluids – including urine, blood, bones, saliva, phlegm, among others – were an integral part of his practice, as were foods that also sustained human growth – figs, butter, eggs, garlic, bread. Measurements were made in proportions of the human body – as much as could be held in two hands, 'two-fingers high' (161v), 'as tall as a man' (168v), a 'handspan' (16r). But the body was more than a tool in production, for the body and the materials of work interpenetrated each other, as when the metalworker inhaled the metals' cold vapours.[53] Both humans and metals had to be tempered and brought into balance by the ingesting of certain substances to counteract undesirable states, such as excess coldness or heat.

The polarities of Aristotelian qualities – hot, dry, wet, and cold – formed a fundamental structuring framework for early modern technical processes and the knowledge that emerged from them, but our study of Ms. Fr. 640 and our examination of other technical writings showed that another set of qualities also informed craft practice: we encountered a persistent paradigm of binary qualities, including hard–soft, brittle–malleable, sour–sweet/soft (*doux*), and fat–lean. The fat–lean binary is common in early modern metalworking and mining texts (and is reportedly still in use by German founders). It appears that the concept of a generative 'fat' emerged out of the practice of agriculture, in which rich, unctuous soils produced bountiful crops, as well as from foundries in which investment and mould materials were made from a buttery unctuous clay produced by long fermentation of clay and wool (as Benvenuto Cellini tells us in the *Two Treatises*). Fat earths were central in mine prospecting, 'stone marrow' being sought in mines; in the seventeenth century, an ambitious chymical practitioner, Johan Rudolf Glauber, and a Leiden professor, Herman Boerhaave, identified fatty earth with the substance out of which metals could be generated, and it remained an object of investigation into the nineteenth century. For seventeenth-century (al)chemical theorists, 'Fatty earth' (*terra pinguis*) would come to be seen as a foundational and transformative element in many material processes, even of interest to Isaac Newton, and informing the concept of 'phlogiston' and then 'oxygen'.[54]

In our experiments in the Making and Knowing laboratory, we came to understand the meaning of fatty sand in sensory terms. In reconstructing paradoxical binders, we struggled with the concepts of fat and lean, as they did not make sense within a modern understanding of salts, much less of fat. The meaning of these terms began to dawn with the grinding and storage of rock salt for the 'Eau Magistra' of fol. 84r. When the researchers returned to the lab a few days later, they found the texture of the salt had become 'sticky'. As they recorded, 'the salt had transformed from a fine, dry powder that poured freely from an outstretched hand to a sticky substance, in which the particles clearly adhered to each other, creating an unexpected sensation of resistance when running one's fingers along the surface. The well-known material of salt unexpectedly possessed a completely different set of properties, and we *felt* the "fattiness" of the rock salt, which suddenly afforded a new workability. This allowed us to apprehend the place of "fatty" in the "science" of the author-practitioner, a rather different perspective than is given by our modern classification of salt as sodium chloride. We thus concluded that these terms, fat and lean, rely very directly upon sensory interaction with materials. Our abstracted modern understanding of this material through its chemical composition differs from the early modern practitioner's, which was a deeper, more intimate knowing through the bodily senses. This intimacy between practitioner and material was close, and sensory testing of materials and ingesting food for medicinal ends were not separated by a great distance – both involved the same types of substances and qualities, including cold, hot, wet, dry, fat, and lean, and both contributed to the intimate connection between practitioner and materials that helped reinforce an understanding of materials in the language of qualities and properties that could be apprehended by the senses.'[55]

In *The Nature and Art of Workmanship*, David Pye regards the effort to transform materials in sculpture, and the representation of that transformation, as part of an aesthetic system: 'perhaps the most constant and delightful aesthetic phenomenon throughout the history of sculpture has been this very expression in hard stone of the properties of soft materials like flesh, hair, and drapery. The stone remains recognizably stone yet the hair is recognizable as hair and the cloth as cloth.' According to Pye, this effort constitutes the essence of skill, or, as he terms it, 'workmanship', as the craftsperson strives to produce objects that display diversity, durability, and equivocality (by which he means a metal showing both its properties of liquid and solid, for example, in a sculpture).[56] To Pye's insight, we can add that the craftsperson also strives to imitate and even compete with the artifice of nature in such an effort. Such imitation and rivalry possessed a long and important genealogy in European craft and art, of course, embodied above all by the enduring employment of Ovid's *Metamorphoses* in discussing and representing natural and artistic process. But, more significantly, we should also see this fascination with transformation as a fundamental part of responding to the challenges that arise in working with the materials of nature, and the exploring, testing, trying by which a practitioner comes to understand the properties and behaviour of these materials. The polarities in metalworking recipes and the efforts to overcome and transform them, then, articulated a mode of work, as well as a framework for working and thinking through the properties of materials and the processes and products they made possible.

Mode of Work: Practice as Investigation of and Experimentation with Materials

This mode of work emerges from human interaction with natural materials – arguably an interaction that has gone on since the dawn of humanity – and can be understood as an essential component of artisanal practice. It was understood as such by sixteenth-century practitioners, who expressed the need for repeated trials of materials.[57] In Ms. Fr. 640, this mode of work is expressed in material terms, for example by the pervasive use of sulfur. In alchemy, sulfur along with mercury is one of the fundamental principles of all metals. But in Ms. Fr. 640, research determined that 'sulfur is used quite practically, as a casting material because its capacity for accepting detail is excellent, and when its deficiencies as a material (its color, luster, or brittleness) make it inappropriate for certain tasks, the author suggests augmenting it with other materials. That sulfur is used in so many other ways, including as an ingredient in pigments and in a variety of other recipes, attests not only to the material's availability and versatility, but perhaps more significantly, its *perceived* versatility. For all of the different uses documented, one can infer that others were tried but did not produce a desired result. The prevalence of sulfur in recipes ranging from coloring to casting to using it to 'boil an egg in cold water without fire' (fol. 35r) provides evidence that the material was both commonly used and tried in a breadth of applications. Even in its uses in processes of casting explored in our reconstructions, one of its purposes appears to have been to experiment with the level of detail one can achieve with a carved pattern or mold.[58] It would seem, then, that, in its ubiquity in trying and testing in this manuscript, sulfur both connoted the idea of experimentation and, as a versatile material, invited further actual trials. Therefore it may not be too speculative to suggest that sulfur metonymically represents the process of trying and assaying, or in other words, of "experimenting" itself.'[59]

11 Wenzel Jamnitzer, *Nautilus Ewer*, c. 1570. Gilded silver, partially enamelled, height 32.5 cm. Munich: Schatzkammer der Residenz. Photo: Bayerische Verwaltung der staatlichen Schlösser, Gärten und Seen.

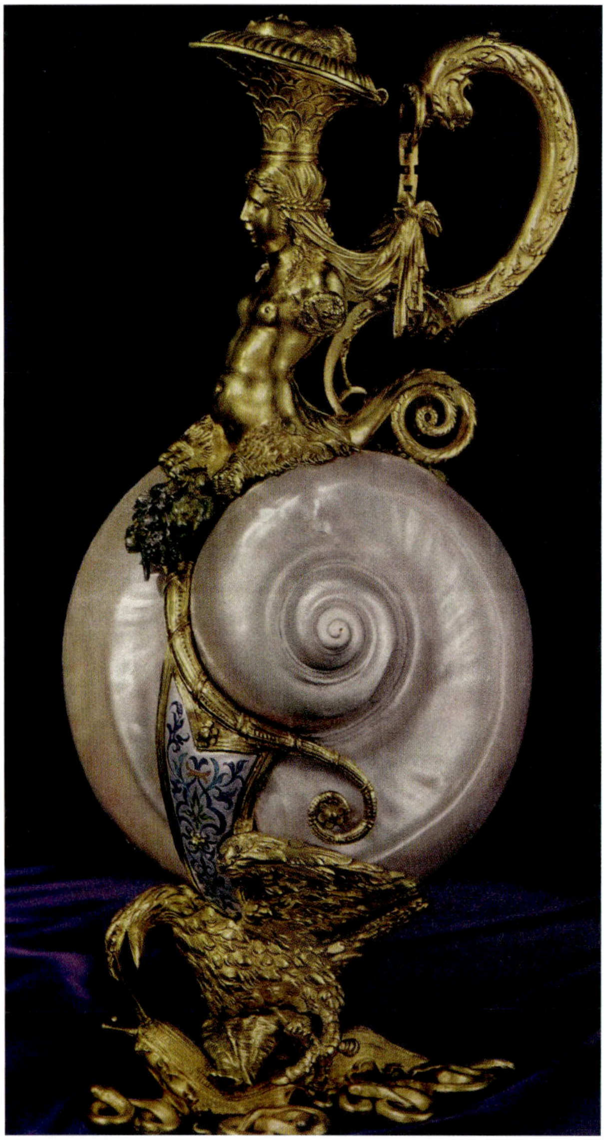

Conclusion

The research carried out by the Making and Knowing project demonstrates that Ms. Fr. 640, and perhaps other early modern recipe books that recount the working of materials and the production of objects, can reveal systems of belief and knowledge, as well as testify to the material, physical, and philosophical engagement of craft with the generative and transformative powers of nature. Ms. Fr. 640 demonstrates that practice (and craft knowledge more generally) was not just productive, but also investigative and 'philosophical'. The author-practitioner's investigation extended seamlessly from exploration of the properties of oyster shells, egg white, metals, sulfur, resin, oils, and all manner of other materials to natural historical observation and experiment. We can see this intersection of craft practice – the moulding, casting, experimenting – with natural historical observation in Ms. Fr. 640 in one further example: the author-practitioner relates where certain animals can be found, their hibernation and feeding habits – even experimenting on snakes to see if they vomit up the live frogs they swallowed – and notes methods for keeping them alive in captivity, and then finally for killing and casting them.[60] This de facto natural historical investigation appears to be represented in the final life-cast objects themselves. Of course, no objects created by the author-practitioner of Ms. Fr. 640 are known to be extant, but he does include instructions for catching, feeding, killing, and casting snakes, as can be found in many extant nature casts, such as Wenzel Jamnitzer's *Nautilus Vessel*

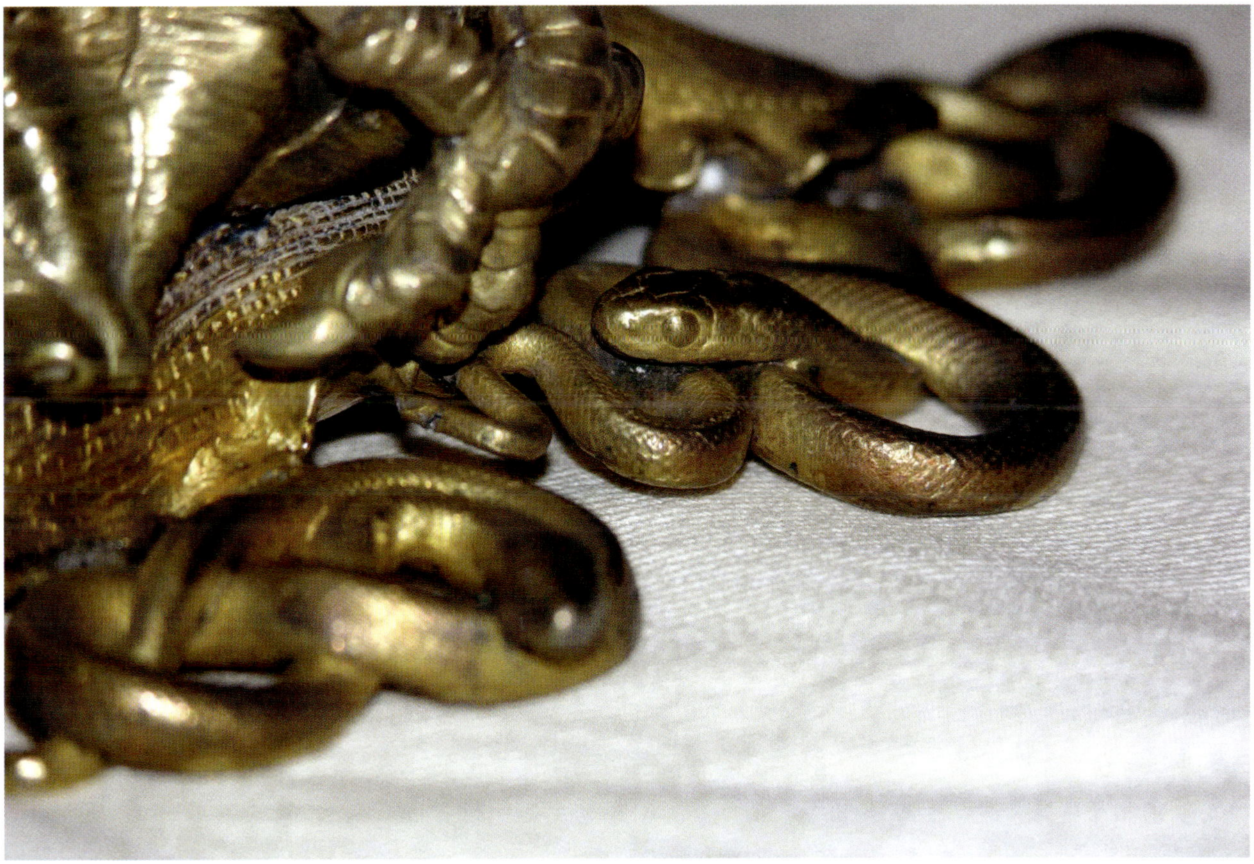

12 Detail of Jamnitzer, *Nautilus Ewer*, showing grass snakes, cast from life. Photo: Pamela H. Smith.

in the Munich Residence Museum (*plate 11*). This fantastical object portrays a mass of common grass snakes at its base, all much the same length and age. I would argue that this object replicates the experience of snake catchers and collectors as they searched out specimens for casting. A film that records the experience of finding a brood nest of snakes conveys this experience vividly (*plate 12*).[61] This codification in metal of practical observation and experience shows clearly the ways in which craft making, such as that articulated in the folios of Ms. Fr. 640, informed and intersected with the emergent modes of scientific thinking and knowing in the sixteenth century.

Over the following centuries, during the period of what is known as 'the Scientific Revolution', the collaboration and experimentation of the craft workshop would go on to be integrated into the practice of the natural sciences. In the course of this integration, the shared origins of art, history, and natural science were obscured. But much can be gained by renewing a dialogue among the arts, history, and natural sciences, and by resuscitating the concept that the investigation of nature and of the human world are deeply entwined, as sixteenth-century natural historians took for granted, and as modern sociologists of science, environmental historians, and climate scientists have recently begun to emphasize.

Appendix

The Making and Knowing Project[62] began in June 2014 with the first of three annual middle French paleography workshops. Working in a collaborative digital space of Google Drive, these workshops teach paleographic skills to about fifteen students already proficient in modern French who then gain practice as they transcribe and translate the manuscript. As they work, they make use of all kinds of reference sources, and their collectively compiled list of dictionaries, texts, and encyclopaedias of early modern techniques and materials is one of the unexpected byproducts of the project. Paleographers annotate the text with comments about sources used and translation choices.

The paleographers' translation forms the basis for a laboratory seminar held each semester, in which students reconstruct some of the recipes in the manuscript – in 2014–15, the courses focused on mouldmaking and metalworking recipes in Ms. Fr. 640;[63] in 2015–16 on colour-making recipes. To reconstruct the recipes, the students must engage in conventional text-, document-, and object-based historical research in order to begin to understand terminology and process, gain a sense of the object that is the goal of the recipe, and try to construct a 'genealogy' for the recipe in order to make clear whether the author-practitioner was drawing upon published (or manuscript) texts. As they work, students take field and laboratory notes, stored on a collective Wiki, in which they record their reconstruction experiments as well as their experience as history students doing hands-on work in a laboratory. In these field notes, as in the entire project, we strive to reflect on the methodology and epistemology of reconstruction – how can we responsibly make use of reconstruction as historical evidence? By the end of each semester, students synthesize their research into annotations on individual recipes in the manuscript which will serve as the final critical commentary of the edition.

The laboratory seminars begin with skill-building exercises, including reconstructing historical culinary recipes (generally not from the manuscript, as it contains very few food recipes), which the class then discusses and (literally) digests. This exercise aims to spur the students into considering the methodological challenges of reconstruction, as well as to formulate a template for a rigorous reconstruction procedure. From culinary recipes, the students moved in 2014–15 to related entries in Ms. Fr. 640, such as bread-moulding. These exercises provided insight into the mechanics of direct and indirect casting as did the subsequent exercises of moulding and casting in cuttlefish bone and producing 'sands' and 'binders' to carry out box moulding. For two weeks in 2014, an 'expert practitioner', Tonny Beentjes, Programme Leader in Metalwork, University of Amsterdam, led the students in sand casting in box moulds, and Andrew Lacey, Sculptor and Independent Scholar, led life-casting in plaster in spring 2015.

For the remaining eight weeks of the semester, students work in groups of two or three on reconstructions of up to three recipes in the manuscript. Aiming for much more than simple accounts of hands-on reconstructions, the students write up their research as an answer to a historical question that their hands-on reconstructions helped to answer. In all cases, the students' annotations must form an integrated text that weaves together in a single narrative both conventional historical research based on documents, texts, and objects with more 'subjective' hands-on historical experiments to make an argument about what research into the recipe reveals. An extensive editing process follows, including two collective revision workshops, and often quite far-reaching revisions by the Making and Knowing Team. At the end of the second semester, a Working Group Meeting is convened, including all students from the two courses, the Making and Knowing team, the two Expert Makers, and a group of scholars, expert in the year's theme, who have had access to the contents of the Wiki and Google Drive throughout the year. This final capstone to the year's research brings both additional expertise into the project, and provides for critical oversight on the student work.[64]

The annotations that formed much of the basis of this article were authored by the Making and Knowing Team (Director Pamela H. Smith; Making and Knowing Postdoctoral Scholars Donna Bilak, Jenny Boulboullé, Joel Klein; Paleography Co-Director Marc H. Smith); students from the 2014–15 Columbia laboratory seminar,

Hist G8906: Craft and Science: Making Objects in the Early Modern World (Emily Boyd, Raymond Carlson, Emogene Cataldo, Giulia Chiostrini, Celia Durkin, Shiye Fu, Sofia Gans, Jordan Katz, Rozemarijn Landsman, Michelle Lee, Caroline Marris, Diana Mellon, Jef Palframan, Stephanie Pope, Jonah Rowen, Julianna Van Visco, Yijun Wang, Zhiqi Zhang); students in the University of Amsterdam MA in conservation and restoration of cultural heritage, metals specialization course (Michaela Groeneveld, Ingeborg Kroon, Elisabeth Kuiper and Marianne Nuij); and a student from the V&A/RCA PhD in History of Design (Maria Alessandra Chessa).

The Making and Knowing Project is collaborating with several institutions, including the Technical Art History Group at the University of Glasgow (Erma Hermens and Mark Richter); Conservation Programmes at the University of Amsterdam (Marjolijn Bol, Tonny Beentjes, Tamar Davidowitz, Ellen van Bork, and Maartje Stols-Witlox); the Victoria & Albert Museum-Royal College of Art Postgraduate Programme in the History of Design (Marta Ajmar); Bard Graduate Center for Decorative Arts, Design History, and Material Culture (Deborah Krohn); the Independent Research Group on Art and Knowledge in Premodern Europe at the Max Planck Institute for the History of Science (MPIWG) (Sven Dupré); the Recipes Project (Elaine Leong) at the MPIWG, and the Colour ConTEXT database compiled by the MPIWG in conjunction with the University of Liège.

A note on authorship and terminology: 'I' refers to Pamela Smith, the principal author of this article, whereas 'we' refers, according to context, to Pamela Smith and Tonny Beentjes, or the Making and Knowing Team and students from Hist G8906 in the 2014–15 academic year.

Notes

The authors gratefully acknowledge the support of Columbia University in the City of New York; the National Science Foundation, Grant #1430843 'The Role of Tacit Knowledge in Experimentation'; the Chemical Heritage Foundation; The Maurice I. Parisier Foundation; The Florence Gould Foundation; and Howard and Natalie Shawn. Thanks also go to Claire Sabel, Miriam Pensack, Tonny Beentjes, Andrew Lacey, Marjolijn Bol, Tamar Davidowitz, Ellen van Bork, Siân Lewis, Marta Ajmar, and the participants in the 2015 Working Group Meeting on Moldmaking and Metalworking.

1 For the development of natural history, see Gianna Pomata and Nancy Siraisi, eds, *Historia: Empiricism and Erudition in Early Modern Europe*, Cambridge, MA, 2005; Brian W. Ogilvie, *The Science of Describing: Natural History in Renaissance Europe*, Chicago, IL, 2006.

2 Pamela H. Smith, *The Body of the Artisan: Art and Experience in the Scientific Revolution*, Chicago, IL, 2004.

3 Benvenuto Cellini, *Trattati dell' Oreficeria e della scultura* (completed by 1565, published 1568), here *Two Treatises on Goldsmithing and Sculpture*, trans. C. R. Ashbee, New York, 1967, 19.

4 Bernard Palissy, *Admirable Discourses on the nature of waters and fountains, either natural or artificial, on metals, salts and salines, on rocks, earths, fire and enamels.... The whole arranged as dialogues, in which are included theory and practice* (1580), trans. Aurele la Rocque, Urbana, IL, 1957.

5 Theophrastus von Hohenheim called Paracelsus, *On the Miners' Sickness and Other Miners' Diseases*, trans. George Rosen, *Four Treatises of Theophrastus von Hohenheim called Paracelsus*, ed. Henry E. Sigerist, Baltimore MD, 1941, 91.

6 For these developments, see Pamela H. Smith, 'Making things: Techniques and books in early modern Europe', in Paula Findlen, ed., *Things*, London, 2013, 173–203, and Pamela H. Smith, 'Craft techniques and how-to books', in Mark Clarke, Bert De Munck and Sven Dupré, eds, *Transmission of Artists' Knowledge*, Koninklijke Vlaamse Academie van Belgie voor Wetenschappen en Kunsten, 2012, 75–84.

7 See http://www.jove.com/.

8 I examine these questions in Pamela H. Smith, *From Lived Experience to the Written Word: Recovering Art and Skill in Early Modern Europe*, in preparation.

9 See for example, William Eamon, *Science and the Secrets of Nature: Books of Secrets in Medieval and Early Modern Culture*, Princeton, NJ, 1994; see the scholarly essays in vol. 3 of Michael of Rhodes, *The Book of Michael of Rhodes: A Fifteenth-Century Maritime Manuscript*, 3 vols, ed. Pamela O. Long, David McGee and Alan M. Stahl, transcription by Franco Rossi, trans. Alan M. Stahl, Cambridge, MA, 2009; and Eleanor Robson, 'Technology in society: Three textual case studies from late bronze age Mesopotamia', in Andrew J. Shortland, ed., *The Social Context of Technological Change: Egypt and the Near East, 1650–1550 BC*, Oxford, 2002, 39–57.

10 Available on Gallica, http://gallica.bnf.fr/ark:/12148/btv1b10500001g/f1.image), with the library title 'Recueil de recettes et secrets concernant l'art du mouleur, de l'artificier et du peintre'.

11 The binding, stamped and gilded with the arms and device of Philippe de Béthune (1565–1649) indicates that the manuscript was part of the bequest of the Béthunes to the royal library in 1662 (registered by Parlement in 1664). Orthography and various dates within the manuscript reveal that it was composed after 1581 and probably before 1600. Watermarks on the endpapers differ from the text block, indicating that the pages were rebound (as does the double set of page numbers). The watermark on the text block is that of the papermaker Nicolas Lebé of Troyes, active c. 1561–96.

12 Study of the *Kunstkammer* began with Julius von Schlosser, *Die Kunst- und Wunderkammern der Spätrenaissance*, 1908; reprint edition: 2 vols, Braunschweig, 1978, and has increased dramatically since the 1980s, especially with publication of Oliver Impey and Arthur Macgregor, eds, *The Origins of Museums: The Cabinet of Curiosities in Sixteenth- and Seventeenth-Century Europe*, Oxford, 1985, and the founding of the *Journal of the History of Collections* in 1989. For a useful introduction, see Thomas DaCosta Kaufmann, 'From mastery of the world to mastery of nature: The *Kunstkammer*, politics, and science', in Thomas DaCosta Kaufmann, *The Mastery of Nature: Aspects of Art, Science, and Humanism in the Renaissance*,

Princeton, NJ, 1993; Martin Kemp, '"Wrought by no artist's hand": The natural, the artificial, the exotic, and the scientific in some artifacts from the Renaissance', in Claire Farago, ed., *Reframing the Renaissance: Visual Culture in Europe and Latin America 1450–1650*, New Haven, CT and London, 1995, 117–96; and Horst Bredekamp, *The Lure of Antiquity and the Cult of the Machine*, trans. Allison Brown, Princeton, NJ, 1995.

13 Most recently on books of secrets, see Elaine Leong and Alisha Rankin, eds, *Secrets and Knowledge in Medicine and Science, 1500–1800*, Aldershot, 2011.

14 Female authorship of BnF Ms. Fr. 640 is not impossible, but appears highly unlikely in light of the orthography, as well as the accounts of travel and the knowledge of other workshops contained in it.

15 Anon., *Der 'Liber illuministarum' aus Kloster Tegernsee*, ed. Anna Bartl, Christoph Krekel, Manfred Lautenschalger and Doris Oltrogge, Stuttgart, 2005; and Anon., *Mediaeval Painters' Materials and Techniques: The Montpellier Liber Diversarum Arcium*, ed. Mark Clarke, London, 2011.

16 Vannoccio Biringuccio, *The Pirotechnia*, first published 1540, trans. C. Stanley Smith and M. Teach Gnudi, New York, 1943.

17 For more on life-casting and BnF Ms. Fr. 640, see Pamela H. Smith and Tonny Beentjes, 'Nature and art, making and knowing: Reconstructing sixteenth-century life-casting techniques', *Renaissance Quarterly*, 63, 2010, 128–79, and for more on our life-casting reconstructions, see Pamela H. Smith, 'In the workshop of history: Making, writing, and meaning', *West 86th: A Journal of Decorative Arts, Design History, and Material Culture*, 19, 2012, 4–31. See most recently, Giulia Chiostrini and Jef Palframan, 'On moulding a rose', http://bit.ly/1hWo6ZI.

18 For a fuller treatment of this, see Pamela H. Smith, 'Between nature and art: Casting from life in sixteenth-century Europe', in Elizabeth Hallam and Tim Ingold, eds, *Making and Growing: Anthropological Studies of Organisms and Artefacts*, Aldershot, 2014, 45–63.

19 Among the many YouTube videos of catching lizards by suspending a slip knot from a stick (as the manuscript advises), see https://www.youtube.com/watch?v=j0pMlAXwuVs.

20 See Smith, 'In the workshop of history'.

21 Smith, 'In the workshop of history'.

22 Francisco Alonso-Almeida, 'Genre conventions in English recipes, 1600–1800', in Michelle DiMeo and Sara Pennell,eds, *Reading and Writing Recipe Books, 1550–1800*, Manchester, 2013, 68–90, calls recipes a 'hive discourse' to indicate this dimension.

23 http://www.makingandknowing.org/.

24 Syllabus for the Spring 2015 course: http://bit.ly/1JOwFSK.

25 For examples of the kinds of evidence reconstruction makes possible, see chapters 1–2 in *Art of the Past: Sources and Reconstructions: Proceedings of the First Symposium of ATSR Study Group*, London, 2005; Maartje Stols-Witlox, 'Sizing layers for oil painting in Western European sources (1500–1900): Historical recipes and reconstructions', in S. Kroustallis and Joyce H. Townsend, *Art Technological Research: Towards a New Discipline. Proceedings of the Second ATSR Symposium*, London, 2008, 147–65; Ken Albala, 'Cooking as research methodology: Experiments in Renaissance cuisine', in Joan Fitzpatrick, ed., *Renaissance Food from Rabelais to Shakespeare: Culinary Readings and Culinary Histories*, Aldershot, 2010, 73–88; and Adelheid Voskuhl, 'Recreating Herschel's Actinometry: An essay in the historiography of experimental practice', *British Journal for the History of Science*, 30: 3, 1997, 337–55.

26 Hasok Chang, 'How historical experiments can improve scientific knowledge and science education: The cases of boiling water and electrochemistry', *Science and Education*, 20, 2011, 317–41. One of the challenges faced by the Making and Knowing Project is the lack of expertise among the individual student researchers on the project. The laboratory seminars do no more than give individual students brief exposure to the experiential knowledge of materials and techniques – really just make them aware of what they do *not* know. It is thus surprising how much the collective nature of the project and the many collaborators with different levels of expertise make up for this lack of expert individual engagement.

27 Deciphering marginalia: Emogene Cataldo and Julianna van Visco, on 'Deciphering Marginalia: Try burnt oysters', http://bit.ly/1PuYjCF; clues to author-practitioner's identity: Rozemarijn Landsman and Jonah Rowen, 'Original Patterns in Ms. Fr. 640', http://bit.ly/1h9c1zO; and Raymond Carlson and Jordan Katz, 'Roses', http://bit.ly/1ER4j2K; techniques outside France: Sofia Gans, 'Knowledge Transmission in Ms. Fr. 640', http://bit.ly/1Ee9tem; learning new techniques: Shiye Fu and Zhiqi Zang, 'Too thin Things', http://bit.ly/1JrUvgH; and Chiostrini and Palframan, 'On Molding a Rose'; failures: Raymond Carlson and Jordan Katz, 'Casting in a Box Mould, fol. 118v', http://bit.ly/1PuYQV6.

28 Rozemarijn Landsman and Jonah Rowen, 'Fol. 92v: Molding a Hollow on One Side and a Relief on the Other (incuse reverse cast medals)' http://bit.ly/1MIJUDu; and Michaela Groeneveld and Marianne Nuij, 'Concerning the Incuse Reverse Casting of Medals', http://bit.ly/1PMVu0y.

29 Elisabeth Kuiper and Ingeborg Kroon, 'What is Spat?', http://bit.ly/1hWKSRf.

30 Experiential knowledge: Caroline Marris and Stephanie Pope, 'Sand with ox foot bone and elm root', http://bit.ly/1NwyAfw; Impalpable: Cataldo and Van Visco, 'Deciphering'; Carlson and Katz, 'Casting in a Box Mould, fol. 118v'; Sofia Gans, 'Calcined Alabaster as Excellent Sand, fol. 83v', http://bit.ly/1V50e4p; and Jenny Boulboullé and Joel Klein, 'Impalpable sand from Toulouse', http://bit.ly/1JfsutE; *s'attaque*: Celia Durkin, 'Molding fruits and animals in sugar', fol. 126r, http://bit.ly/1fBue85.

31 Mustard: Diana Mellon, 'Excellent Mustard, fol. 84r', http://bit.ly/1KHSP3O; millet consumption: Giulia Chiostrini and Jef Palframan, 'Making Millas, fol. 20r', http://bit.ly/1U2zqoy; and sand for hourglasses: Caroline Marris and Stephanie Pope, 'Powder for hourglasses, fol. 10r', http://bit.ly/1ER6Nyj.

32 As in the fluid-text edition of Herman Melville's Typee manuscript: http://rotunda.upress.virginia.edu/melville/default.xqy.

33 As, for example, in the innovative digital essay 'Kafka's wound' by the writer Will Self: http://thespace.lrb.co.uk/.

34 Fol. 150r.

35 For a full treatment of this, see Pamela H., Smith, 'Making as knowing: Craft as natural philosophy', in Smith, Amy R. W. Meyers and Harold J. Cook, eds, *Ways of Making and Knowing: The Material Culture of Empirical Knowledge*, Ann Arbor, MI, 2014, 17–47; and Pamela H. Smith, 'Vermilion, mercury, blood, and lizards: Matter and meaning in metalworking', in Ursula Klein and Emma Spary, eds, *Materials and Expertise in Early Modern Europe: Between Market and Laboratory*, Chicago, IL, 2010, 29–49.

36 Maria Alessandra Chessa, 'Imitation Coral', fol. 3r, http://bit.ly/1U2zuVe. See also the video of reconstructing the 'Imitation Coral' recipe on fol. 3r in the Making and Knowing laboratory: https://vimeo.com/129811219.

37 John Gage, 'Colour words in the high Middle Ages', in Erma Hermens, ed., *Looking through Paintings: The Study of Painting Techniques and Materials in Support of Art Historical Research* (Leids Kunsthistorisch Jaarboek XI), Baarn, The Netherlands, 1998, 35–48, 39.

38 See Pamela H. Smith, 'Itineraries of materials and knowledge in the early modern world', in Anne Gerritsen and Giorgio Riello, eds, *The Global Lives of Things*, London, 2015, chapter 1. For photographs of British Library, MS Egerton 1821, fols. 1v-2r, see https://www.bl.uk/catalogues/illuminatedmanuscripts/record.asp?MSID=8389&CollID=28&NStart=1821.

39 See for example, the information in vol. 3 of Michael of Rhodes, *The Book of Michael of Rhodes: A Fifteenth-Century Maritime Manuscript*. Further study about the use of ratios and proportions and its effects for conceptions of number in the early modern period is merited.

40 Mellon, 'Excellent Mustard'.

41 The process of 'extension' in experimentation to which Hasok Chang alludes. For evidence in the manuscript, see Cataldo and Van Visco, 'Deciphering'; Shiye Fu and Zhiqi Zang, 'Too thin Things'; Sofia Gans, 'Calcined Alabaster'; Kuiper and Kroon, 'What is Spat?'

42 Fol. 90r alludes to this searching out materials in nature.

43 'Les fondeurs prenent de la racine de jeune orme quand il est en sabe, & le font bouillir en vin ou pour mieulx vinaigre, et en font provision pour tout l'an dans un barriquet'. BnF Ms. Fr. 640, fol. 87v.

44 Deanna F. Curtis, Curator of Woody Plants at the New York Botanical Garden, suggested via email that an elm hybrid cultivar known as 'Pioneer' (binomial *Ulmus x hollandica*) might be more suitable for our purposes: as the Pioneer hybrid is formed by the crossing of two European species, *Ulmus glabra* (or Wych Elm) and *Ulmus minor* (the Smooth-leaved Elm), it probably more closely approximated the elms from which our author-practitioner would have sourced his roots. See

Caroline Marris and Stephanie Pope, 'Sand with ox foot bone and elm root'.

45 Email communication, dated 10 May 2015, from Joel T. Fry, Curator, Bartram's Garden, John Bartram Association, Philadelphia. As Dr Fry made clear, the commercial uses of plants are still considered in hybridization; but they are generally not part of the taxonomic system.

46 'Having thus moistened the sand in order to give it a nice hold, though it still came apart easily...' ['Ayant ainsy humecte le sable facon quil faisoict bonne prise sesmiant...'], fol. 118v. See Emogene Cataldo and Julianna van Visco, 'Eau Magistra': Investigating Binders for Sandcasting', and Cataldo and van Visco, 'Deciphering Marginalia: Try Burnt Oysters, fol. 84v', http://bit.ly/1NPGNIJ.

47 Cataldo and van Visco, 'Eau Magistra'; 'Deciphering'; and Marris and Pope, 'Sand with ox foot bone and elm root'.

48 Alessio Piemontese, *De Secreti*, also describes the perfect sand for casting as 'very soft, as if impalpable'. Alessio Piemontese, *Secreti del Reverendo Donno Alessio Piemontese*, Venice, 1555, 206. See also, Carlson and Katz, 'Casting in a Box Mould, fol. 118v'.

49 See Christoph Krekel and Manfred Lautenschlager, 'Bearbeitung von Glas, Edelstein, Bein und Horn', in Anna Bartl, Christoph Krekel, Manfred Lautenschalger and Doris Oltrogge, *Der 'Liber illuministarum' aus Kloster Tegernsee*, Stuttgart, 2005, 673–78. An excellent example of these kinds of recipes that attempt to bring about opposite states can be found in *Künstbüchlin gerechten gründlichen gebrauches aller kunstbaren Werckleüt*, Augsburg, 1535.

50 Emogene Cataldo and Julianna van Visco, 'Wax and Tallow: A Material Investigation', http://bit.ly/1NPGMo7.

51 Rozemarijn Landsman and Jonah Rowen, 'Concerning the uses of sulfur for casting', fols. 140v and 12r, http://bit.ly/1WOsb2a.

52 Yijun Wang, 'Powder of ox bone and rock salt, fol. 89r', http://bit.ly/1I8RYai; and Gans, 'Calcined Alabaster'.

53 Ms. Fr. 640 contains a prophylactic against metal fumes: 'In the morning, take a piece of thin toast with butter, neither antimony nor any other vapors will harm you. Or put half a pig's bladder in front of your face' (fol. 79v). For more on this, see Pamela H. Smith, 'Making as knowing: Craft as natural philosophy'.

54 For more on this, and all pertinent citations, see Pamela H. Smith, 'The matter of ideas in the working of metals in early modern Europe', in Christy Anderson, Anne Dunlop and Pamela H. Smith, eds, *The Matter of Art: Materials, Practices, Cultural Logics, c. 1250–1750*, Manchester, 2015, 42–67.

55 Michele Goun Lee, with Pamela H. Smith, 'Eau Magistra, fol. 84v', http://bit.ly/1U2zWCK.

56 David Pye, *The Nature and Art of Workmanship*, Cambridge, 1968, 47.

57 For example, Biringuccio, *Pirotechnia*, 114.

58 Cellini recommends that 'you will do well to make wax impressions from time to time, while you are cutting, to see how you are getting on'. *The Two Treatises*, 73–4.

59 Landsman and Rowen, 'Concerning the uses of sulfur'.

60 Another example of the intersection between natural historical and craft interests can be found in the author-practitioner's terms for the techniques of life-casts – he calls the channels by which metal enters the mould 'veins', and the tubes by which ashes can be blown out of the mould after burnout and by which air escapes as metal replaces it in the pour 'breathing holes'. See Chiostrini and Palframan, 'On Molding a Rose'.

61 https://www.youtube.com/watch?v=W6sxEf997iM. See the sequence at 57 seconds into the video. I thank Joosje van Beenekom, Senior Metals Conservator, Rijksmuseum, Amsterdam, for informing me of this film.

62 For further information on the Making and Knowing Project, see http://www.makingandknowing.org/. For photographs of much of the work done in the Project, see its Flickr site https://www.flickr.com/photos/128418753@N06/albums.

63 Syllabus for the Spring 2015 course: http://bit.ly/1JOwFS.

64 The first Working Group Meeting on Moldmaking and Metalworking in BnF Ms. Fr. 640 was held 27–30 May 2015. Meeting Programme at http://bit.ly/1U2Ab0Q.

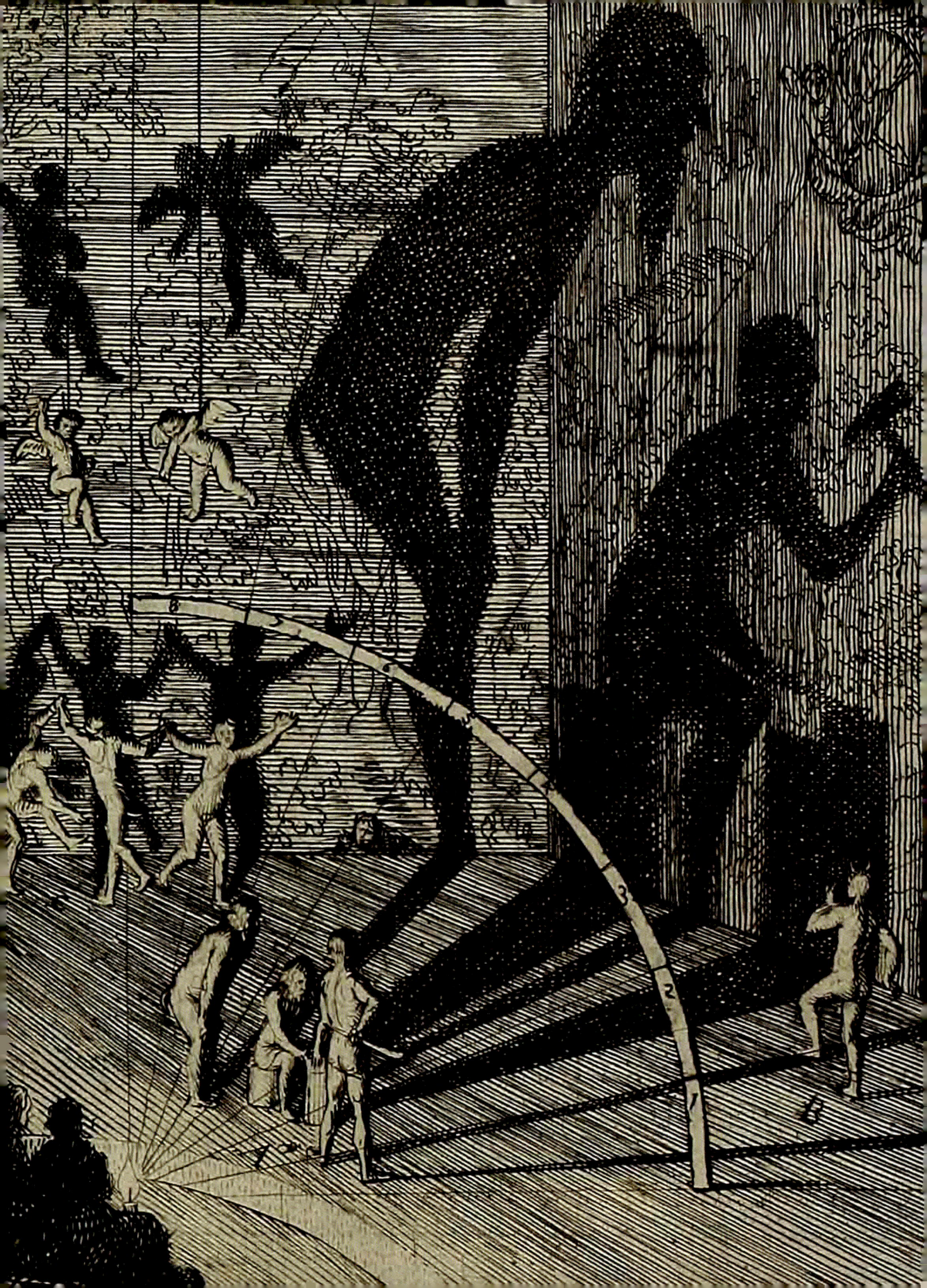

Chapter 3
Works in Progress: Painting and Modelling in Seventeenth-Century Holland

Jan Blanc

Even though I say that chance is everything, I anticipate what chance will look like through my own aesthetic sense of line and framing. I've studied photography and art long and hard. I know Caravaggio and Goya or any number of people in my eye socket. ... It's not that I'm out photographing and thinking, 'Oh my God, I see a Botticelli.' No, I see a woman at a fashion show. But things happen in life that bear a kind of a bell, a chime of recognition.[1] (Larry Fink, 2014)

Recalling the important works that have punctuated his career, and also the artistic principles that guided him, the photographer Larry Fink recently observed that all perception is informed by models shaped in the memory of the observer. Fink is not seeking to defend the point of view of an aesthete who, like Marcel Proust's Swann, believes that life imitates art, and who likes to 'see his wife as a creation of Botticelli', buying her 'a fabulous blue and pink oriental scarf, ... because it precisely resembled that worn by the *Madonna of the Magnificat*'.[2] Rather it is a question of emphasizing, with Ernst Gombrich, that every glance is constructed from pre-existing patterns acquired through education and socio-cultural inculcation, rather than a blank, innocent gaze capable of directly and purely imitating nature.[3]

Yet art historians of the Dutch Golden Age have noted that there is a significant difference between what Netherlandish painters called studies *naer het leven* (after life), based on the direct observation of nature, and those *uyt den geest* (from memory), drawing on the treasures of accumulated visual experience. It is largely on the basis of this distinction that, following in the wake of Svetlana Alpers's path-breaking work, Dutch painting has often been analysed in opposition to Italian painting and in connection with seventeenth-century developments in the new science of observation.[4] Philosophers such as Francis Bacon, René Descartes and Galileo Galilei held that nature should be interpreted on the basis of a dialogue between the inductive accumulation of facts and the deductive construction of arguments, in which the image and the visual analysis of nature henceforth occupied a predominant place. Art historians following Alpers have contended that Dutch artists sought to establish similar procedures, prioritizing the direct observation of natural phenomena and pondering the optical conditions in which such observation is possible.

Johannes Vermeer has come to occupy a crucial position at the heart of such a defence of a specific Dutch identity and even of a distinct Dutch visual culture within the territory of the old Netherlands.[5] In order to show that his paintings have played 'their exemplary role in the definition of the art of describing',[6] Svetlana Alpers

Detail from Samuel Van Hoogstraten, *La Danse des ombres*, c.1678 (*plate 10*).

Art and Technology in Early Modern Europe *Edited by Richard Taws and Genevieve Warwick*.

emphasizes two particular aspects of his work. First of all she reminds us – and we shall return to this later – that Vermeer probably used a camera obscura in the design of a number of his pictures. This choice is seen as characterizing the approach of Dutch painters, who, according to Alpers, often favoured an 'optical image' of the visible world over a 'perspectival image'.[7] Moreover, Alpers accords special importance, among Vermeer's works, to the *Art of Painting*,[8] emphasizing the emblematic character of what she calls 'the mapping impulse in Dutch art'.[9] The map hanging on the wall in the background of that work bears a legend that, as Daniel Arasse has also shown, features the Latin word *descriptio*, which is held to exemplify the eminently descriptive aspect of Vermeer's art – and also of Dutch painting as a whole.[10]

The object of this article is not to undertake a detailed examination of these analyses, which have already been the subject of a major critical review.[11] It is rather to qualify or even problematize them firstly by looking again at the distinction, which is more complex than might be imagined, between the concepts *naer het leven* and *uyt den geest*. These two ways of working in fact embody two different forms of the same modelling of nature, which are then translated into the artistic terminology of the seventeenth century. Finally, the example of Johannes Vermeer highlights the key role played by models, whether material or mental, in the development of Dutch painting of the seventeenth century.

Naer Het Leven versus *Uyt Den Geest*

The Dutch painters of the sixteenth and seventeenth centuries used the term *naer het leven* to denote any form of imitation based on direct observation of a model. Karel van Mander explains, for example, that Barocci

> painted many excellent portraits from life (*nae t'leven*), but also ... numerous charming pictures of the Virgin in oil, almost always using the face of his female servant, sometimes depicting her with her child in order to copy their gentle and graceful way of sitting.[12]

The Dutch painters used *uyt den geest*, on the other hand, to refer to studies executed from the resources of their memory and broader visual culture:[13]

> After having sketched out his thoughts in chalk for his apprentices' sake, [Frans Floris] then had them execute his painting on the basis of these ideas, telling them to: 'Adapt this or that head in this place' for he always had a large number of heads prepared ready on panels. This way, his apprentices would acquire a certain boldness and facility, and would then venture to paint their own canvases and even to compose and paint a picture out of their own head (*uyt den gheest*).[14]

This distinction could lead one to suppose that for Rembrandt's contemporaries there existed a more direct and 'realistic' way of perceiving and representing the world (*naer het leven*) that contrasted with a more learned approach based on models provided by the great masters (*uyt den geest*). However, I would like to demonstrate that this was far from the case, firstly because theoretical texts from the sixteenth or seventeenth centuries do not support this interpretation, but also because it is important not to confuse a study *after life* with a study *drawn from nature*. Svetlana Alpers rightly cites the inventory of Rembrandt's possessions as including 'a small book filled with statues drawn by Rembrandt from nature'

(*Een dito [boeckie] vol Statuen Van Rembrandt nae't leven geteckent*).[15] By contrast, drawings such as those made by the engraver Pieter Quast from figures drawn *after life* by Roelandt Savery as well as the engravings he made from the compositions derived by Salomon Savery from the same drawings by his uncle and godfather (*plate 1*) cannot be considered studies *naer het leven*.[16] The reason is that these drawings were not executed directly before their natural model even though they are as detailed and precise as Savery's sketches. What distinguishes a drawing *naer het leven* from a drawing *uyt den geest* is not, therefore, the degree

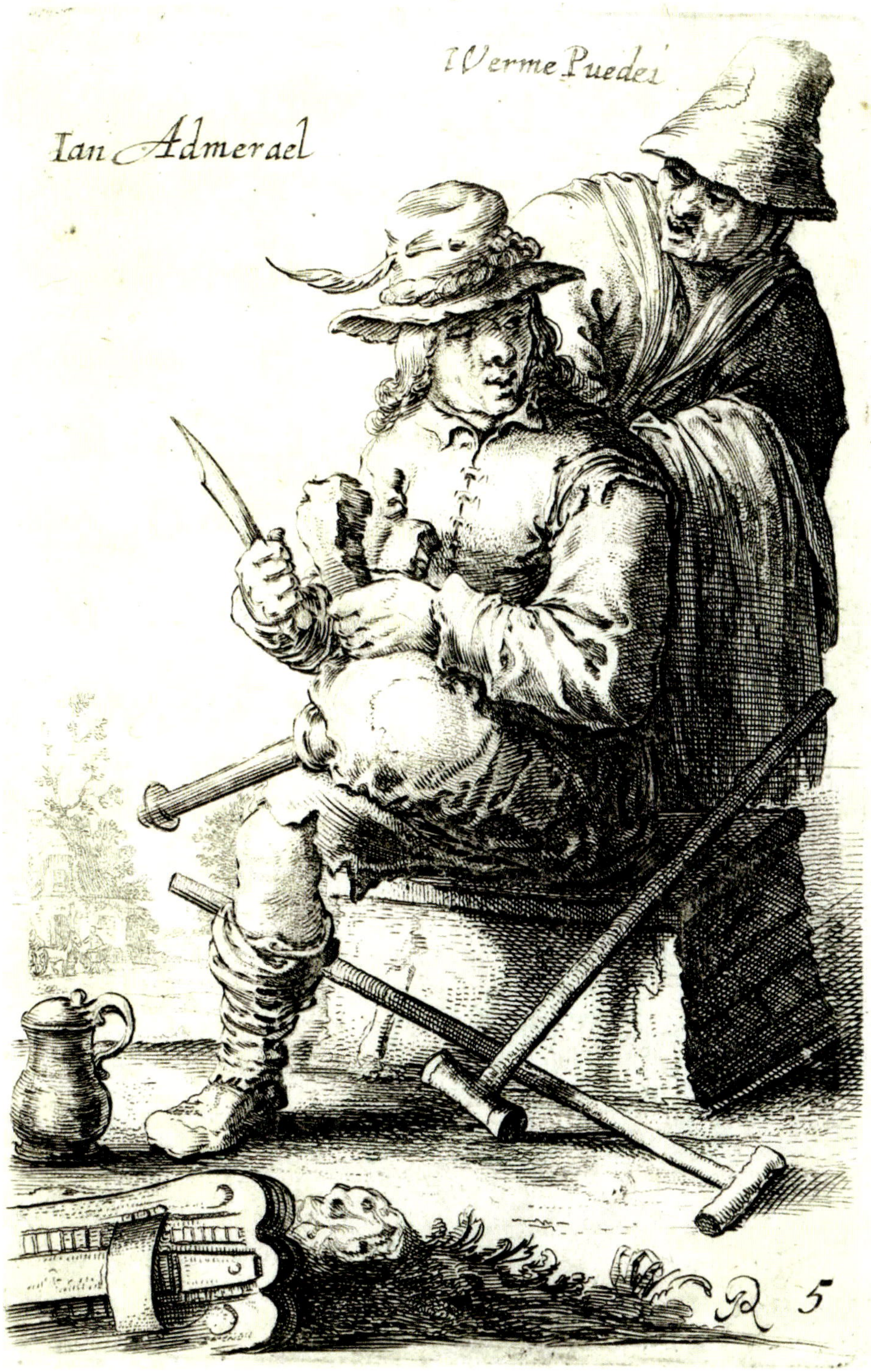

1 Pieter Quast after Salomon Savery, *A Male Beggar (Ian Admerael) and Another Beggar (Werme Puedes)*, c.1634–40. Etching, 13.7 × 8.9 cm. London: British Museum, Department of Prints and Drawings (inv. D,7.65). Photo: © The Trustees of the British Museum.

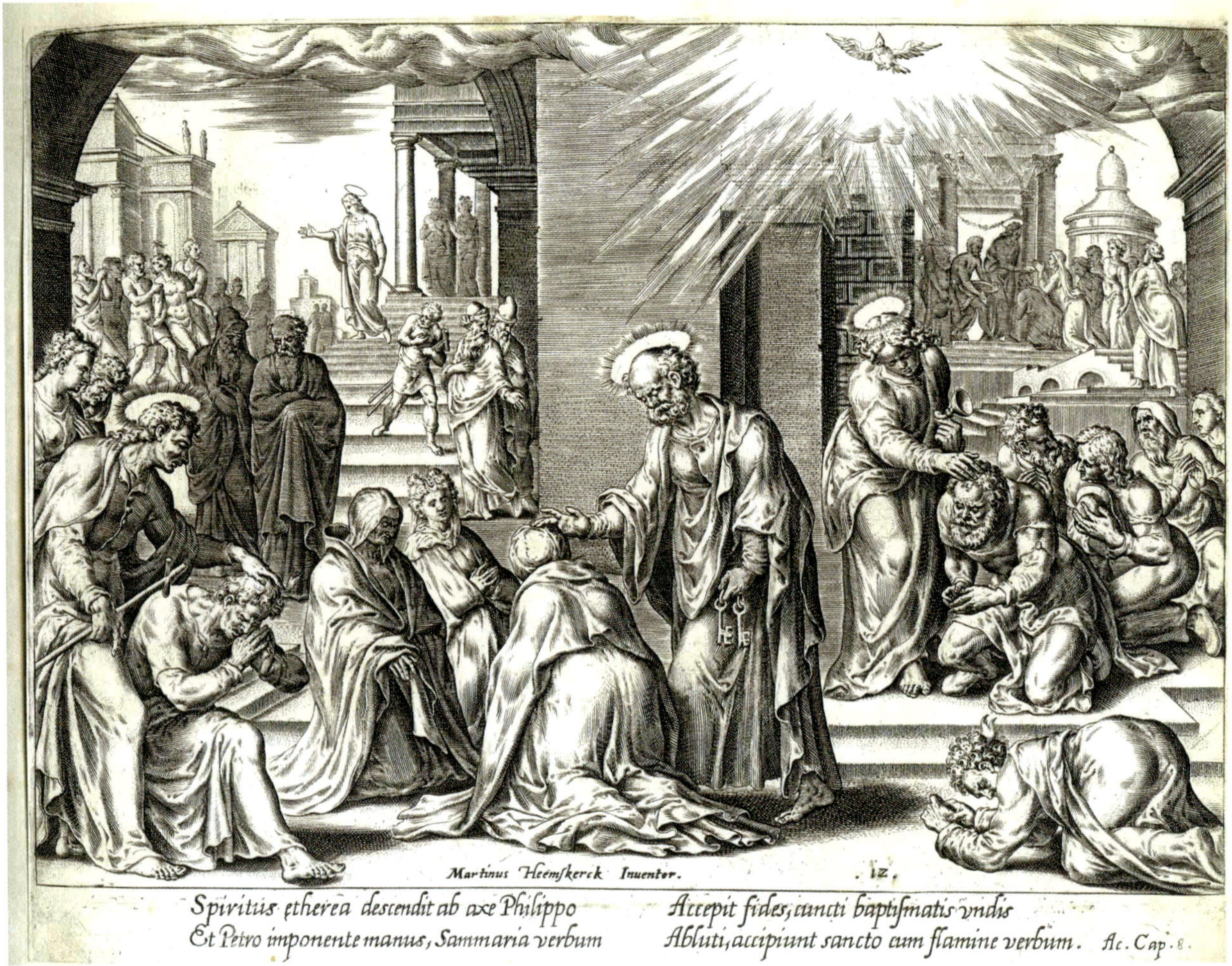

2 Unknown engraver after Jan van der Straet, *The Descent of the Holy Ghost*, 1582. Engraving, 21.0 × 27.0 cm. London: British Museum, Department of Prints and Drawings (inv. 1954, 0813.12.12). Photo: © The Trustees of the British Museum.

of faithfulness of the 'copy' to the 'original', but the manner of execution. The one should be designated a 'first-hand' drawing and the other a 'second-hand' drawing in the sense in which Antoine Compagnon uses the term when describing the workings of quotation.[17]

Both these graphic practices clearly have advantages and disadvantages. As stressed by van Mander, nature offers the enormous advantage of being an omnipresent model, available and agreeable to study at all times. It is a place of repose that awaits those who, 'eager to learn and know more, [have] dulled [their] senses almost to the point of going blind and losing consciousness'.[18] It is also a real landscape in which all the rules of art are applied and can be studied in action by painters: 'Let us go and see the beauty that is to be found outside, / where the pipes are freely played! There shall we observe views aplenty / from which we can paint landscapes on linen cloth or panels of hard oak from Norway.'[19] But by plucking a figure from the 'mind' (*ingenium*, *geest*) rather than from nature itself, a painter knows that he is drawing on the source of the best works. What he loses in diversity, he gains in perfection, by calling upon the authority of the greatest masters.[20] Describing the series of the *Acts of the Apostles*, begun by Maarten van Heemskerck and completed by Johannes Stradanus (*plate* 2), who on this occasion 'did many other things that furnish us with sufficient indications and proofs of his resourcefulness and intelligence, both in the composition of histories as well as in the placing and arranging of his figures in all the different circumstances of art', van Mander praises the ingeniousness of the iconographical and formal choices made by the engraver, who works 'out of his head or imagination' (*uyt zijnen geest oft inventie*).[21]

For all this, in none of these passages does van Mander claim that a study *naer het leven* is more 'realistic' or 'faithful' to nature than a study *uyt den geest* for he knows perfectly well that in both cases the eye immediately encodes and models what it sees.

Vision as Code

In order to emphasize the importance of the modelling process in perception and artistic practice, Dutch painters of the seventeenth century put forward arguments that are similar to those developed by Ernst Gombrich in *Art and Illusion*. For them, pure imitation of the visible world is quite simply impossible because it passes through the filter of the mind. It is for this reason that van Mander, whose ideas were taken up by Dutch theoreticians of the seventeenth century, makes the mind (*geest*), rather than nature (*natuer*), the origin of art. Thus at the beginning of his life of Jan Gossaert he declares that 'the initial seeds of the art of painting reside in the mind (*geest*) or in thought (*ghedacht*), which manifest themselves in the form of internal imaginings (*inwendighe inbeeldinghen*) before being irrigated by the hand and brought to completion'.[22]

In this respect, a representation of nature *naer het leven* is no less modelled than a representation *uyt den geest*: it is simply that the models and the effect they have on the viewer differ. The landscapes of Pieter Bruegel the Elder (*plate* 3) strike van Mander as perfect objects of nature. This was not only because he believed them to have been observed and studied *naer het leven*, but also because

3 Johannes and Lucas van Doetechum after Pieter Bruegel the Elder, *Alpine Landscape*, c. 1555/57. Etching and engraving, 32.3 × 43.0 cm. Washington, DC: National Gallery of Art, Rosenwald Collection (inv. 1964.8.411). Photo: © National Gallery of Art.

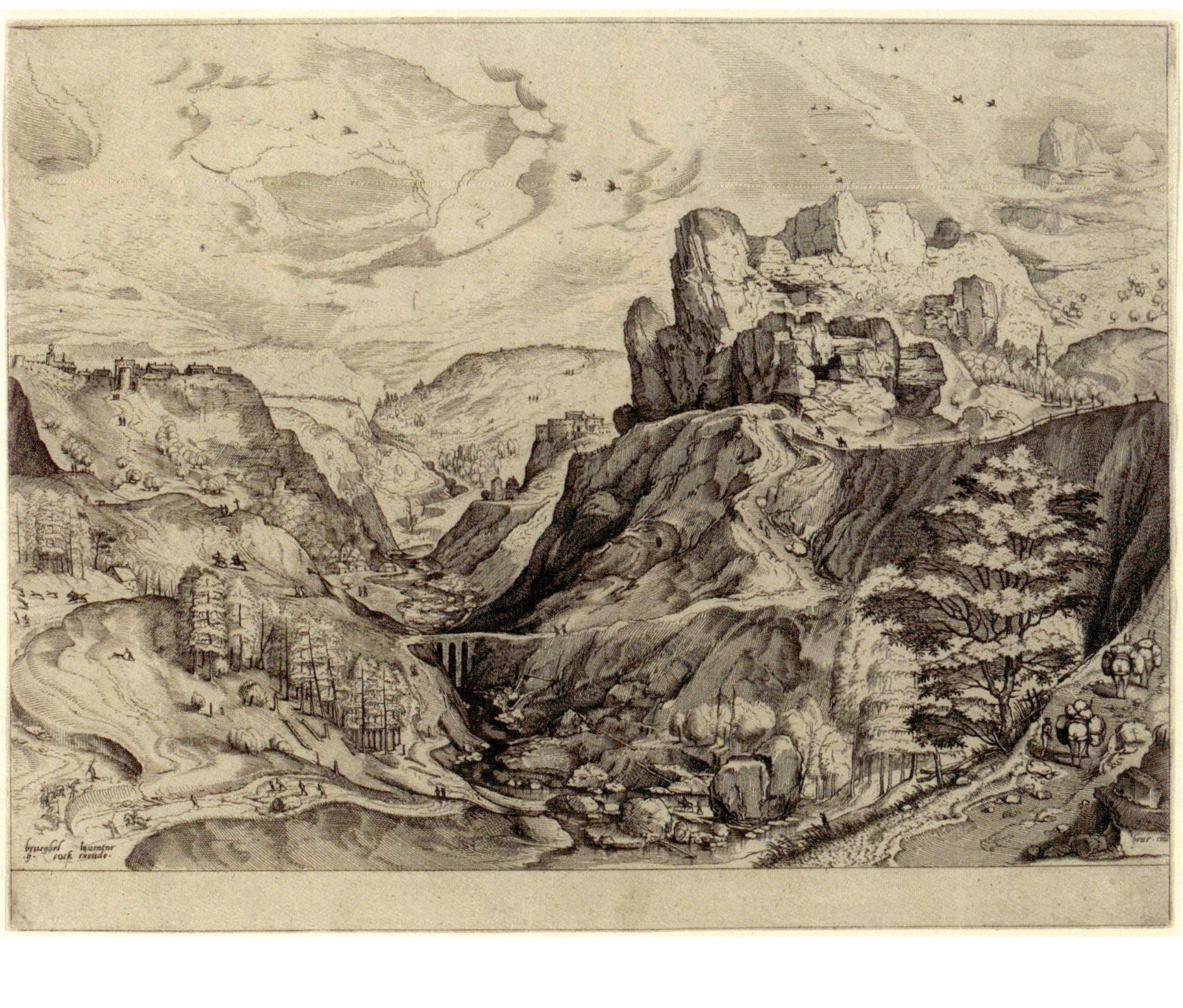

they were seen and retranscribed by the mind of an artist familiar with the types to which the idea of nature shared by the viewers of his works corresponded:

> On his travels, he depicted numerous views from life (*nae t'leven*) so well that it was said he had swallowed all the mountains and rocks while in the Alps and upon his return home had spat them out onto canvases and panels, so true (*eyghentlijck*) could his imitations of these and other parts of nature be.[23]

If nature, as observed in the Alps, constituted Bruegel's basic model, this model was itself the object of a double series of modellings likened by van Mander, in the sixteenth-century tradition of the poetic metaphors of mimesis and innutrition, to two organic

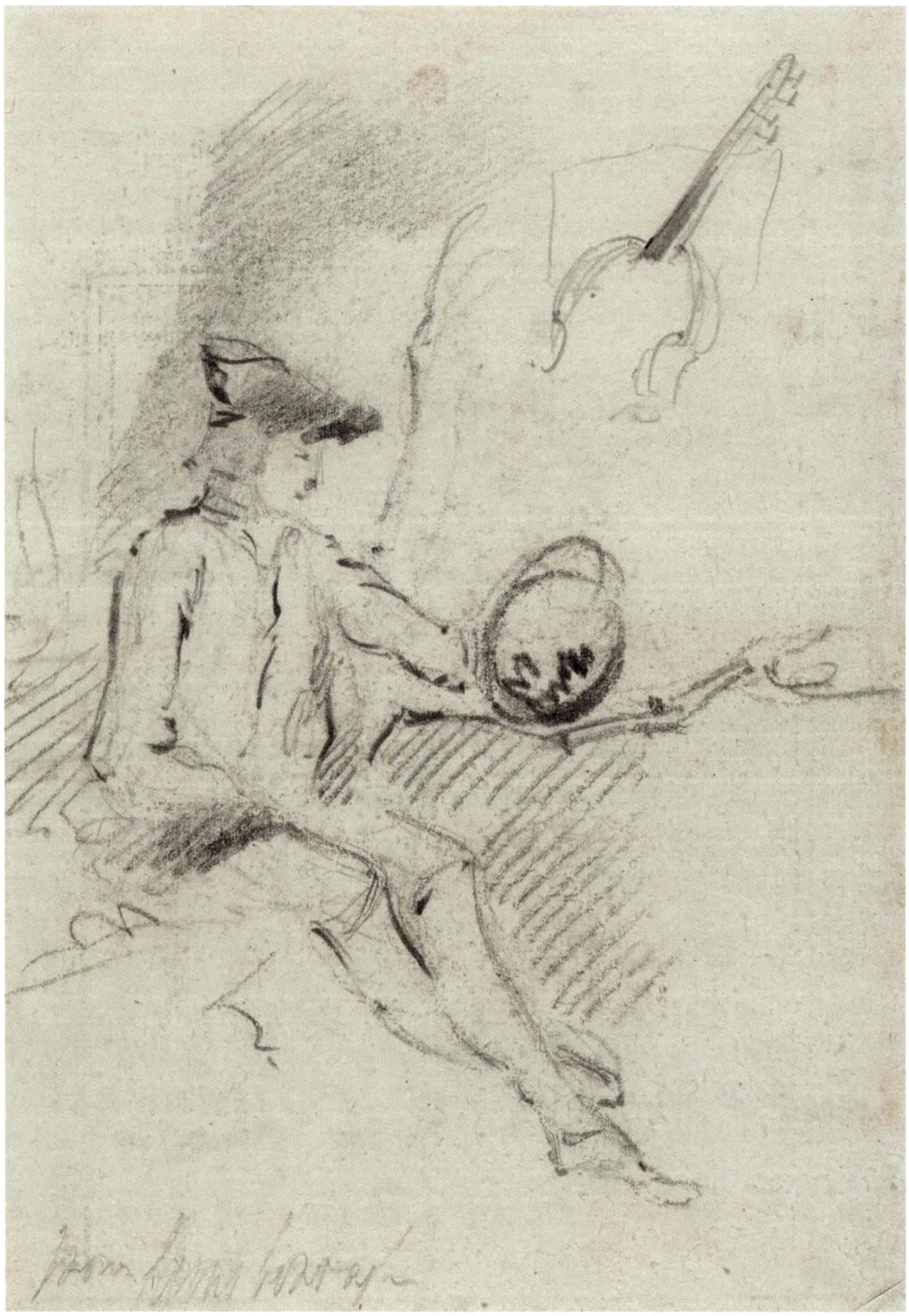

4 Copy after Thomas Gainsborough, *Study of a Man Sketching Holding a Claude Glass*, c. 1750. Graphite on slightly textured, cream woven paper, 24.1 × 17.1 cm. New Haven: Yale Center for British Art, Paul Mellon Collection (inv. B1975.4.24). Photo: © Yale Center for British Art.

processes. Firstly *ingurgitation*, by means of which Bruegel initially *swallowed* the Alps, thereby performing the first two transformations of the visual material (sectioning as a result of biting, and breaking down as a result of mastication). Secondly *regurgitation*, whereby Bruegel 'spat out' this same food, thus reassembling the objects seen.[24]

Two main lessons can be learned from the analysis of this famous passage. First of all it can be observed that the categories of 'realism' and 'naturalism', still regularly used by art historians today, would appear to be completely unsuited to explaining how painters of the seventeenth century conceived of their relationship with nature. For van Mander, it went without saying that the works of the most admired Netherlandish landscape painter of the second half of the sixteenth century seemed singularly faithful to nature not because they were a perfect reflection of their natural source but because they made use of stereotypes and models of nature that were particularly easy to understand and remember. Furthermore, after reading these lines it is difficult to persevere in claiming that Dutch painting of the seventeenth century encouraged use of the camera obscura and the practice of cartography and drawing *naer het leven* in order to cultivate an optical image of nature. This is evident from treatises on painting such as the *Introduction to the Academy of Painting* (*Inleyding tot de hooge schoole der schilderkonst*, 1678) by Samuel van Hoogstraten. According to this former student of Rembrandt, a camera obscura offers the advantage of showing 'the overall or general aspect that a truly natural painting should possess', which can also be seen 'in miniature in glasses and small mirrors, which, although the drawing is somewhat deformed, display the entirety and economy of its colour range in all its purity'.[25] The 'glasses' and 'small mirrors' (*glazen en spiegeltjes*) described here do not refer to those whose use and representations are analysed in the article by Genevieve Warwick in this volume. They correspond to another type of mirror known as 'black mirrors' or 'Claude glasses' in tribute to the French painter Claude Lorraine, who at that time made regular use of them, as did landscape painters such as Thomas Gainsborough (*plate* 4) in the eighteenth century. These small convex tinted mirrors gave artists access to a transformed perception of nature. The image they could see in them distorted the forms and colours of nature by accentuating the contrasts between the planes and reducing objects to their main hues.[26]

Paradoxically, the same modificatory principles govern the way maps work, according to Samuel van Hoogstraten. In this connection Svetlana Alpers quotes the Dutch theoretician ('How wonderful a good map is in which one views the world as from another world thanks to the art of drawing'[27]). Pondering the 'mapping impulse' in Dutch art of the seventeenth century, she first of all reveals its highly conventional dimension, emphasizing in particular the symbolic and mimetic codes that have governed the production of maps since the fifteenth century. She then stresses that these same conventions seem to her to be marked by a more visual and graphic conception than that which marks the approach of the Italian painters, who are more idealist but also more influenced by the textual model. In particular, she recalls that van Hoogstraten's observations on maps are integrated into a passage in which the Dutch painter and theoretician links drawing 'to letters formed in writing, to planning war maneuvers, to medicine, astronomy, natural history, and geography'.[28] It nevertheless seems to me that on this point her analysis needs to be pushed a little further. When van Hoogstraten asserts, in the passage referred to by Alpers, that drawing is like 'a second kind of writing' (*als een tweede wijze Van Schrijven*), he is clearly alluding to the well-known ambivalence of the Greek *graphein*, which means both 'to write' and 'to draw'. Thus van Hoogstraten does not content himself with defending the universally necessary character of the art of drawing any more than he promotes the 'optical image' of the visible world to the detriment of the 'perspectival image'. He seeks above all to show that the art of painting, like that of writing, is only superficially mimetic, and that its primary function is to represent nature via a collection of to some extent arbitrary signs, analogous to those used in forming letters and words when writing. He confirms this, moreover, by likening the art of painting to an art of speaking 'in a hieroglyphic manner' (*op een Hierogliphische wijze*).[29] Asserting, as he does, that a 'good map' shows the world 'as from another world' this enables van Hoogstraten to emphasize that, like a camera obscura or a black mirror, a map acts as a veritable prosthesis that modifies our 'vision of the world'.[30]

Borrowing from the vocabulary of current scientific modelling theory, we could therefore affirm, in support of the analyses proposed above, that the representation of nature, whether mental – resulting from observation – or material – giving rise to a drawing – is always a modelling operation and furthermore that this modelling is based not on simple analogies between the 'source' and the 'model' but on a compression and selective presentation of the observed visual information.[31]

Models and Modelling

It is probable that Dutch painters of the seventeenth century were not merely aware of the considerable importance of the modelling processes in the observation and representation of nature, but were also greatly interested in the plastic and formal resources that could be derived from such reflections.

As evidence I would point, in the first place, to the words used by artists to describe the modelling of their works. As with the English word 'model', which evolved along similar lines, the Dutch term model appeared in its vernacular form in the final third of the sixteenth century. It was characterized from the outset by a double ambiguity. This ambiguity relates to the word's two main meanings. The first, based on the vocabulary of logic from scholastic and neo-Platonic philosophy, is that of the model as a mental construction. Thus in 1578 the Ptolemaic system was described as 'a new Theoricke, or model of the world, shewing that the earth resteth not in the Center of the whole world, but only in the Center of thys our mortal world'.[32] The second meaning of the word model, occuring later and more frequently in the Dutch language, denotes the entire set of drawings realized within the context of a construction, structure or project and then, by specialization, a small-scale model of a projected work.[33] Karel van Mander, for example, refers to the 'models and drawings' (*modellen en teyckeningen*) that Leonardo da Vinci 'made every day'.[34] Half a century later, the historiographer Dirck Van Bleyswijck describes having seen in Delft the 'model of the great tower of the Nieuwe Kerk in the state in which it found itself prior to the well-known fire of 1536' (*het model van den grooten nieuwe Kercks Toorn, hoedanigh die was voor den vermaerden brandt van 't Jaer 1536*).[35]

In such uses, the word model, in English as well as in Dutch, thus simultaneously denotes the mental concept and the concrete materialization of a project. But it also presents another ambiguity that relates to its etymology. Some authors associate the origins of the word model with the Italian *modello* and the notion of a preparatory study for a work. Others insist on an architectural origin, tracing the word to the Latin *modulus*, used by Vitruvius to designate the diameter of the column, which, repeated variously within a building, forms the basis for the harmony and proportions of the whole.[36] One of the texts in which this hesitation is most evident is Leon Battista Alberti's *De pictura*. In the initial, Latin, version of *De pictura*, intended for a select band of learned readers capable of understanding this language, Alberti used the word *modulus* to describe the 'models' used by painters:

> When we have an *istoria* to paint, we will first think out the method (*modis*) and the order (*ordine*) to make it most beautiful; we will make our models (*modulos*) of all the *istoria* and every one of its parts first of all; we will call our friends to give advice about it.[37]

However, in the Italian translation of his text, which was probably addressed more particularly to artists, Alberti chooses the word *modello*, which he links to the term 'concept' (*faremo nostri concetti e modelli di tutta la storia*).[38] Thus this etymological indecisiveness does not, I believe, stem merely from specifically philological issues. It also reveals a hesitation over problems relating to the functions and definitions of the concept of the *model*. But when the definition of model is associated with the notion of *modulus*, this concept is inscribed into the process of *communication* by means of which the engineer discusses matters with his associates. He presents them with a model corresponding to the project in its entirety, just as the *modulus* reflects the construction of a harmonious and proportionate relationship between the individual parts and the whole (*commodulatio*). The function on which Alberti insists, while in all probability addressing learned readers or architect colleagues, is that of the exchange of knowledge and information permitted and even facilitated

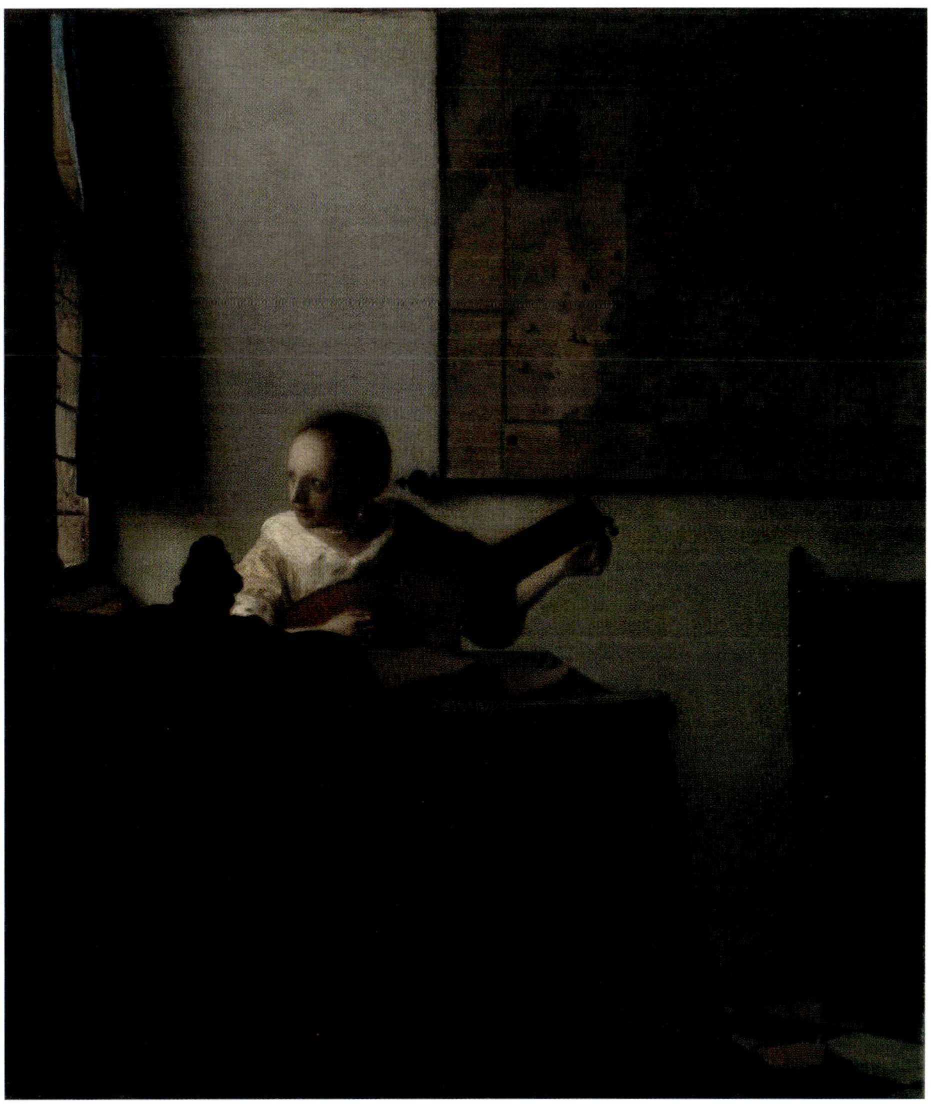

5 Johannes Vermeer, *Woman with a Lute*, c. 1662–75. Oil on canvas, 51.4 × 45.7 cm. New York: Metropolitan Museum of Art, Bequest of Collis P. Huntington (inv. 25.110.24). Photo: © Metropolitan Museum of Art.

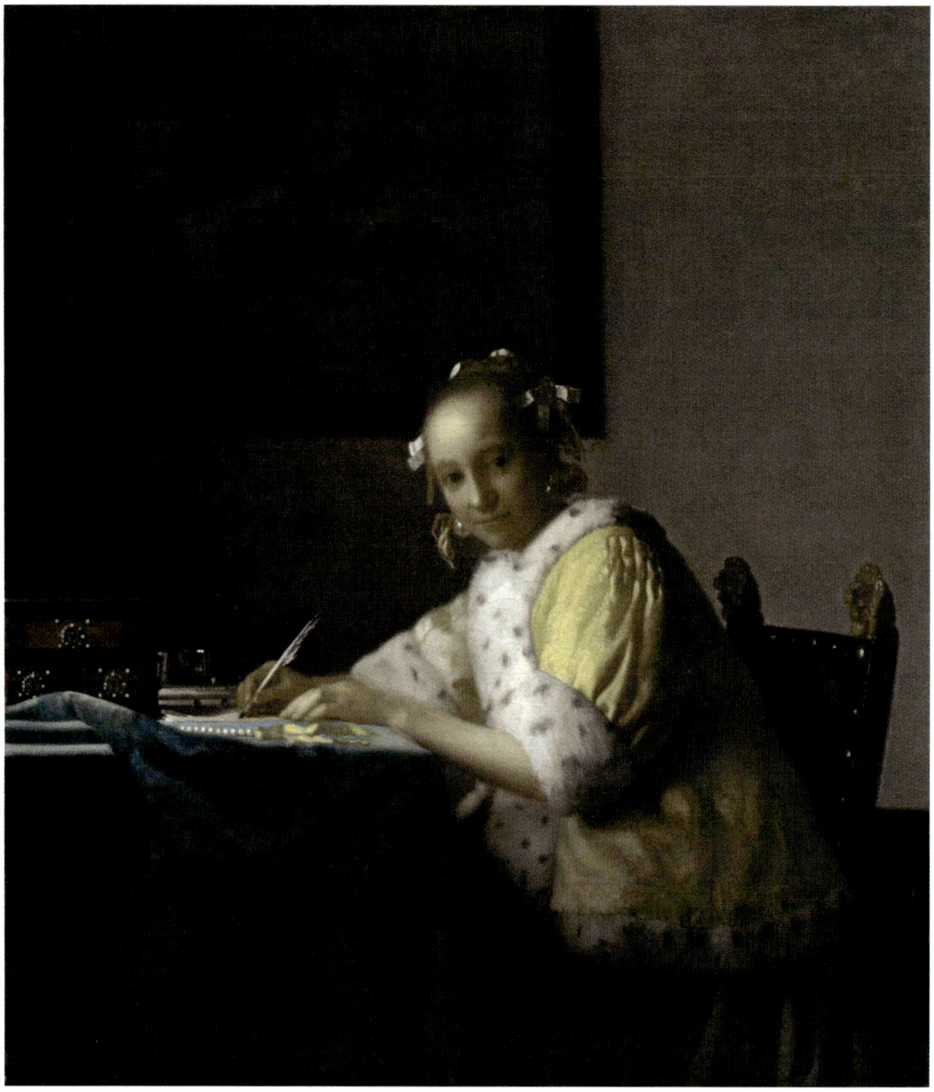

6 Johannes Vermeer, *A Lady Writing*, c. 1670–72. Oil on canvas, 45 × 39.9 cm. Washington, DC: National Gallery of Art, Gift of Harry Waldron Havemeyer and Horace Havemeyer, Jr (inv. 1962.10.1). Photo: © National Gallery of Art.

by the model while also accelerating it. When associated with the notion of *modello*, however, this same idea becomes more the actual locus of a construction and of a presentation. The *construction* regards the genesis of a work of art, which is only made possible by the prior and concomitant fabrication of one or more models. The *presentation* concerns the potential work that the artist submits beforehand to the scrutiny of his patrons.[39]

More widely, this hesitation also relates to problems of definition. Denoting alternately an object (*modello*) and part of an object (*modulus*), the model raises the question of the relationship between the preparatory phases in the genesis of a work and the execution of the work in the strict sense. Alberti was himself aware of this and warned architects of the danger of regarding their maquettes as works of art in their own right:

> I must not omit to observe, that the making of curious, polish'd Models, with the Delicacy of Painting, is not required from an Architect, that makes it his Business by charming the Eye and striking the Fancy of the Beholder, to divert him from a rigorous Examination of the Parts which he ought to make, and to draw him into an Admiration of himself.[40]

For Alberti, it is clear that architects' models should be understood as sketches, not 'too delicate and neat, but plain and simple, more to be admired for the Contrivance of the Inventor, than the Hand of the Workman', as if it were necessary to avoid the work becoming a prisoner of the model that rendered it possible or, worse, the model itself becoming a work of art in its own right, emancipating itself from the preparatory tasks of which it is a part.[41]

Vermeer and the Art of Modelling

If Alberti insists on the danger of models and modelling processes assuming an autonomous position in the genesis of works of art, it is no doubt because he judged it likely that artists were indeed confronted by such a situation. Possibly this was due to the proliferation of, and increasing role played by, the physical models (maquettes) conceived by Renaissance architects and their painter (the *schizzi*, *disegni*, *modelli* and *cartoni* under Giorgio Vasari's classification) and sculptor (*bozzetti* and *modelli*) colleagues.[42]

As my scope is confined here to Dutch painters of the seventeenth century, I will do no more than emphasize the role played by those artists who made frequent use of such modelling techniques in the enrichment and diversification of these techniques, based on the example of one of the most striking and important exponents of the practice, Johannes Vermeer. He was among the Dutch painters of his day who devoted the most attention to the issue of modelling and its use in the genesis of paintings.[43] The place of the camera obscura in his work has been the subject of numerous studies, including recent ones, and does not need to be developed further here, any more than does his use of perspective, which enabled him to organize the disposition of the various elements within his paintings with great precision.[44] Here I will emphasize above all the use Vermeer made of the space in his own studio. Philip Steadman has shown definitively how Vermeer undertook a careful modelling of his pictures within his normal workspace. Having set up his studio on the first floor of the large house in which he lived with his mother-in-law, wife and children, he would first of all select a precise viewpoint facing into one of the corners of his studio, either with the help of a camera obscura or through the careful positioning of his chair and easel. He would then arrange the space delimited by his field of vision and fill it with furniture and objects belonging to the household. The characters, meanwhile, were probably 'played' by a mannequin dressed by the painter and moved from place to place within the studio space in order to represent the different figures that were ultimately to inhabit the painting, thereby explaining why the same clothing can be seen in a number of paintings (*plate* 5 and *plate* 6).

7 Hendrik Danckerts, *View of St James's Park, Seen from the Eastern End of the Canal and Looking Directly Along it Towards a Distant Building, Probably Tart Hall*, c.1660–79. Pen and brown ink, over graphite, with grey, brown and blue-green wash, on paper, 28.2 × 90 cm. London: British Museum, Department of Prints and Drawings (inv. 1874,0808.106). Photo: © The Trustees of the British Museum.

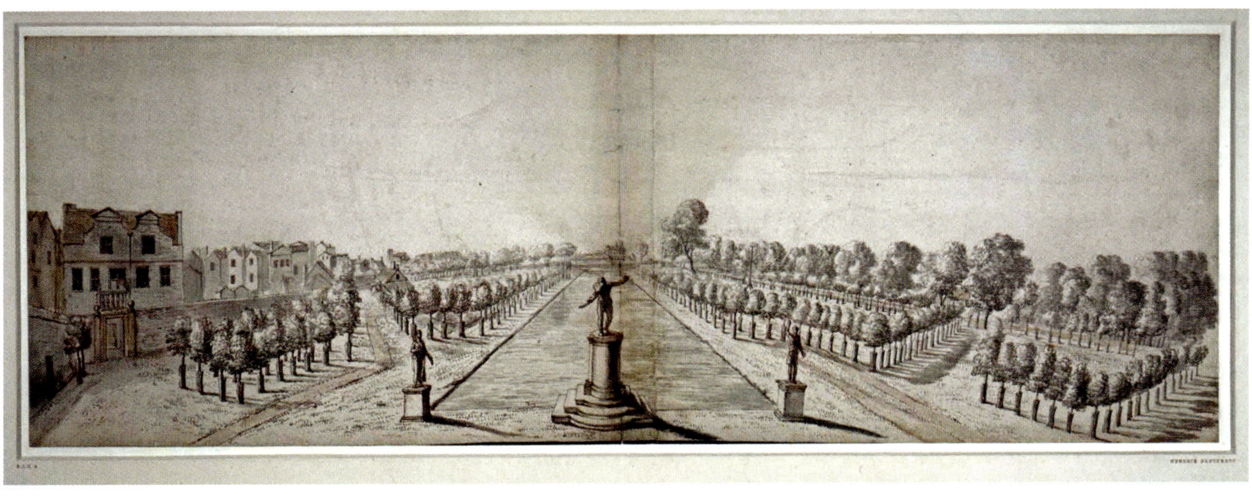

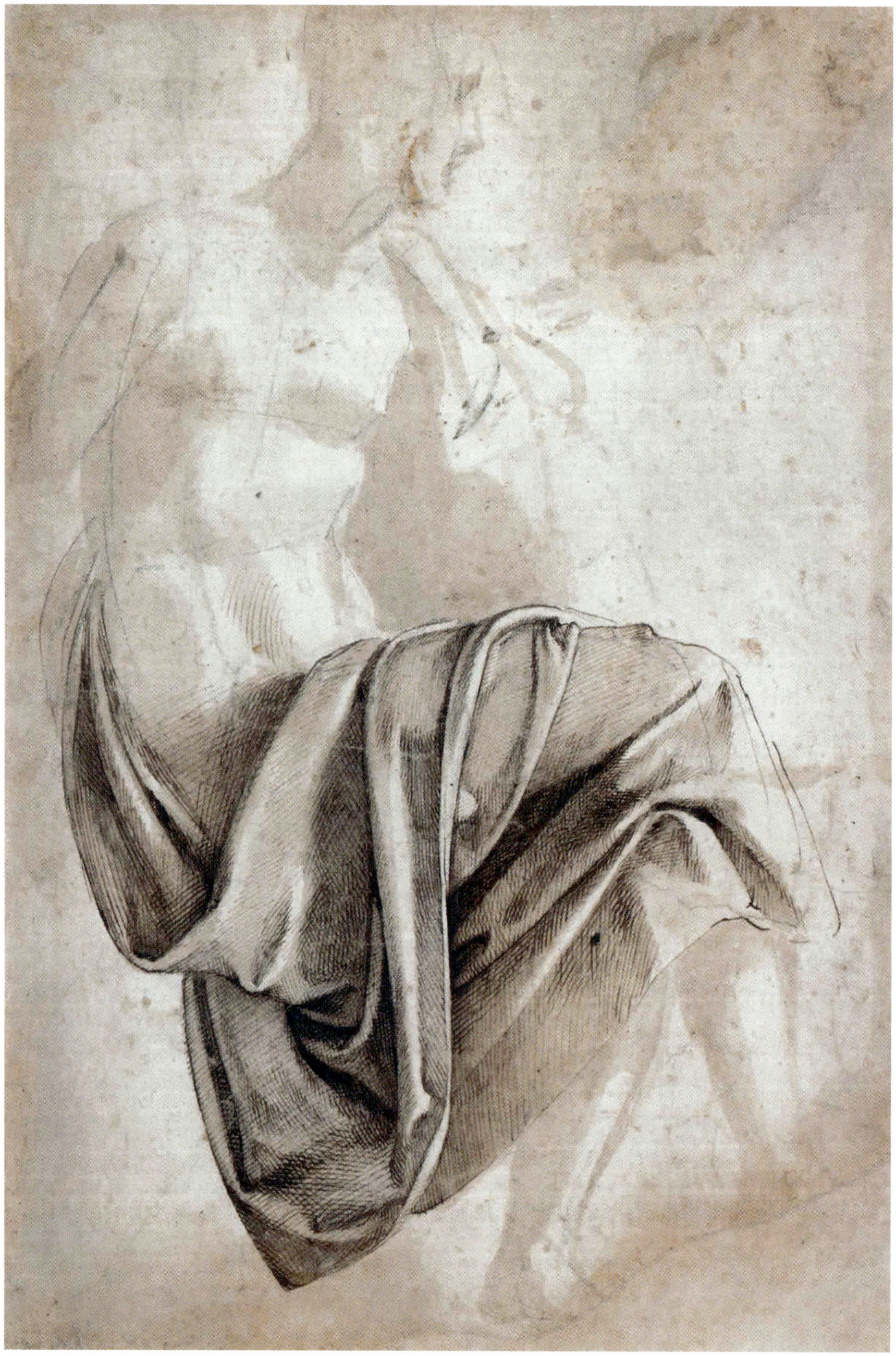

8 Michelangelo, *Study for the Drapery of the Erythraean Sibyl, Seated to Right with Legs Crossed*, c.1508–12. Brown wash and pen and dark brown ink over a black chalk under-drawing on paper, 38.4 × 25.8 cm. London: British Museum, Department of Prints and Drawings (inv. 1887,0502.118). Photo: © The Trustees of the British Museum.

This method is often presented as extremely unusual in the seventeenth century, but it is so only up to a point. Other seventeenth-century Dutch artists are known to have used the camera obscura to define the limits and visual characteristics of a space. One such is the landscape painter Hendrick Danckerts, some of whose drawings reveal that he also used it (*plate* 7). Articulated mannequins (*manichino*), small 'wooden mannequins' (*figuretta di legname*) and clay or wax models are also known to have been used by painters and sculptors since the fifteenth century and became common in the sixteenth and seventeenth centuries. They were even recommended by Italian theoreticians of painting and sculpture as an aid to the study of foreshortening, shadows and drapery (*plate* 8).[45] Neither were life-size mannequins rare. According to

Giorgio Vasari, Fra Bartolommeo 'caused a life-size figure of wood to be made, which moved at the joints' in order to clothe it 'with real draperies, from which he painted most beautiful things, being able to keep them in position as long as he pleased, until he had brought his work to perfection'.[46] And it is known that various of Vermeer's colleagues, such as Cornelis van der Voort, Gerard ter Borch, Jan van Hoogstraten and Aert de Gelder, also made regular use of this technique.[47]

Neither is the idea of modelling the design of a painting by means of a maquette new.[48] It seems this practice developed from the second half of the sixteenth century onwards. Carlo Ridolfi writes that Tintoretto was in the habit of constructing small-scale models of his compositions, which he then filled with wax figurines and illuminated with candles in order to study the effects of light and shadow.[49] According to various contemporary accounts, Nicolas Poussin did the same.[50] And there is also evidence of the practice among Vermeer's contemporaries, as Samuel van Hoogstraten reports in 1678:

> In the past, I saw in The Hague an experienced painter, Samuel Smits, who was in the habit of initially modelling in wax miniature that which he wanted to compose (*ordineeren*). In his studio I saw all kinds of boards supporting small, indistinct figures in crude relief. Here was Tiresias, sitting in judgement over Juno and her husband as they argued. There were Pan and Apollo competing to see who was the better musician. Elsewhere, Narcissus was gazing at himself in a crystal-clear spring, enamoured of his own reflection. The little gowns were made of painted white paper. This was first moistened and then folded in a generous and attractive manner. The trees were made from twigs of oak. The ground and grottoes were intelligently constructed of moss and the water was a mirror.[51]

Vermeer did not, then, initiate entirely new methods of composition and invention. He did, however, innovate by combining one method with another. I believe it is even possible to claim that he was one of the few painters of the seventeenth century to prepare the configuration of his paintings by initially constructing what could be termed a 'life-size maquette'. It is likely that he was not the first to proceed in this way, as the following recommendation of Roger de Piles, written seven years before the death of Vermeer, appears to describe the precise method of the Delft painter:

> For what remains, you are to have a Lay-man almost as big as the life, for every Figure in particular, besides the natural Figure before you, on which you must also look, and call it for a witness, which must first confirm the thing to you, and afterwards to the Spectators as it is in reality. You may make use of these Models with delight, if you set them on a Perspective Plan, which will be in the manner of a Table made on purpose. You may either raise or let it down according to your convenience; and if you look on your Figures through a hole so contriv'd, that it may be mov'd up and down, it will serve you for a point of Sight and a point of Distance, when you have once fix'd it. The same hole will further serve you to set your Figures in the Ceiling and dispos'd upon a Grate of Iron-wire, or supported in the Air by little Strings rais'd at discretion, or by both ways together. You may joyn to your Figures what you see fitting, provided that the whole be proportioned to them; and in short what you your self may judge to be of no greater bigness than theirs. Thus, in whatsoever you do there will be more of truth seen,

> your work it self will give you infinite delight, and you will avoid many doubts and difficulties which often hinder you, and chiefly for what relates to linear perspective, which you will there infallibly find, provided that you remember to proportion all things to the greatness of your Figures and especially the points of Sight and of Distance.[52]

For Roger de Piles, as for Johannes Vermeer, the fictive scene, consisting of the space (the maquette) and the figures (the mannequin), fulfils similar functions to the small-scale model that artists, and more particularly architects, have made of their works. It enables the painter to anticipate, evaluate and simulate his picture before it is painted with the help of visual experimentation and the spatial arrangement of its principal elements, thereby no doubt dispensing with the long and tiresome task of making preparatory drawings.[53]

But this method also offers the possibility of discovering new viewpoints, surprising lighting effects and unexpected framing, organized according to a nature that even when modelled, remains at all times infinitely rich and varied in keeping with Karel van Mander's motto as recalled at the beginning of this article. A model such as the maquette constructed and used by Vermeer does not, any more than a camera obscura or a geographical map, reveal the world as it is but rather as it could be. The model functions at the same time like a piece of 'simulation software', such as might be spoken of in the field of technology or virtual reality today, or as a 'vision machine', to borrow an expression coined by Paul Virilio, in that it enables the painter to learn to see nature better by looking differently and discarding visual habits, thereby allowing him to work *naer het leven and uyt den geest* at the same time.[54]

It is known, however, that Vermeer also constructed his picture space by continuing to modify the composition of his paintings until their completion, suppressing figures or motifs (which, incidentally, invalidates the hypothesis put forward by Philip Steadman that the Delft painter painted his pictures from start to finish using a camera obscura) and employing perspective to organize his visual material (*plate* 9). For Vermeer, the 'life-size maquette' method was therefore part of a different enterprise to that described by Roger de Piles, which was focused on facilitating the genesis of a work. This can be compared to the method of Vermeer's contemporary, Samuel van Hoogstraten. According to Arnold Houbraken, his former master, van Hoogstraten had set up a small theatrical stage in the attic of his home in Dordrecht. In order to stimulate their spatial and expressive imaginations, van Hoogstraten encouraged his apprentices to stage and perform plays there, which their parents and friends could attend.[55] It is no doubt to these attempts at modelling that van Hoogstraten's famous 'dance of the shadows' (*schaduwendans*) testifies, a visual scenographic experiment on projected shadows proposed and in his treatise the *Inleyding* (*plate 10*). The considerable amount of time devoted by van Hoogstraten to this kind of installation (best described as a 'stage design' in the primary sense of the term[56]), or by Vermeer to his life-size maquettes, would no doubt have merited criticism under the traditional concept of the *modello*. As we will recall from Leon Battista Alberti's criticisms, this was originally meant to display 'the Contrivance of the Inventor' rather than 'the Hand of the Workman'. It would seem, however, that these painters were not interested in using these models merely in the elaboration of their works; they also used them as an important tool in boosting the prestige of their savoir-faire. This was conceived both as a glorification of their technical dexterity (their *handeling*, as it was called at the time) and as a celebration of their gaze, combining both their perceptive *savoir-voir*, to borrow an expression from Giovani Paolo Lomazzo, and intellectual (what might be called *savoir-concevoir*) qualities.[57]

No doubt it was in this sense that the words of Pieter Teding van Berkhout are to be understood. After visiting Vermeer's studio on 14 May and 21 June 1669, the art lover recorded in his journal, which he kept in French, that he had admired the 'curiosities' of the Delft painter along with the 'specimens of his art, the most extraordinary aspect of which is their perspective'.[58] These laudatory observations paint the picture of a painter 'curieux' in the sense in which the word was understood in the seventeenth century, that is to say 'with a desire to learn, to see good things, the marvels of art and nature'. But the wealthy magistrate from The Hague was also speaking of a curious art, falling within the sphere of what were then known as 'the curious sciences, familiar to only a few, which possess their own peculiar secrets, such as chemistry and part of optics, which reveals extraordinary phenomena with mirrors and lenses', and which on that account 'contains many remarkable things known to few men'.[59] This is an art in which beauty does not merely reside in perfectly mastered technical skills, but also in the ideas conceived of and presented by the artist.

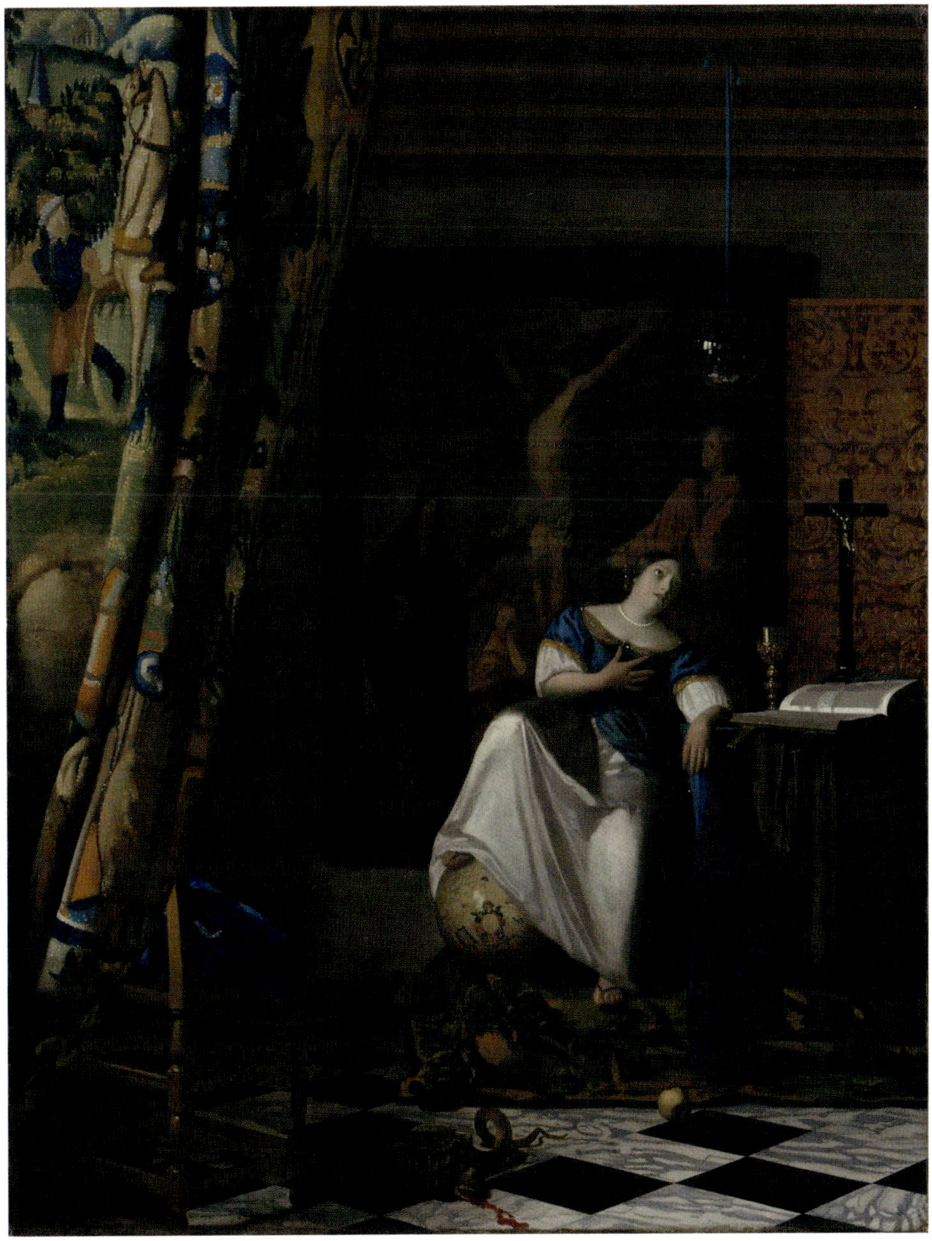

9 Johannes Vermeer, *Allegory of the Catholic Faith*, c. 1670–72. Oil on canvas, 114.3 × 88.9 cm. New York: Metropolitan Museum of Art, The Friedsam Collection (inv. 32.100.18). Photo: © Metropolitan Museum of Art.

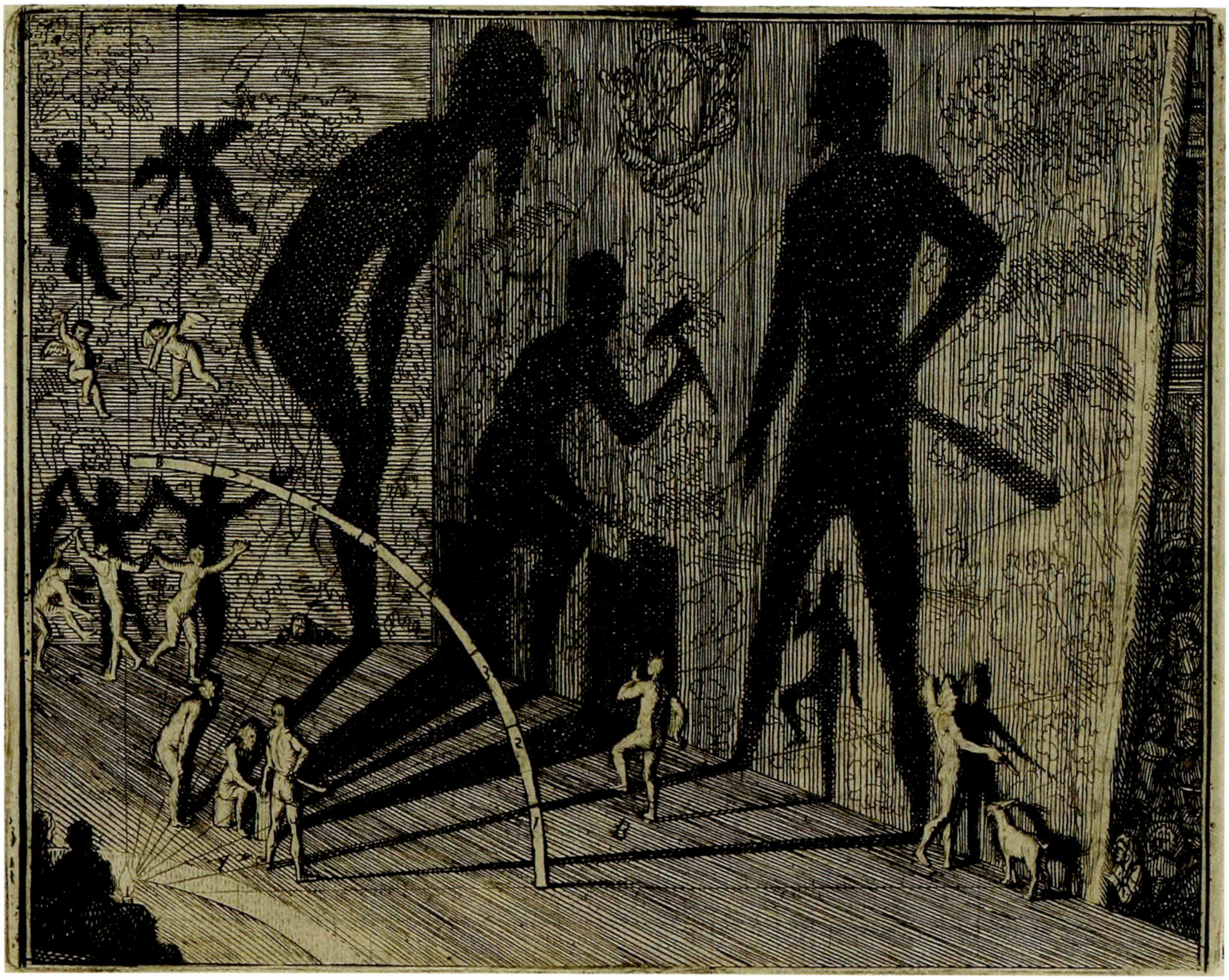

10 Samuel Van Hoogstraten, *La Danse des ombres*, c.1678, from *Inleyding tot de hooge schoole der schilderkonst*, Rotterdam: François Van Hoogstraten, 1678. Etching, 9.7 × 12.6 cm. Photo: © Rijksmuseum, Amsterdam.

Although it is clearly delicate to make Vermeer the emblem of seventeenth-century Dutch painting, the evidence demonstrates that his keen interest in – if not obsession with – different ways of modelling his paintings – the camera obscura, mannequins, maquettes, perspective – stem from a broader preoccupation shared by a number of his contemporaries. Starting from this point, I believe it is necessary to reassess the way that seventeenth-century Dutch artists could have perceived the differences between a work made 'from life' and 'from memory'. Svetlana Alpers is quite correct in emphasizing the importance to many seventeenth-century Dutch painters of glorifying the savoir-faire element of their art, which they regarded as a liberal art notwithstanding its craft aspects.[60] But this interpretation needs to be qualified by demonstrating at the same time that this glorification by no means involved a rejection of, or distancing from, the intellectual dimension of artistic practice. In this context, a model can be understood as a place where theories and practices combine and intertwine. Consisting of a long and slow phase of preparation during which the pictorial scene gradually takes shape, before the results of this modelling are geometrically arranged by means of perspective, the method of conception and composition analysed in the case of Vermeer brings together what Francis Bacon called the experimental and the rational faculties. If it is comparable to the experimental model praised by the British philosopher, this is precisely because the painter makes technical and visual experimentation the locus of a form of invention that differs in no way, as has been shown, from some of the definitions found from the middle of the sixteenth century onwards in Italian art theory. It is from the direct experience of the pictorial space, modelled by the maquette, the mannequin and the scene-setting, that ideas about composition spring forth. And it is thanks to the regulating capacity of geometrical perspective that this experimentation assumes form and meaning within the confines of the painted composition. Vermeer wanted this preparatory work to be careful and thorough, but also recognizable to the viewer, as suggested by the importance accorded to the figure of the painter in the *Art of Painting*, and also the ironic intrusion into *The Concert* of his own easel. This is because the Dutch painter was making the modelling both its *mise en valeur* (contribution to the finished work) and its *mise en abyme* (self-reflection in the finished work), a key criterion in the evaluation of his paintings. Once again, this taste for the art of modelling is to be found in both Dutch and Italian practice. Indeed it is strikingly apparent in the seventeenth-century Netherlandish case, and in other related areas too, such as dolls' houses (*poppenhuisen*).[61] Giorgio Vasari explains that Filippo Brunelleschi was also 'for ever making, on the slightest occasions, designs (*disegni*) and models (*modelli*)', often 'with his own hand' (*di sua mano*), not only of his buildings, but also of the machines used on his building sites.[62] Other artistic practices that were especially popular in southern Europe, such as *sacri monti* or the arranging of polychrome sculptures into tableaux, derive from the same ambition to model a coherent visual space.[63] In basing my argument on the example of painters who shared this new concept of artistic modelling, I hope to have demonstrated that for them, the genesis of their paintings could no longer be reduced simply to the particular space and time in which they were conceived and painted. Progressively claiming independence from the works for which it was meant to be merely preparation, the task of modelling enabled these artists to think of the genesis of their works as an art form in its own right. This is how they played a role in the prehistory of the notion of 'work in progresss', which thus makes its entrance alongside them, and on an equal footing, into the field of Western art.

Notes

1 Larry Fink, *On Composition and Improvisation*, New York, 2014, 90.

2 Marcel Proust, *A la recherche du temps perdu*, Paris, 1999, 489. The author is referring here to Sandro Botticelli's *Madonna del Magnificat* (c. 1480–81, Florence, Galleria degli Uffizi).

3 Ernst Hans Gombrich, *Art and Illusion: A Study in the Psychology of Pictorial Representation*, London, 1960.

4 Svetlana Alpers and Paul Alpers, 'Ut Pictura Noesis? Criticism in literary studies and art history', *New Literary History*, III: 3, 1972, 437–58; Svetlana Alpers, 'Art history and its exclusions: The example of Dutch art', in Norma Broude and Mary D. Garrard, eds, *Feminism and Art History: Questioning the Litany*, New York, 1982, 183–99; Svetlana Alpers, *The Art of Describing: Dutch Art in the Seventeenth Century*, Chicago, IL, 1983. See also Robert D. Huerta, *Giants of Delft: Johannes Vermeer and the Natural Philosophers: The Parallel Search for Knowledge During the Age of Discovery*, Lewisburg, PA, 2003; Claudia Swan, *Art, Science, and Witchcraft in Early Modern Holland: Jacques de Gheyn II (1565–1629)*, Cambridge, 2005; Eric Jorink, ed., *Art and Science in the Early Modern Netherlands*, Zwolle, 2011.

5 Svetlana Alpers, 'Picturing Dutch culture', in Wayne Franits, ed., *Looking at Seventeenth-Century Dutch Art: Realism Reconsidered*, Cambridge, 1997, 57–67. These aspects have since been substantially qualified or even challenged. See in particular Christopher Brown, *The Dutchness of Dutch Art*, Amsterdam, 2002; Thomas DaCosta Kaufmann, ed., *L'Art flamand et hollandais: Belgique et Pays-Bas*, 1520–1914, Paris, 2002.

6 Alpers, *The Art of Describing*, 222.

7 Alpers, *The Art of Describing*, 30–2, 240–1.

8 Vienna, Kunsthistorisches Museum, inv. GG_9128.

9 Alpers, *The Art of Describing*, 165–8.

10 Daniel Arasse, *L'Ambition de Vermeer*, Paris, 2001, 76–140.

11 Ivan Gaskell, *The Oxford Art Journal*, VII: 1, 1984, 57–60; Eddy de Jongh, *Simiolus*, XIV: 1, 1984, 51–9; Jeroen Stumpel, *The Burlington Magazine*, CXXVI, 1984, 580–1; Kim H. Veltman, *Kunstchronik*, XXXVII, 1984, 262–7; Jan Bialostocki, *The Art Bulletin*, LXVII, 1985, 520–6; Josua Bruyn, *Oud-Holland*, XC, 1985, 155–60; Anthony Grafton and Thomas DaCosta Kaufmann, *Journal of Interdisciplinary History*, XVI: 2, 1985, 264–5; Gregory Martin, *Apollo*, CXXI, 1985, 66.

12 Karel Van Mander, *Het Schilder-Boeck*, Haarlem, 1604, fol. 186v.

13 Alpers, *The Art of Describing*, 40–1. See also Claudia Swan, 'Ad vivum, naer het leven, From the Life: Defining a mode of representation', *Word & Image*, XI: 4, 1995, 353–75; Robert Felfe, 'Naer het leven: eine sprachliche Formel zwischen bildgenerierenden Übertragungsvorgängen und ästhetischer Vermittlung', in Claudia Fritzsche, Karin Leonhard and Gregor J. M. Weber, eds, *Ad Fontes! Niederländische Kunst des 17. Jahrhunderts in Quellen*, Petersberg, 2013, 165–95.

14 Van Mander, *Het Schilder-Boeck*, fol. 242v. Regarding these issues, see Alpers, *The Art of Describing*, 40–1.

15 Cornelis Hofstede De Groot, *Die Urkunden über Rembrandt (1575–1721)*, The Hague, 1906, 204.

16 Jurij I. Kuznetsov, 'Pieter Quast copied Savery's "Naer het leven"-drawings', in J. Buryn et al., eds, *Album amicorum J. G. Van Gelder*, The Hague, 1973, 215–18.
17 Antoine Compagnon, *La Seconde main ou le travail de la citation*, Paris, 1979.
18 'Dat ghy staersichtich schier stomp hebt versleten / U sinnen, leer-lustigh om meer te weten' (Van Mander, *Het Schilder-Boeck*, 'Grondt', VIII, 1, fol. 34r°).
19 'En gaen sien de schoonheyt, die daer is buyten, / Daer ghebeckte wilde Musijckers fluyten, / Daer sullen wy bespieden veel ghesichten, / Die ons al dienen om Landtschap te stichten / Op vlas-waedt, oft Noorweeghsch' hard' eycke plancken' (Van Mander, *Het Schilder-Boeck*, 'Grondt', VIII, §3, fol. 34v°. Author's emphasis).
20 Hessel Miedema, *Kunst, kunstenaar en kunstwerk bij Karel Van Mander: een analyse van zijn levensbeschrijvingen*, Alphen aan den Rijn, 1981, 1–3.
21 'Oock neffens Hemskercken begonnen wercken der Apostelen het heel ghevolgh, en veel meer ander dinghen, welcke die ons ghenoeghsaem aenwijs doen, en ghetuyghnis gheven zijnen overvloedigen cloecken geest, soo in't ordineren der Historien, als in't stellen, en toemaken zijner beelden, met alderley omstandicheden der Const' (Van Mander, *Het Schilder-Boeck*, fol. 267v°). For the second quotation, see Van Mander, *Het Schilder-Boeck*, fol. 296v°.
22 'De Schilder-const, die haer eerste baringhe heeft in den geest, oft ghedacht door inwendighe inbeeldinghen, aleer sy met der handt opghevoedt, en ter volcomenheyt ghebracht can worden' (Van Mander, *Het Schilder-Boeck*, fol. 225r°).
23 'In zijn reysen heeft hy veel ghesichten nae t'leven gheconterfeyt, soo datter gheseyt wort, dat hy in d'Alpes, wesende, al die berghen en rotsen had in gheswolghen, en t'huys ghecomen op doecken en Penneelen uytghespogen hadde, soo eyghentlijck con hy te desen en ander deelen de Natuere nae volghen' (Van Mander, *Het Schilder-Boeck*, fol. 233r°).
24 Terence Cave, *The Cornucopian Text: Problems of Writing in the French Renaissance*, Oxford, 1985.
25 'Zoo zietmen hier wat gros of generael een recht natuerlijke Schildery behoorde te hebben. Maer dit zelve is ook te zien in verklein glazen en spiegeltjes, die, hoewelze de reykening wat verbuigen, het gros der koloreeringe en houdinge zuiverlijk vertoonen' (Samuel van Hoogstraten, *Inleyding tot de hooge schoole der schilderkonst*, Rotterdam, 1678, 263).
26 Arnaud Maillet, *The Claude Glass: Use and Meaning of the Black Mirror in Western Art*, New York, 2004.
27 Alpers, *The Art of Describing*, 141.
28 Alpers, *The Art of Describing*, 142.
29 Hoogstraten, *Inleyding tot de hooge schoole der schilderkonst*, 346. On this point, see also Thijs Weststeijn, 'From hieroglyphs to universal characters: Pictography in the early modern Netherlands', *Nederlands Kunsthistorisch Jaarboek*, LXI, 2011, 238–81. This notion of the map's symbolic rather than mimetic function is, moreover, ancient. It can be found, for example, in Richard Mulcaster, who wrote in 1581: 'Whose use [of drawing] all modelling, all mathematikes, all manuaries do finde and confesse to be to so notorious and so needefull' (Richard Mulcaster, *Positions Wherin those Primitive Circumstances be Examined, which are Necessarie for the Training Up of Children, Either for Skill in their Booke, or Health in their Bodie*, London, 1581, 35) and in the poet George Gascoigne, who states a wish to 'tel some reasonable worde, / Of Hollandes state, the which I will present, / In Cartes, in Mappes, and eke in Models made' (George Gascoigne, *The Posies*, London, 1575, 173).
30 Regarding this question, see in particular Arnaud Maillet, *Prothèses lunatiques : les lunettes, de la science aux fantasmes*, Paris, 2007.
31 Franck Varenne, 'Modèles et simulations dans l'enquête scientifique : variétés traditionnelles et mutations contemporaines', in Franck Varenne and Marc Silberstein, *Modéliser & simuler: épistémologies et pratiques de la modélisation et de la simulation*, vol. 1, Paris, 2013, 11–51, 27.
32 Leonard Digges, *A Prognostication Everlastinge of Righte Good Effecte Fruictfully Augmented by the Auctour*, London, 1578, 'To Rdr. M', n. p.
33 Euclid, *The Elements of Geometrie of the Most Auncient Philosopher Euclide of Megara*, London, 1570, preface, fol. ciijv.
34 Van Mander, *Het Schilder-Boeck*, fol. 112r°.
35 Dirck van Bleyswijck, *Beschryvinge der stadt Delft, betreffende des selfs situatie, oorsprong en ouderdom*, Delft, 1667, 128.
36 'Soit que l'on fasse un Tétrastyle, un Hexastyle ou un Octostyle, une de ces parties sera le module, qui n'est autre chose que la grosseur d'une Colonne' (Vitruvius and Claude Perrault, *Les Dix livres d'architecture de Vitruve* (1684), Liège, 1996, III, ii, 78). In 1598, Richard Haydocke wrote in his English translation of Giovanni Pietro Lomazzo's *Trattato della pittura* that 'Vitruvius measureth this order by models' (Giovanni Paolo Lomazzo, *A Tracte Containing the Artes of Curious Paintinge, Carvinge & Buildinge*, Oxford, 1598, I, 89).
37 Leon Battista Alberti, *On Painting* (1435), New Haven, 1970, III, 61, 95.
38 Alberti, *On Painting* (1435), 267.
39 Philippe Fleury, 'Vitruve et le métier d'ingénieur', *Cahiers des études anciennes*, XLVIII, 2011, 7–34.
40 Leon Battista Alberti, *The Ten Books of Architecture* (1485; 1755 Leoni Edition), New York, 1986, II, i, 22.
41 Alberti, *The Ten Books of Architecture*, II, i, 22.
42 Giorgio Vasari, *Le vite de' più eccellenti architetti, pittori, et scultori italiani, da Cimabue, insino a' tempi nostri* (1550), Turin, 1986, 60–2. On the use of maquettes in architecture, see Andres Lepik, *Das Architekturmodell in Italien 1335–1550*, Worms, 1994; Bernd Evers, ed., *Architekturmodelle der Renaissance: die Harmonie des Bauens von Alberti bis Michelangelo*, Munich, 1995. On models sculpted during the Renaissance, see in particular Johannes Myssok, *Bildhauerische Konzeption und plastisches Modell in der Renaissance*, Münster, 1999. On preparatory drawings, see above all Lizzie Boubli, ed., *Savoir-faire: la variante dans le dessin italien au XVIe siècle*, Paris, 2003; Lizzie Boubli, *L'Atelier du dessin italien à la Renaissance: variante et variation*, Paris, 2003; Eric Pagliano, ed., *L'Atelier de l'oeuvre: dessins italiens du Musée Fabre*, Ghent, 2013.
43 For a broader discussion of the issues broached here, see Jan Blanc, *Vermeer: la fabrique de la gloire*, Paris, 2014, 58–171.
44 Philip Steadman, *Vermeer's Camera: Uncovering the Truth Behind the Masterpieces*, Oxford, 2001; Wolfgang Lefèvre, ed., *Inside the Camera Obscura: Optics and Art under the Spell of the Projected Image*, Berlin, 2007. See also Christopher Heuer, 'Perspective as process in Vermeer', *Anthropology and Aesthetics*, XXXVIII, 2000, 82–99; Kobayashi-Sato Yoriko, 'Vermeer and his thematic use of perspective', in A. Golahny, M. M. Mochizuki and L. Vergara, eds, *In His Milieu: Essays on Netherlandish Art in Memory of John Michael Montias*, Amsterdam, 2006, 209–18.
45 Antonio Averlino Filarete, *Trattato di architettura* (1499), Milan, 1972, 2 vols, XXII–XXIV; Giovanni Battista Armenini, *On the True Precepts of the Art of Painting* (1587), New York, 1977, II, v, 164–70; Giovanni Paolo Lomazzo, *Idea del tempio della pittura* (1590), Florence, 1974, 92; Giovanni Pietro Bellori, *The Lives of the Modern Painters, Sculptors and Architects* (1672), Cambridge, 2005, 323.
46 Giorgio Vasari, *Lives of the Painters, Sculptors and Architects* (1568), London, 1996, 2 vols, I, 679.
47 Arnold Houbraken, *De groote schouburgh der nederlantsche konstschilders en schilderessen*, Amsterdam, 1718, 3 vols, II, 169, III, 207–8; E. F. van der Grinten, 'Le Cachalot et le mannequin. Deux facettes de la réalité dans l'art hollandais du seizième et du dix-septième siècle', *Nederlands Kunsthistorisch Jaarboek*, XIII, 1962, 156–79.
48 Avigdor Arikha, 'De la boîte, des figurines et du mannequin', in Pierre Rosenberg and Louis-Antoine Prat, eds, *Nicolas Poussin: 1594–1665*, Paris, 1994, 44–7.
49 Carlo Ridolfi, *Vita di Giacopo Robusti detto il Tintoretto, celebre pittore cittadino venetiano*, Venice, 1642, 8.
50 Antoine Le Blond de la Tour, *Lettre du sieur Le Blond de La Tour à un de ses amis, contenant quelques instructions touchant la peinture*, Bordeaux, 1669, 38–41; Joachim von Sandrart, *Academie der Bau-, Bild- und Mahlerey-Künste von 1675: Leben der berühmten Maler, Bildhauer und Baumeister*, Munich, 1925, 184, 258.
51 Hoogstraten, *Inleyding tot de hooge schoole der schilderkonst*, 192.
52 Charles-Alphonse Dufresnoy, Roger de Piles and Richard Graham, *De Arte Graphica: The Art of Painting*, London, 1695, 146–7.
53 Hélène Vérin, *La Gloire des ingénieurs: l'intelligence technique du XVIe au XVIIIe siècle*, Paris, 1993, 43–8.
54 Paul Virilio, *The Vision Machine*, Bloomington, IN, 1994.
55 Arnold Houbraken, *De groote schouburgh der nederlantsche konstschilders en schilderessen*, Amsterdam, 1718, 3 vols, II, 162–3.
56 Vitruvius and Perrault, *Les Dix livres d'architecture*, I, ii, n. 7, 10.

57 'Sapere vedere' (Giovanni Paolo Lomazzo, *Trattato dell'arte de la pittura: ne' quali si contiene tutta la theorica e la prattica d'essa pittura*, Milan, 1584, 262–3).
58 Blanc, *Vermeer: la fabrique de la gloire*, Paris, 2014, 87.
59 Antoine Furetière, *Dictionnaire universel, contenant généralement tous les mots français tant vieux que modernes, et les termes de toutes les sciences et des arts*, The Hague, Rotterdam, 1690, 3 vols, art. 'Curieux'.
60 Alpers, *The Art of Describing*, 100, 114–15.
61 Jet Pijzel-Dommisse, *Het Poppenhuis van Petronella de la Court*, Utrecht, 1987; Karin Braamhorst and Marlies Visser, *Het poppenhuis van Petronella Oortman*, Amsterdam, 2014.
62 Giorgio Vasari, *Lives of the Painters, Sculptors and Architects* (1568), London, 1996, 2 vols, I, 346, 349.
63 Timothy Christopher Verdon, *The Art of Guido Mazzoni*, New York, 1978; Maria Luisa Tibone and Antonella Bo, eds, *Sindone e Sacri Monti: l'arte e la società, lo spettacolo e la devozione*, Turin, 1998; Gabriele Landgraf, *Die Sacri Monti im Piemont und in der Lombardei: zwischen Wirklichkeitsillusion und Einbeziehung der Primärrealität*, Frankfurt, 2000.

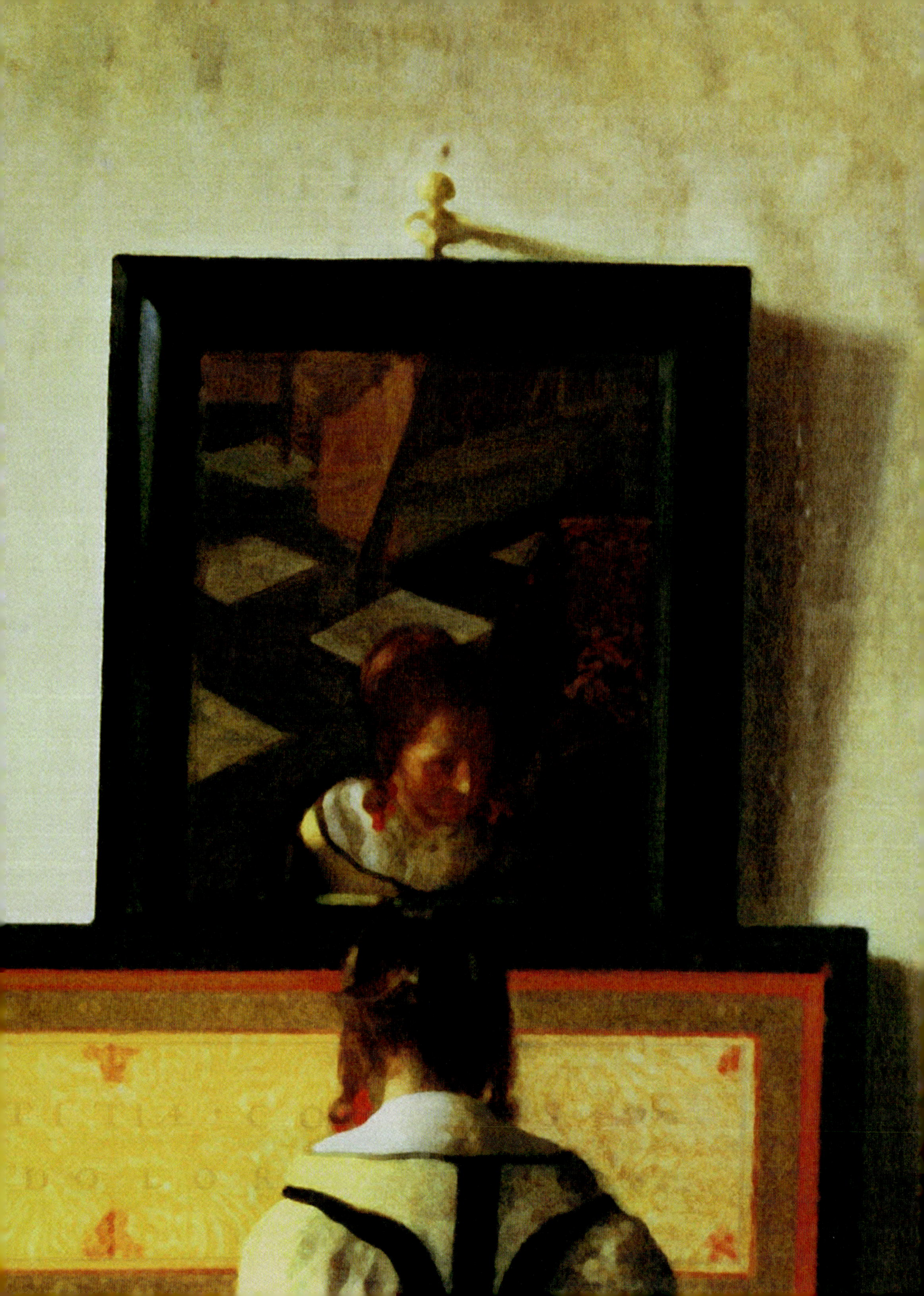

Chapter 4
Looking in the Mirror of Renaissance Art

Genevieve Warwick

Detail from Johannes Vermeer, *Music Lesson*, 1662–65 (*plate 5*).

Art and Technology in Early Modern Europe *Edited by Richard Taws and Genevieve Warwick*.

In his 1568 biography of the Venetian painter Giorgione, Giorgio Vasari described a now-lost or possibly fictive painting by this artist composed of a series of mirrored effects. Apparently it depicted a male nude seen from behind:

> at his feet, a limpid stream of water bearing his reflection. To one side was a burnished cuirass that the man had taken off, and this reflected his left profile since the polished surface of the armour revealed everything clearly. On the other side was a mirror reflecting the other profile of the nude figure. This was a very fine and fanciful idea . . . . greatly praised and admired for its beauty and ingenuity.[1]

While the Venetian critic Paolo Pino had twenty years earlier identified the figure as a St George clad in armour, by 1590 G. P. Lomazzo would alter the identity of the figure yet again, transforming it into a female as an allegory of painting: 'Giorgione represented Painting as a nude, reflected in a stream ... with a mirror behind her to show her from the back, and a polished cuirass at her side that reflected her in profile.'[2] Whether fiction or the description of a lost work of art, the conception fits well into a historical trajectory of pictorial, as well as critical, concerns of the period. For at this moment Renaissance art saw an efflorescence of paintings concerned precisely with the effects of mirrored reflection, represented by mirrors themselves and in a further range of reflective surfaces – armour, pools of mirroring water, and decorative objects such as glass vessels and jewels.[3] As Lomazzo's transformation of the figure's identity into an allegory of *Pittura* makes manifest, the mirror reflection in Renaissance art was to become an emblem of the art of painting itself.

Indeed, the depiction of mirrors within Renaissance painting may be said to have articulated a pictorial declaration of art's very powers of illusion. Inset mirror images served to remind the viewer that a painting, however persuasive in its imitation of nature, was in fact a flat surface covered with coloured pigments. As such, the inset mirror within the image acted as an '*abrégé du tableau*' – a painting in reduction – even as it underscored the analogy between the two.[4] The mirror within the painting was nothing other than the representation of painting within itself, and a metonymy of its own art. It thus acted as a reflexive 'doubling' of art, a return of painting on itself. This *mise-en-abyme*, or modality of self-reflection within the work famously defined Michel Foucault's view of early modern cultural production. In the mirror-within-the-painting Foucault recognized what he termed 'the enchantment of the double', which in his view constituted the structuring sign of early modernity. Such emulation of the mirror was, for Foucault, not only a literary preoccupation of the period (the 'play within the play'), but also a lodestone of its mimetic art, manifest above all in '*les miroirs-tableaux*', those works that explicitly displayed their conceptualization of painting as specular.[5] Further, as Lucien Dällenbach famously argued in a study of the specular double in literature, the painted mirror image in early modern art may be understood as a representation of the art of painting itself.[6] In this regard it was a recollection, within the illusion of the painted surface, of what André Chastel would term 'the scenario of painting's own production', that is, a depiction of the art of art.[7]

During this same period, the mirror as object became an increasingly common workshop tool, with a range of uses within artistic practice. Chief among these was as an instrument of mimesis. In the pursuit of an ever-more accurate pictorial representation of nature, Renaissance artists turned to the growing supply of mirrors as a means of reflecting, and so pictorializing, the world around them. In the studio, the mirror served as a practice-based device to facilitate the transposition of the visible world of three dimensions into the flat plane of a painting. Its reflected image offered a simulacrum of painting's field of visible observation to the painter's study of his subject, already translated into a two-dimensional, pictorial format. Moreover, it rendered the scene in reduced scale, just like a painting. Thus the mirror reflection was both an instrument and a model of the visible world for the painter's brush.[8] The increasingly prevalent workshop use of the mirror as a mimetic tool may thus be tied to what has been analysed as a greater mechanization of artists' methods from the fifteenth century on.[9] This is linked to the changing paradigms of visual imitation that scholars identify as 'Renaissance', in which painting was admired for its proximity to a mirror-reflected field of vision. In this context the mirror was but a technology of imitation, to be sure, yet its visual effect was recognized as an equivalent to painting itself.

This essay therefore takes a double focus, between making and viewing, to argue that a Renaissance art of mimetic resemblance was predicated on a deeply worked approximation between painting and the specular image. The analysis proceeds by singling out particularly pronounced historical manifestations of the mirror analogy in Renaissance and early modern painting, weaving them together with accounts of pictorial experiments with mirror reflections and their elaboration in critical and theoretical treatises of the period. It moves from Brunelleschi's lost but celebrated reflection-painting at the door of Florence Cathedral (*c.* 1413–20) to the great Galerie des Glaces at Versailles (1678–84) to argue for an early modern elision of the picture plane with the mirror, which drew together the surface of a painting with the technological means of its own production.

Mirrors and Mimesis

The pictorial trope of the mirror's reflective surface as the sign of painting was, in a Renaissance historiography of art, famously tied to the developing technology of perspective and optics in the story of Brunelleschi's formative experiment at the door of the Florentine cathedral. Standing inside the Duomo looking out at the Baptistery, Brunelleschi painted a panel of the scene framed by the doorway. The panel's square shape – reported to have been a half *braccio* along each side – was designed to correspond in reduced scale to the view of the Baptistery as seen through the Cathedral's double doors, and to match the size of a mirror he used to reflect the scene back to himself. Brunelleschi clearly conceived of the painted panel and the mirror as a pair.[10] To further the correlation between the panel and the mirror reflection as parallel representations of the view itself, Brunelleschi burnished the sky of his panel with silver. In this way the panel's upper section was also a reflective surface which, when placed inside the Duomo's doorway facing outward, would reflect the actual sky with its passing clouds and changes of atmospheric effects. With the mirror in front of the panel and facing towards it, its reflective surface mirrored the painting of the Baptistery and its doubled reflection of the actual, moving, sky. At the centre of his panel Brunelleschi made a small peephole. When the viewer stood inside the Cathedral doors facing out at the view of the Baptistery, as Brunelleschi had done when painting it, and peered through the panel's peephole from the back, s/he saw the panel's reflection in the facing mirror, made to look just like the view itself.

In this pictorial experiment, Brunelleschi's claim was that painting could perfectly reproduce the visible world like a mirror reflection. His motive was to demonstrate the representational accuracy of a mathematically derived perspective system in painting. He deployed mirror technology to advance that claim through a visible demonstration of it. It allowed the viewer to compare the view through the doorway with the painted view seen through the peephole, and to find them fully equivalent. The overriding historical importance of the early fifteenth-century development of perspectival rendering in painting has inclined the scholarly literature to subsume Brunelleschi's great experiment wholly within the narrative of one-point perspective. While acknowledging the central importance of the historic development of perspective for Renaissance art, my interest here is in Brunelleschi's instrument, the mirror, and its role as a technology of mimesis in the story of Renaissance painting. For the claim that painting could represent the fiction of space on a two-dimensional surface was made by analogy with the mirror.

The history of the glassed mirror as a material object is long and diffuse, marked by incremental technological changes in the history of glass production and in the application of different combinations of liquid metals to a glass surface.[11] Early Renaissance mirrors were made from blown glass which was then silvered and cut to form half-globes. The means of making even small flat panes of glass which could then be silvered into mirrors did not become prevalent until the early sixteenth century, and large sheets of flat glass awaited technological developments in glass-making in the mid to late seventeenth century.[12] Yet a much longer tradition of small, flat reflecting surfaces of polished metal, both circular and square, constituted the mirrors of the ancient world, as was known across the Renaissance. Steel gave a particularly clear and colourless reflection, and could be tempered into sheets of considerable size. These metal mirrors might also be silvered to increase the reflectivity of the surface finish, as was seemingly the case with Brunelleschi's panel in its reflection of the sky. Thus the metaphor of the flat squared surface of a painting as mirror-like that Brunelleschi appealed to in his view of the Baptistery was well known across the period. The Renaissance ushered in a newly insistent approximation between painting and the mirror, yet this drew on a much longer history of their analogy and comparison.

The typology of painting as a mirror already present in Brunelleschi's experiment deepened with the increased production of mirrors during the fifteenth and sixteenth centuries. This progressed alongside a growing diffusion of oil paint, whose reflective surface also served to heighten the approximation between a painting and a mirror. The early use of oil sought to form a glistening pictorial surface out of light-reflecting glazes, much like the brilliance of a mirror's sheen. This new-found specularity of painting was prescient above all in the work of Jan Van Eyck, as Vasari acknowledged, noting its

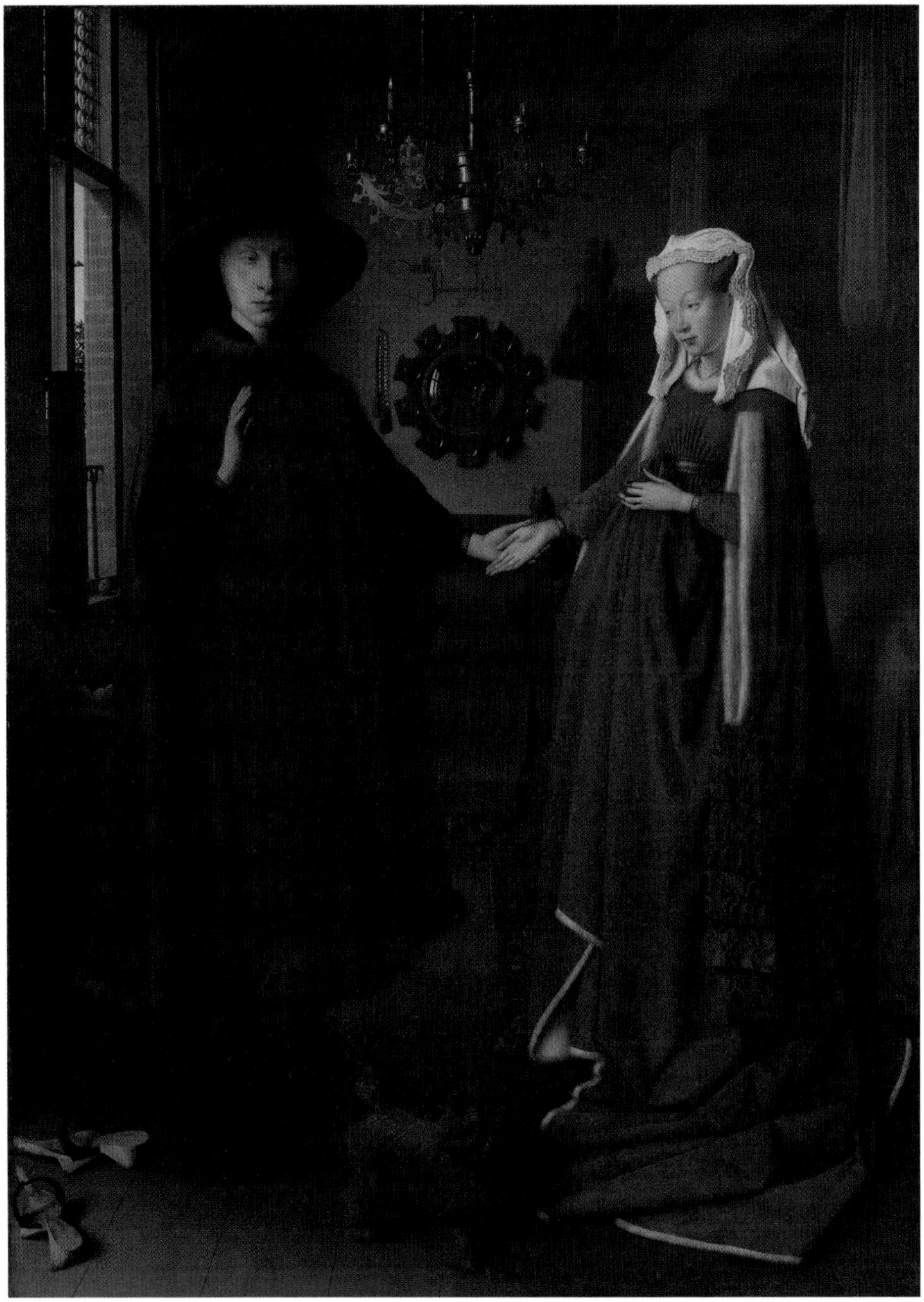

1 Jan van Eyck, *Arnolfini Double Portrait*, 1434. Oil on panel, 82 × 60 cm. London: National Gallery. Photo: Scala, Florence.

heightened mimetic possibilities particularly in the rendering of lustre and gleam.[13] It is epitomized in his *Arnolfini Double Portrait* (1434) whose celebrated perspectival centre concludes in a convex mirror of tondo form. The painted mirror 'reflects' conversely between its pictorial and viewing spaces (*plate 1*).[14] This small circular mirror rises towards its centre, typifying the blown glass manufacture of the early Renaissance. We see the couple from the front in the painting, while the mirror reflection behind shows them in reduction from the back. In the depths of the mirror we also see another pair of figures facing them, which is somehow strangely us, for they stand in what must be our position as viewers. At the same time, it must also have been Van Eyck's position as he painted the work, so conflating the painting's production with its new-found specularity. A metonymy of painting's art as founded in the study of light and reflection, the Arnolfini mirror is surrounded by a succession of objects that amplify this theme: glass beads, a brass chandelier, the small glass roundels of the mirror's frame, all studies in the painter's art of reflection. The detail of the mirror may be seen as an artistic reflection, both literally and metaphorically, on what painting is. Yet it must also have been the means of the painted mirror's making, Van Eyck surely studying from a mirror reflection in order to paint it. This is further corroborated by the use of oil as a binding medium, so deepening the identification between the pictorial and the specular. Here Van Eyck apparently plays with the nature of the illusion of art, fusing viewer space with the fictive space of the painted mirror. This most celebrated of mirror images within painting thus reflects on its own status as pictorial representation. The motif of artistic self-reflection is further amplified in Van Eyck's subsequent depiction of a tiny figure standing at an easel, thought to be himself as he painted, in the shield of St George in his *Virgin and Child with Canon van der Paele* (1436).[15] It is noteworthy that painters and mirror-makers belonged to the same guild in Van Eyck's Bruges, providing an easy point of transfer between their respective crafts of illusion. It is perhaps also significant that the Dutch term for shield – *schilder* – was commonly used to refer to painting, as Carel Van Mander's 1604 book on painting titled *Het Schilderboeck* would later make manifest, and which might underlie Van Eyck's representation of himself painting in the shield of his St George.

According to some scholars the Arnolfini double portrait may have had a painted cover, also attributed to Van Eyck, of a bathing female nude reflected in an inset mirror to show her figure from the side and the front. Now lost, perhaps separated from the double portrait by Arnolfini's wife herself before her death, the image of this possible painted cover for the portrait is known from copies (*plate 2*).[16] Here a small convex mirror shaped of blown glass, as in the Arnolfini portrait, mirrors the figures as if in a globe. This reflected image was, perhaps, like Giorgione's lost work of reflections, designed to 'confound the sculptors' by showing the figure from multiple viewpoints as if in the round. For the sculptors had claimed this as proper to their art alone in the great debate of '*paragone*' or comparison of the arts that structured much Renaissance art criticism.[17] Conversely the painters argued that their art required greater knowledge in order to represent the effects of space, volume, light fall and reflection in a two-dimensional medium. Thus, in his lost work depicting multiple reflecting surfaces with which this essay began, Giorgione used the motif of the mirror to represent the art of painting. For the surface of a painted canvas was just like a mirror, as Brunelleschi had demonstrated, reflecting back to the viewer all that is visible. Thus the painted mirror image was, in this sense, a painting within a painting, both the model and the instrument of its art.

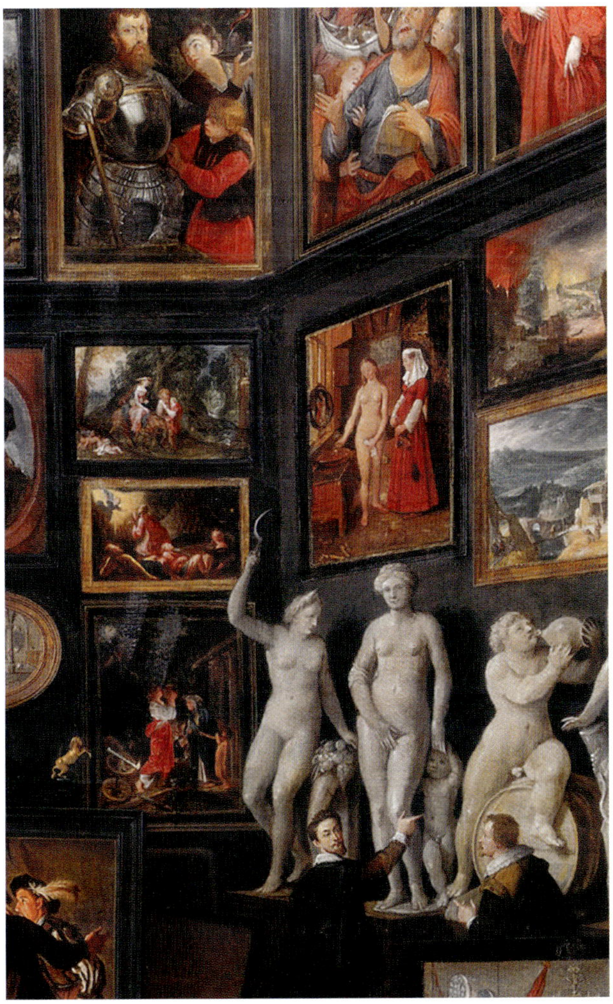

2 Detail of Willem van Haecht, *Gallery of Cornelis van der Geest*, 1628, showing the panel of a female at her toilet linked to the lost cover of the Arnolfini double portrait. Oil on panel, 100 × 130 cm. Antwerp: Rubenshuis. Photo: Scala, Florence.

As Van Eyck painted his Arnolfini portrait, Leon Battista Alberti was writing his great theorization of the art of painting, *De Pictura* of 1435. Indelibly bound up with the development of one-point perspective as the new structuring logic of the picture plane, Alberti's treatise famously conceived of painting as an open window 'through which I see what I want to paint'.[18] This analogy

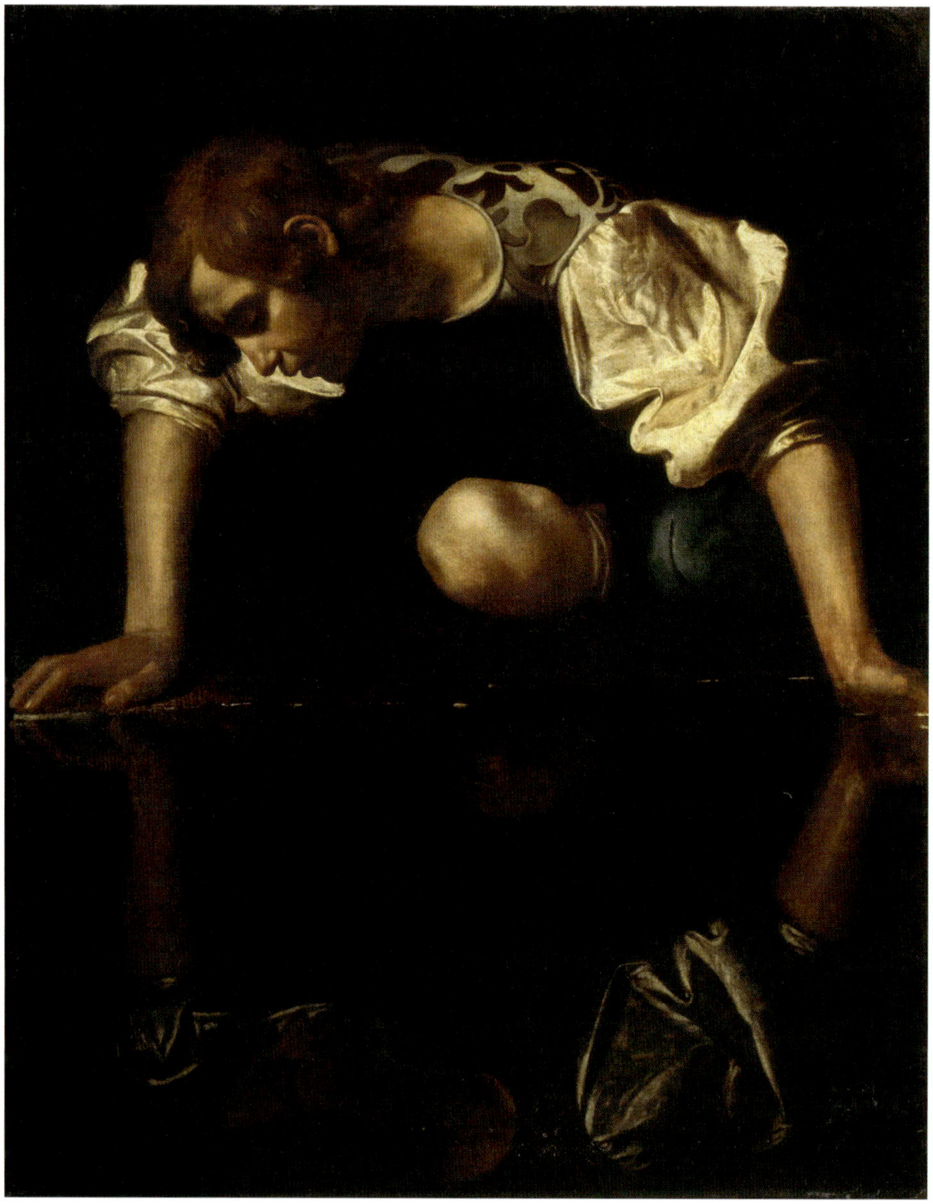

3 Caravaggio, *Narcissus*, c. 1597–99. Oil on canvas, 110 × 92 cm. Rome: Galleria Nazionale d'Arte Antica, Palazzo Barberini. Photo: Scala, Florence.

between painting and an open window is already at play in the depiction of the window within the Arnolfini portrait, and repeated in the reflection of the same window within the painting's inset mirror. The simile is also at work in Brunelleschi's conceptualization of painting as the view through a doorway, transposed into the reduced size of a mirror. The widespread adoption of the window motif within the Renaissance, as an Albertian paradigm of the nature of painting, would become a parallel to the metaphor of the mirror and an equivalent definition of painting's art.

At the same time, like his friend Brunelleschi to whom Alberti dedicated the Italian edition of his book, Alberti also understood the picture plane in terms of the technology of its making. He theorized the plane as what he termed the 'intersection' between the viewer and subject depicted, which he characterized as a perspectival grid standing at a midpoint between, and intersected by, on the one side the viewer's rays of vision and on the other the converging lines of pictorial space. In his practice he advocated the use of a framed grid or squaring device to embody his conceptualization of the intersection, which in effect became the technological means of conceptualizing a work of art. Standing between the artist's eye and the object to be depicted, this grid was just like a window. By its means the artist could frame the composition, using it to transpose her/his perception of the world into two dimensions. S/he constituted this workshop-based device out of a transparent veil of very finely woven cloth like muslin, marked out by parallel lines in a larger thread:

> Nothing can be found, so I think, which is more useful than that veil which among my friends I call an intersection. It is a thin veil, finely woven, dyed whatever colour pleases you and with larger threads [marking out] as many parallels as you prefer. This veil I place between the eye and the thing seen . . . . This veil can be of great use to you . . . . The veil will greatly aid you in learning how to paint . . . .[19]

Like Brunelleschi's mirror, Alberti's intersection was both an instrument of artistic production, and a virtual simulacrum of the work of art in the making. The conflation of the picture plane with the workshop instruments of a mimetic art heightened the proximity between painting as representation and the processes of its own production. The inset mirror, as both a metaphor and an instrument of the art of painting, directly articulated this correlation.

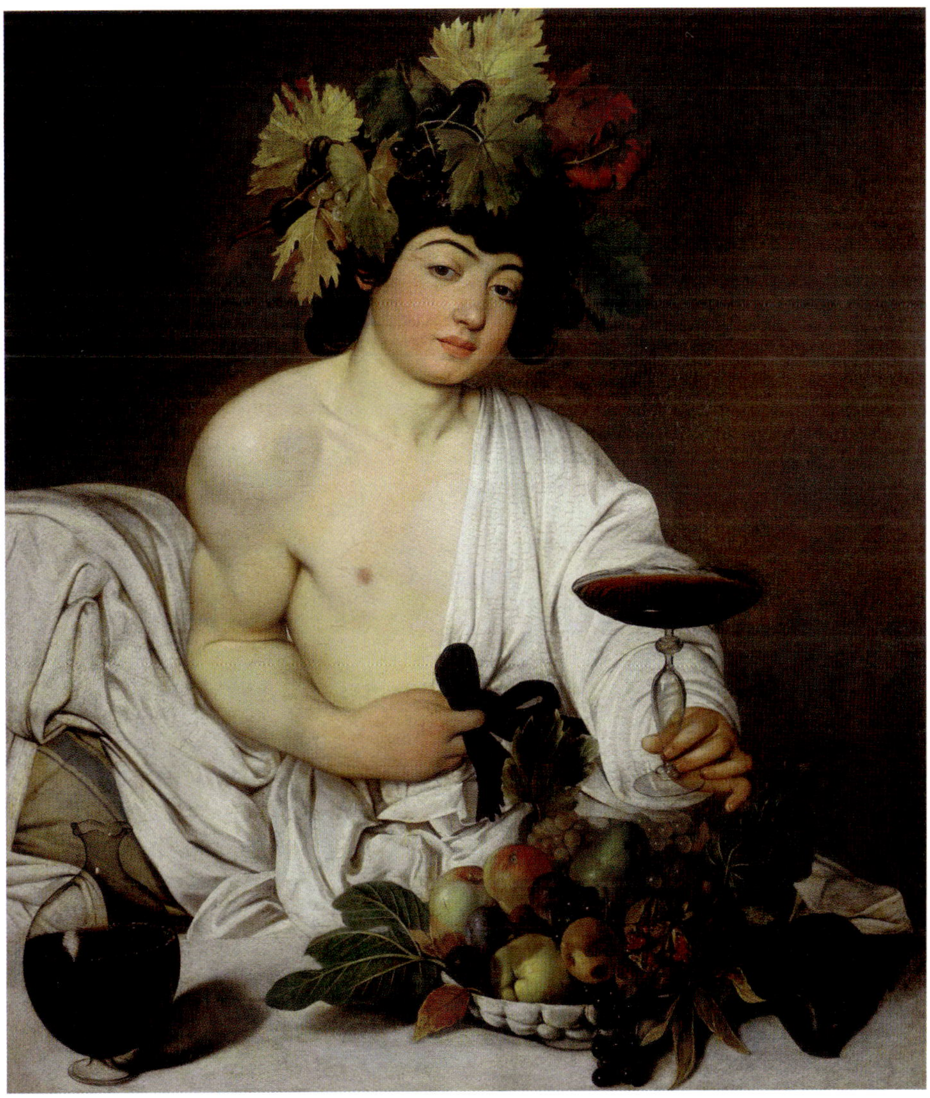

4 Caravaggio, *Bacchus*, c. 1595. Oil on canvas, 95 × 85 cm. Florence: Galleria degli Uffizi Gallery. Photo: Scala, Florence.

Elsewhere, like Brunelleschi, Alberti also conceived of painting as a mirror, for he traced the invention of painting to the myth of the beautiful youth Narcissus who fell in love with his own reflection in a watery pool. Becoming so enamoured that he reached out to embrace this image of himself, he was pulled into the water which, in Ovid's poetic version of the story in the *Metamorphoses*, was the origin of the narcissus plant. In Alberti's retelling, the myth refers instead to the origins of painting, where the boy's reflection was the sign of art:

> I say among my friends that Narcissus who was changed into a flower, according to the poets, was the inventor of painting. Since painting is already the flower of every art, the story of Narcissus is most to the point. What else can you call painting but a similar embracing with art of what is presented on the surface of the water in the fountain?[20]

Long after Alberti, Caravaggio would pictorialize the myth of Narcissus and his embrace of the mirror reflection in the surface of the shimmering pool. The canvas in question represents both the boy and the reflection he gazes at (*plate 3*).[21] The love that Narcissus sought was more than an image; it was a mimetic translation of his

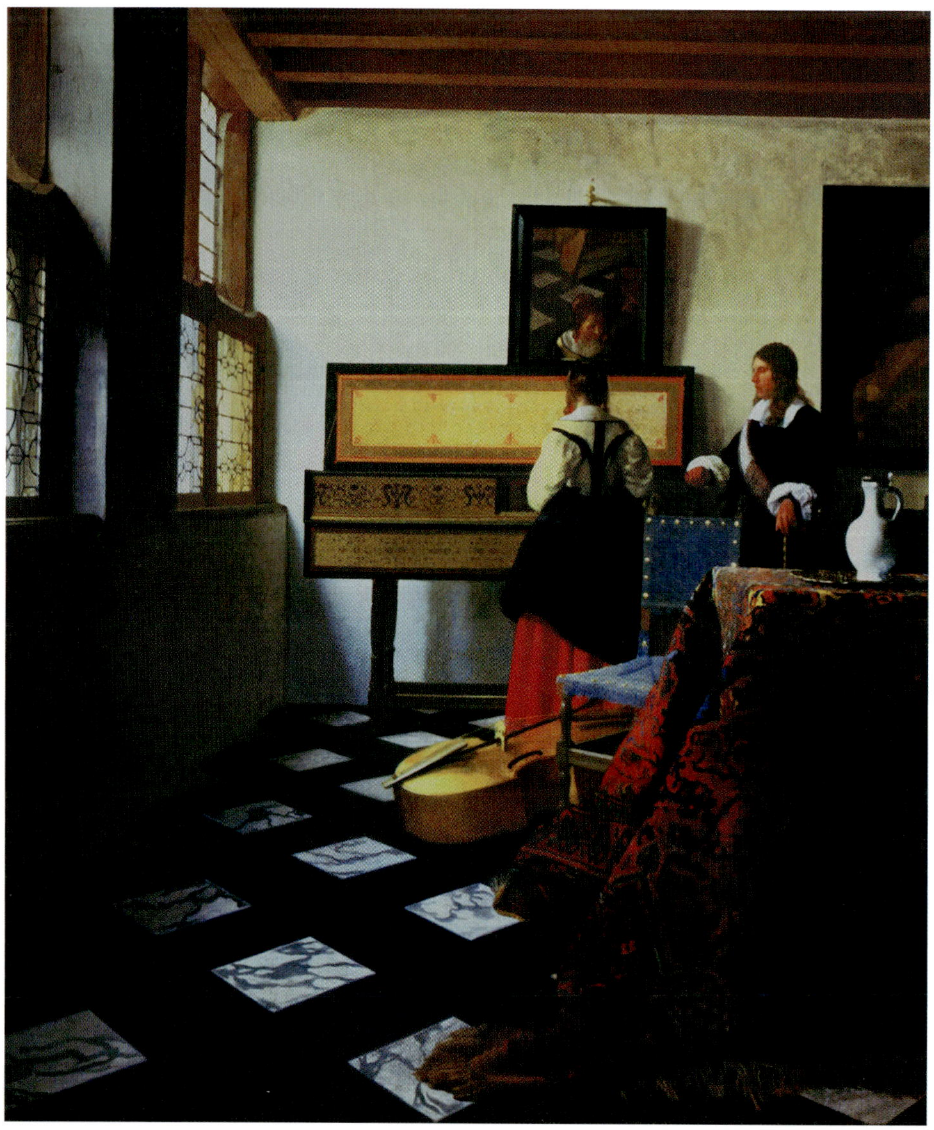

5 Johannes Vermeer, *Music Lesson*, 1662–65. Oil on canvas, 74.6 × 64.1 cm. London: Royal Collection. Photo: Scala, Florence.

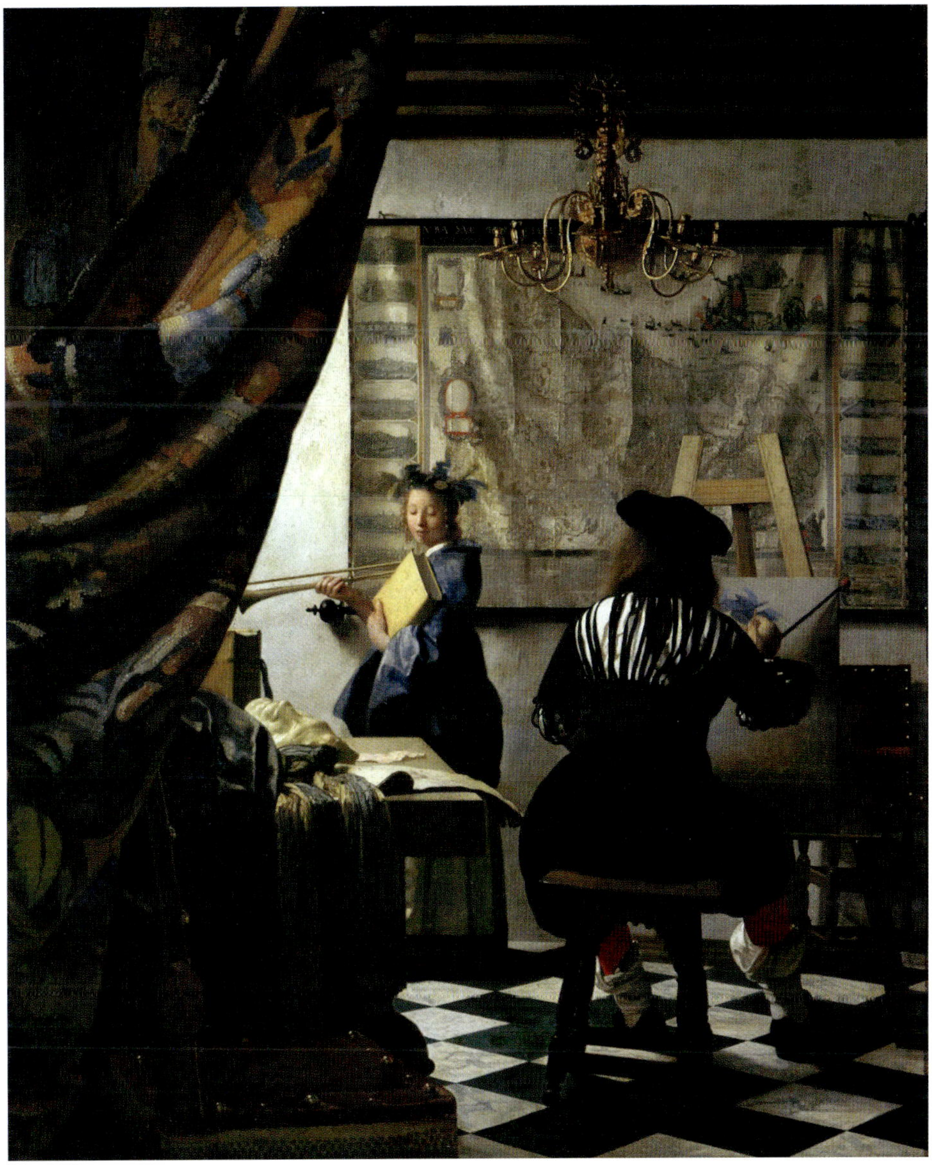

6 Johannes Vermeer, *Allegory of Painting*, 1665–68. Oil on canvas, 130 × 110 cm. Vienna: Kunsthistorisches Museum. Photo: Scala, Florence.

three-dimensional self onto a two-dimensional surface. In this lay its art as well as its lure – it looked just like him, yet was in fact a composition of glistening ribbons of colour on the flat surface of the water, as if a painting. The mirroring effects within Caravaggio's work stand as an emblem of this painter's art of realism, advancing the claim of this artist who famously purported to 'paint what I see'.[22]

In terms of both representation and of process, Caravaggio's *Narcissus* was bound up with the analogy of painting as a mirror. Like Brunelleschi's panel, Caravaggio's *Narcissus* would also appear to have been conceived through the use of mirrors as the technology of its artistic production. Here the artist seemingly worked from a model placed before two flat mirrors situated with one perpendicular to the other. One mirror was laid flat to capture the reflection of the boy in the 'water', the other upright for the depiction of the youth leaning over the 'pool'. Among his otherwise meagre possessions listed in an inventory of Caravaggio's property dating to 1605 were two mirrors, one large and one convex.[23] Giovanni Baglione, an artist himself and one of Caravaggio's early biographers, observed that this artist commonly painted his figures by studying his own reflection in a mirror. This testimony regarding the use of mirrors within Caravaggio's working practice is one that scholars of the period acknowledge as widespread.[24]

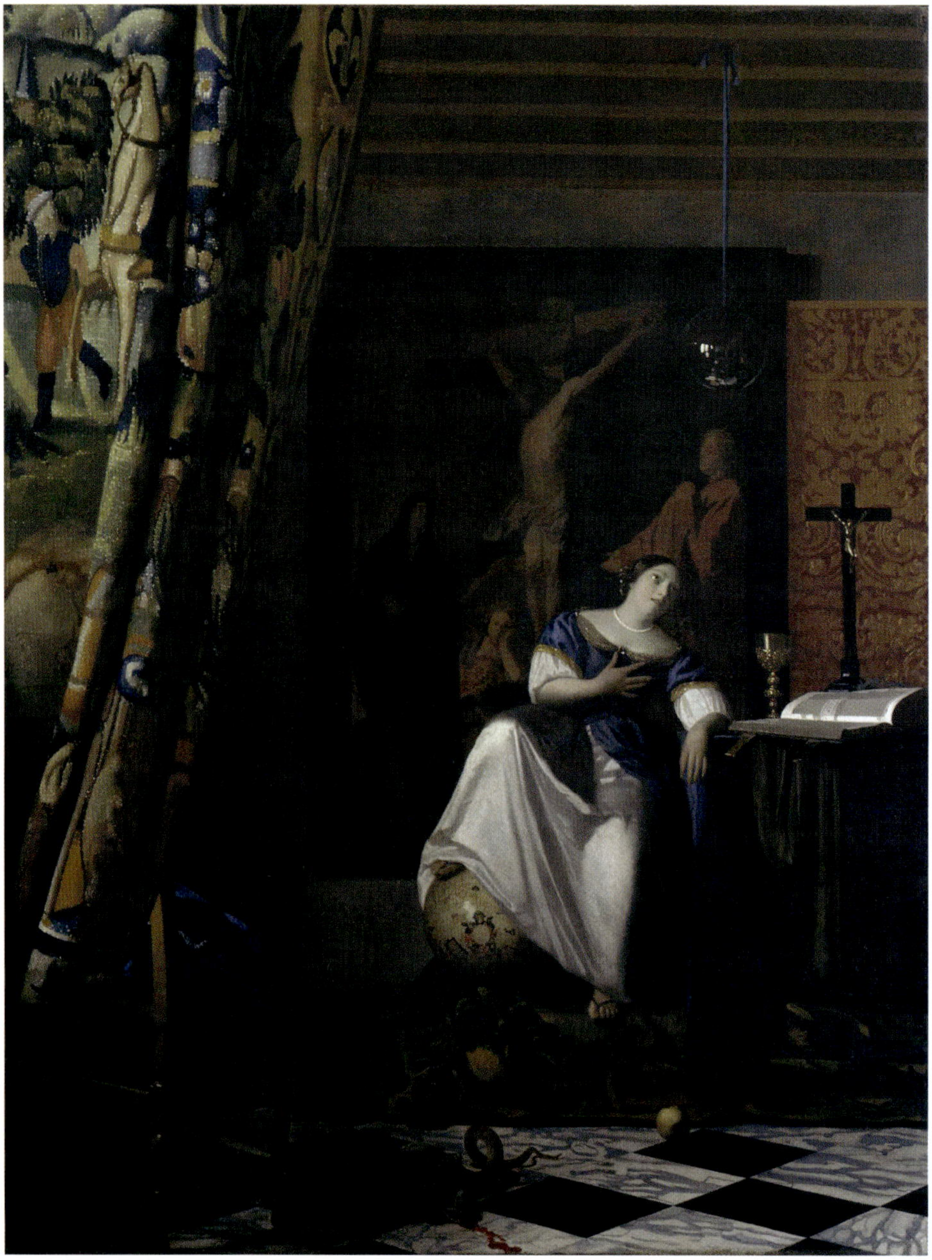

7 Johannes Vermeer, *Allegory of Faith*, c. 1670–72. Oil on canvas, 114.3 × 88.9 cm. New York: Metropolitan Museum of Art. Photo: Scala, Florence.

Since the cleaning of Caravaggio's Uffizi *Bacchus* in 1922 the fine detail in the reflections depicted in the carafe have become clearly visible (*plate* 4). Mina Gregori here observed 'the reflected head . . . of a male who wears a contemporary costume with a white collar; there also seems to be a painting seen from the back, as though on an easel'.[25] Both versions of Caravaggio's *Boy bitten by a lizard* similarly suggest the reflection of a window in the carafe of water, with the intimation of the room in which the work was painted, as in the Arnolfini mirror image of the room depicted in the painting. In Caravaggio's attentive art of visible realism, he seemingly rendered a reflection of himself and his surroundings as he worked on the Uffizi *Bacchus*. Through the prism of the reflective glass carafe, Caravaggio pictorialized his own artistic process in front of a mirror whose reflection he could then transpose onto the canvas. This suggestion furthers the intimation of a heightened proximity between painting and specularity, both in representation but also, crucially, in painting's depiction of its own production. The reflected images in *Boy bitten by a lizard* also suggest the painter's studio, lending further credence to Caravaggio's use of the mirror as a technology for translating visual observation into art. Thus Alberti's story of the origins of painting in the pool of Narcissus had its practice-based counterpart. In the studio, the mirror was the artist's instrument for transposing the visible world into the two-dimensional surface of painting. In representation, the mirroring surface inset within the painting became the locus for the reflected depiction of its own production.

The study of mirror reflection in the art of Vermeer is most clearly defined in his *Music Lesson* (plate 5). The mirror that hangs above the virginals reflects not only the face and bust of the woman who plays the instrument but also, at the top of the image, the legs of the artist's easel. Thus the reflection intimates the painter's presence. Vermeer's *Allegory of Painting* and his *Allegory of Faith*, like the *Music Lesson*, also pictorialize the 'scenario of their own production', albeit in very different ways. This is explicit in the *Allegory of Painting* (plate 6) where the painter sits at his easel in the act of painting the model before him; while in the *Allegory of Faith* (plate 7) we also see the artist at work before a canvas, but here in the miniaturizing reflection of the blown globe of clear glass that hangs from the timbered ceiling.[26]

The practical use of the mirror as a mimetic aide in painting surely adopted by Vermeer was one that Leonardo had much earlier counselled in the notebooks he kept throughout his working life.[27] With the intention of assembling them as a series of treatises on various aspects of his art, these included a discussion of the mirror as a simulacrum of painting, but also of its instruction:

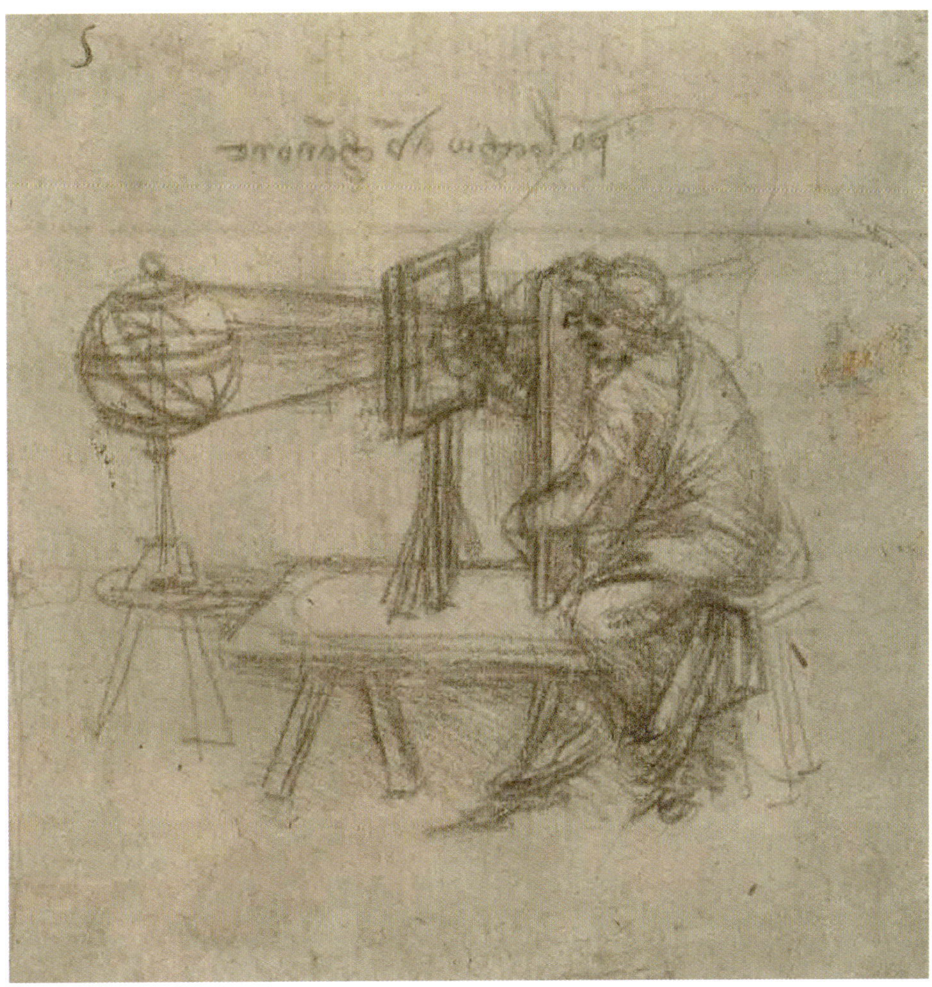

8 Detail of Leonardo, page of sketches including bellows for raising water and a draughtsman drawing after an astrolabe through a perspectograph, *Codex Atlanticus*, f. 5r, c. 1480. Silverpoint on prepared paper, 297 × 198 mm. Milan: Ambrosiana. Photo: Veneranda Biblioteca Ambrosiana.

> How the mirror is the master of painters: You should take the mirror as your master, that is a flat mirror, because on its surface things in many ways bear a resemblance to a painting. That is to say, you see a picture which is painted on a flat surface showing things as if in relief: the mirror on a flat surface does the same. The picture has but one surface and the mirror the same.[28]

Leonardo so fused painting with the reflection of the mirror that elsewhere he urged the painter to 'liken his mind to its surface'.[29] Across his notes, made over many years, his thoughts encompassed a wide-ranging and varying analogy between the mirror and the canvas. Most profoundly he understood the mirror as a parallel form to painting in its ability to render the illusion of all visible things in its reflecting sheen. Vasari had noted Leonardo's extensive experiments with mirrors as part of his interest in science and technology, and also his extraordinary use of reversed or 'mirror' writing throughout his notebooks made over some twenty-five years.[30] In his artistic practice and scientific experiments, as in his writing, the mirror served Leonardo as an instrument and as a model to transpose the world into the visual field.

Elsewhere, Leonardo touched on the related issue of artistic memory, predicated on imitation, which he understood as a compendium of 'mirror' reflections. That is, he advocated training the artist's visual memory through repeated copying. This he advised through the use of an 'exemplar traced on to a thin flat plane of glass. Place this on top of the drawing you have done without the exemplar, and note carefully where the tracing does not match up with your drawing.'[31] Leonardo's glass exemplar was made from a flat pane, like a window or a sheet mirror, on which he could trace the outlines of what he saw, to compare with his drawing on paper. A sketch among his notes in the Codex Atlanticus illustrates the technological means of this form of artistic translation (*plate* 8). Here an artist draws after an armillary sphere by looking through a transparent pane of glass. By this means he transposes the perception of the third dimension into two-dimensional representation.[32] Collectively, Leonardo's notes on the mirror and the glass pane theorize their conceptual equivalence with painting in relation to their practical use as technologies of imitation. This in turn drew on his larger interests in optics, the science of reflected light, and the relationship between reflection and vision.[33] Throughout his practice and in his notes, Leonardo thus drew together by analogy the window with the mirror, likening them both to painting.

The Picture Plane

As Leonardo was collecting his notes for an intended treatise on painting, Dürer was preparing his multi-volume work on artistic measurements, the *Underweysung der Messung* (1525).[34] This treated the geometry of figural forms in art, and specifically the representation of complex volumes, in the two-dimensional medium of drawing. Aptly described as a revolving door in its accommodation of theory with practice, it familiarized 'coopers and cabinet-makers with Euclid and Ptolemy … [and] professional mathematicians with what may be called workshop geometry'.[35] As with Brunelleschi's mirror, Alberti's grid, and Leonardo's glass exemplar, which Dürer surely knew, this technological apparatus was the means for translating the appearance of volumes into two dimensions. At the close of the volume he included a plate depicting an artist drawing after a three-dimensional object through the use of various glass pane instruments. In the first illustration, a draughtsman draws the outline of a vase onto an upright sheet of glass framed like a picture, or indeed a window, using also a sighting device to keep the position of the viewing eye constant (*plate* 9). Like a preparatory drawing, the outline on the glass forms the first stage in the artistic translation of the world into the flat field of a painting. As with Leonardo's description and sketch of his glass exemplar, Dürer's illustrations visibly demonstrate the use of a range of optical devices in Renaissance artistic practice to facilitate the translation of observed volume and space into the illusion of drawing or painting. This would culminate in the extensive range and development of optical and perspectival devices described by Galileo's friend and fellow artist-mathematician Ludovico Cigoli in his *Trattato pratico*, which illustrated the growing complexity and elaboration of artists' technologies originating in Dürer's glass and grid.[36] In this vein Dürer's second illustration is particularly instructive. It shows a draughtsman seated at a table with a large sheet of squared paper before him on which he has begun to draw (*plate* 10). However, his eyes do not follow his hand at this moment; rather, he looks through a sighting device onto an upright framed grid surely similar to Alberti's 'intersection', whose squares correspond to those on the artist's sheet of paper below. Looking through the grid to what lies beyond it, we/the artist observe a life-size female nude awkwardly recumbent across the further half of the table, seemingly a sculpture. Clearly, the artist is using the framed grid to establish within its squares where each part of this female anatomy lies, in order to transpose it onto the squares of his paper. Again, the framed viewing device – in this instance a grid – acts as the method of artistic translation from three into two dimensions. As Dürer stressed in his notes, the framed grid enabled the artist to alter the scale, rendering the imitation either larger or smaller, as wished.[37] Here Alberti's celebrated metaphor of the picture plane as a 'window' or intersection becomes a working instrument.[38] The workshop technologies of Dürer's craft are thus fully entwined with the broad conceptualization of painting as a window or mirror, as Brunelleschi had demonstrated, and Alberti and Leonardo had theorized.[39] What we witness in Dürer's prints is the practice-based, technological counterpart to a Renaissance theory of mimesis.

Elsewhere Alberti advocated the use of a mirror as a controlling/corrective device, as Leonardo later also did, understanding it as a kind of surrogate or heightened reproduction of the work under judgment:

> A good judge for you to know is the mirror. I do not know why painted things have so much grace in the mirror. It is marvellous how every weakness in a painting is so manifestly deformed in the mirror. Therefore things taken from nature are corrected with a mirror.[40]

If Alberti's story of Narcissus represented painting's claim to the mimetic powers of the mirror, the use he advocated within practice was related in concept but different in its practical orientation. Here he commended the mirror as an instrument of comparison and judgment, using it to reflect back the painting in progress so as to make adjustments and corrections. Renaissance workshop technologies such as the mirror were commonly deployed as a means of verifying, even authenticating, a mimetic paradigm of art.[41] Alberti's concern was to test through comparison with the mirror reflection the depiction of relief in painting, rendered through the judicious application of white and black to give the illusion of highlight and shadow and so of volume. Leonardo also advocated using the mirror this way, but he integrated it more profoundly into the ongoing development of a painting, deploying it to reflect

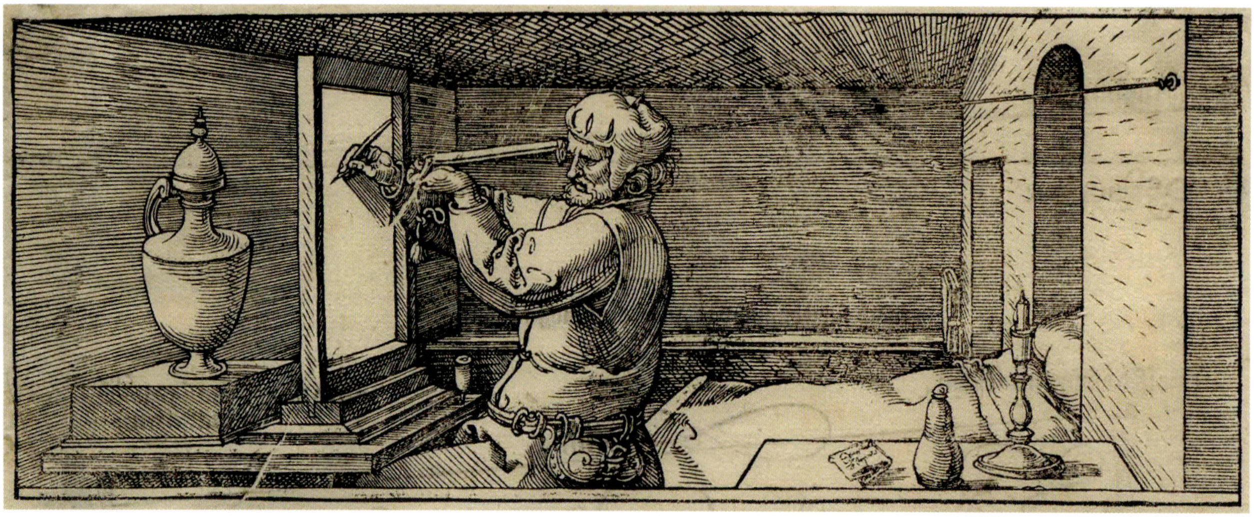

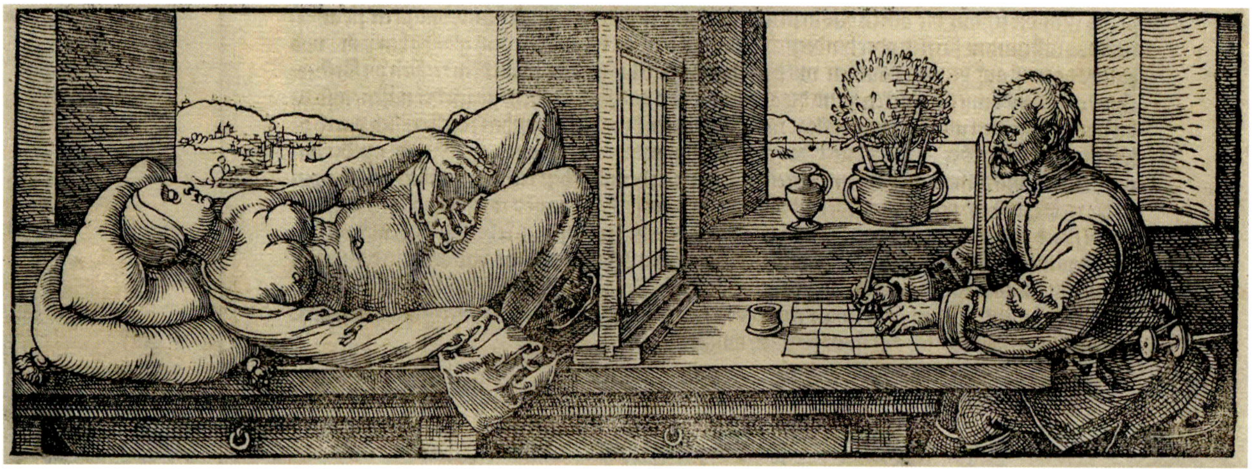

9 Albrecht Dürer, illustration from *Underweysung der Messung mit dem Zirckel und Richtscheyt*, Nuremberg, 1538. Photo: © Trustees of the British Museum.

10 Albrecht Dürer, illustration from *Underweysung der Messung mit dem Zirckel und Richtscheyt*, Nuremberg, 1538. Photo: © Trustees of the British Museum.

the subject of representation in order to transcribe it onto the canvas in much the same way that Alberti's and Dürer's artists worked from what they saw through the squaring device. In Leonardo's practice the mirror would increasingly come to take the place of the grid. Further, the dimming reflection of the Renaissance mirror offered the artist a reduction of local colour into a unifying scale of tonal gradations. This specular shift in means and technology may be joined to Leonardo's broader reconceptualization of perspective as atmospheric as well as linear, and so of the art of painting itself.[42]

The Self-Portrait

The conjoined functions of the squaring device with the mirror in the years around 1500 take on a particular significance with the historical emergence of self-portraiture. Here the use of the mirror as the technology of a mimetic process was clearly instrumental to the development of the genre. Dürer's and Parmigianino's early self-portraits brilliantly seed this historical trajectory. These works have long been recognized as pivotal not only for the history of art but for the definition of Renaissance culture itself.[43]

Dürer's earliest self-portrait drawing as a child, to which, forty years later, he ascribed a date of 1484 (*plate 11*), inaugurated a sequence of portrayals of himself in both drawing and painting that span his career. In his mature annotation of the drawing Dürer ascertained not only its status as a self-portrait, but also that the technology of its making was a mirror reflection: 'This I fashioned after myself out of a mirror in the year 1484 when I was still a child. Albrecht Dürer.'[44] The drawing itself is the size of a Renaissance mirror, further testament to the implied equivalence between them. By this means Dürer sought to ensure the drawing's transmission to historical memory, as Joseph Koerner has argued, in an act of *gedächtnus* long subsequent to the drawing itself. For the destiny of this drawing by a child was not foreseen at the moment of its making; it may rather be linked to a portrait of the same year of Dürer's father, the Nurenberg goldsmith.[45] While Dürer the Elder holds a small silver statuette as the emblem of his art, the young artist's hand is instead held

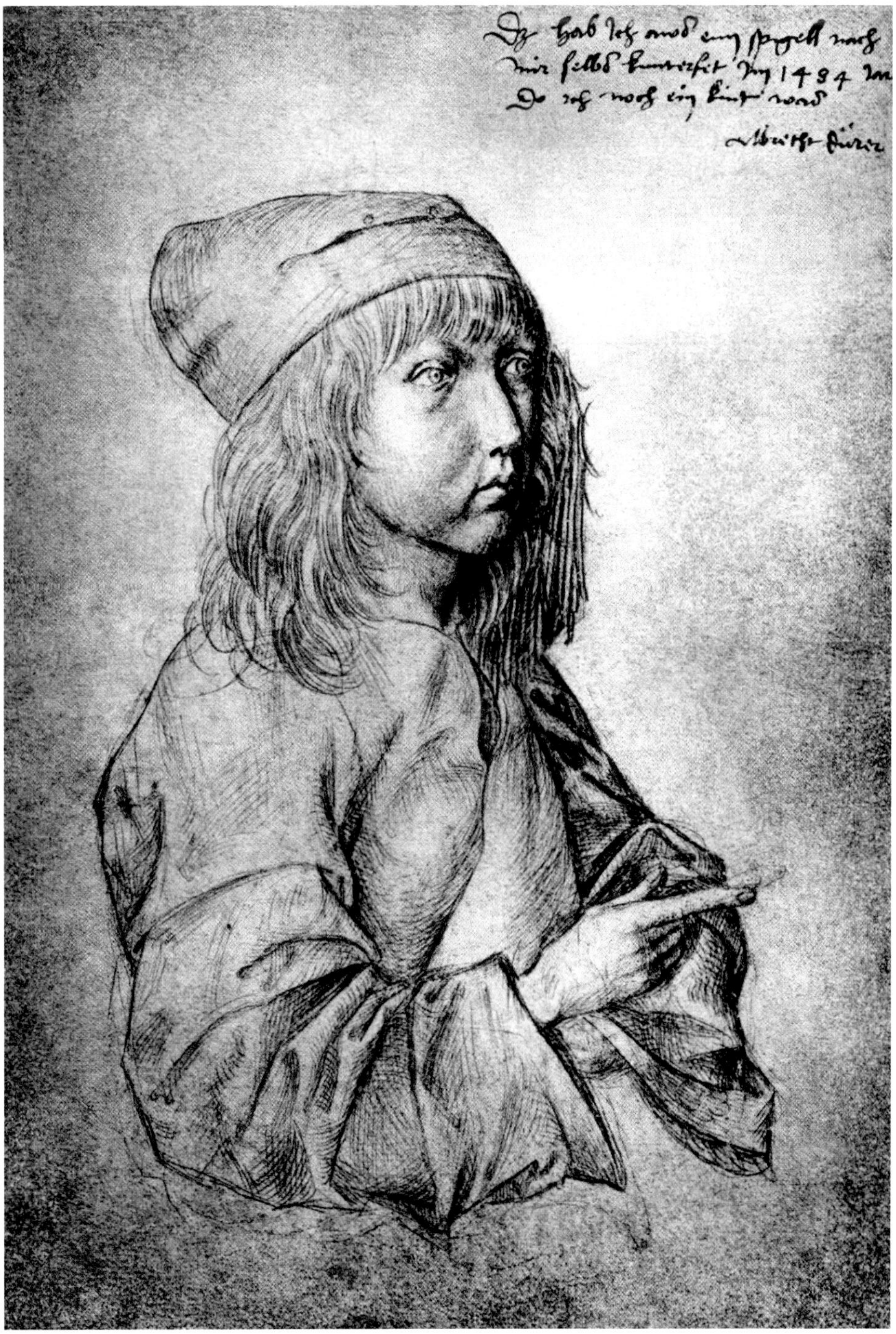

11 Albrecht Dürer, *Self-Portrait Aged 13*, 1484. Silverpoint on prepared paper, 27.5 × 19.6 cm. Vienna: Albertina. Photo: Scala, Florence.

in what is surely an act of 'sizing'. With his fingers he gauges the dimensions of a detail of his body as seen in the mirror to capture it accurately in his drawing.

Parmigianino's early self-portrait of 1525 (*plate 12*) is, like the Dürer drawing, the work of a young adolescent, dating to the same moment as Dürer's later inscription of 1484. Vasari tells us that the young Parmigianino painted it as an artistic introduction to papal circles in a display of skill in the observation of reflections.[46] Here Parmigianino rendered his image on a convex panel he had fabricated to the same dimensions as the convex mirror in which he studied his face, so reinforcing the visual equivalence between them. The image meticulously observes the distortions of the mirror's reflection caused by the rise of the blown glass. This is particularly pronounced in the lozenge-shaped window on the upper left, recognizably that of a reflection in a spherical mirror; and in the exaggerated size of the artist's hand in the foreground. Thus the image conflates the mirror, which constitutes the technology of the painting's production, with the material form of the panel on which it is painted. In a doubled play with this conceptualization of the painting as a mirror image, it also severs the two. To the right side of the painting we see the edge of a tondo-shaped frame, surely the mirror from which Parmigianino, according to Vasari, had worked. This has the effect of recalling to mind the temporal processes of the painting's production, for we seem to witness the artist studying the canvas on which he paints having just turned from looking in the mirror to observe what he means to depict. The painting is both the reflection in Parmigianino's round mirror, and the canvas that

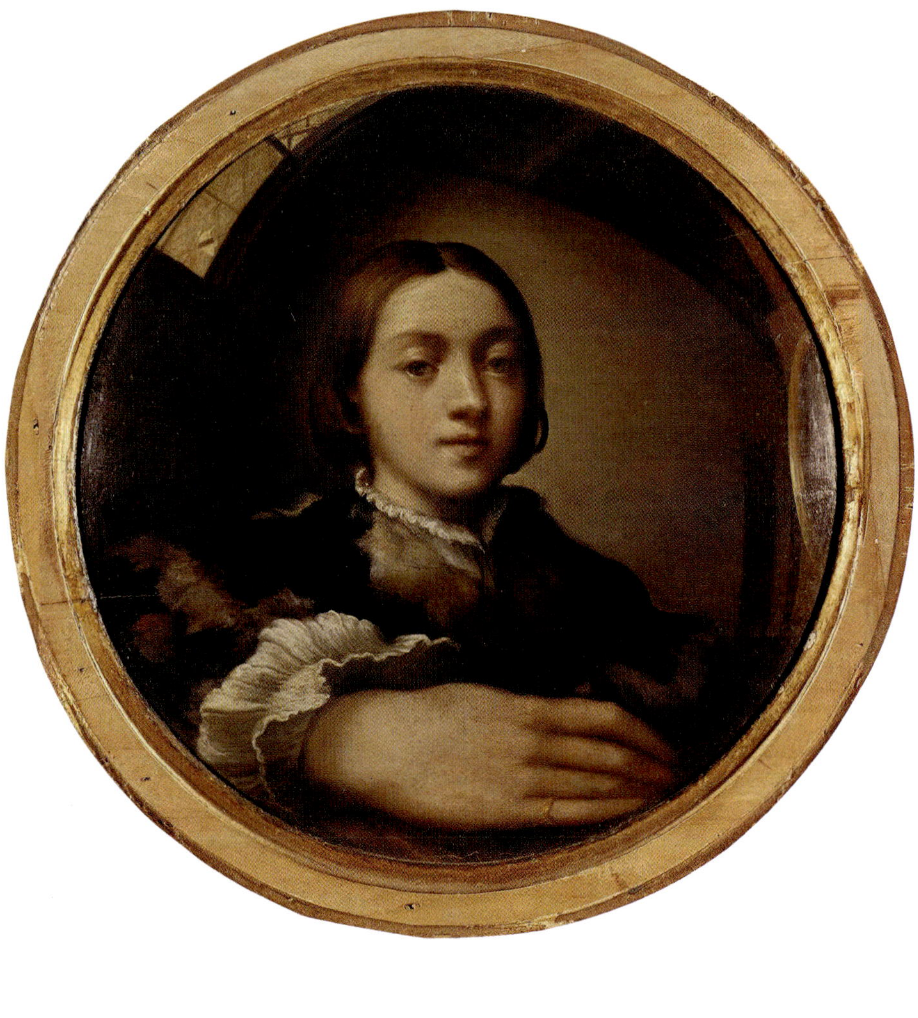

12 Parmigianino, *Self-Portrait*, c. 1525. Oil on convex panel, diameter 24.4 cm. Vienna: Kunsthistorisches Museum. Photo: Scala, Florence.

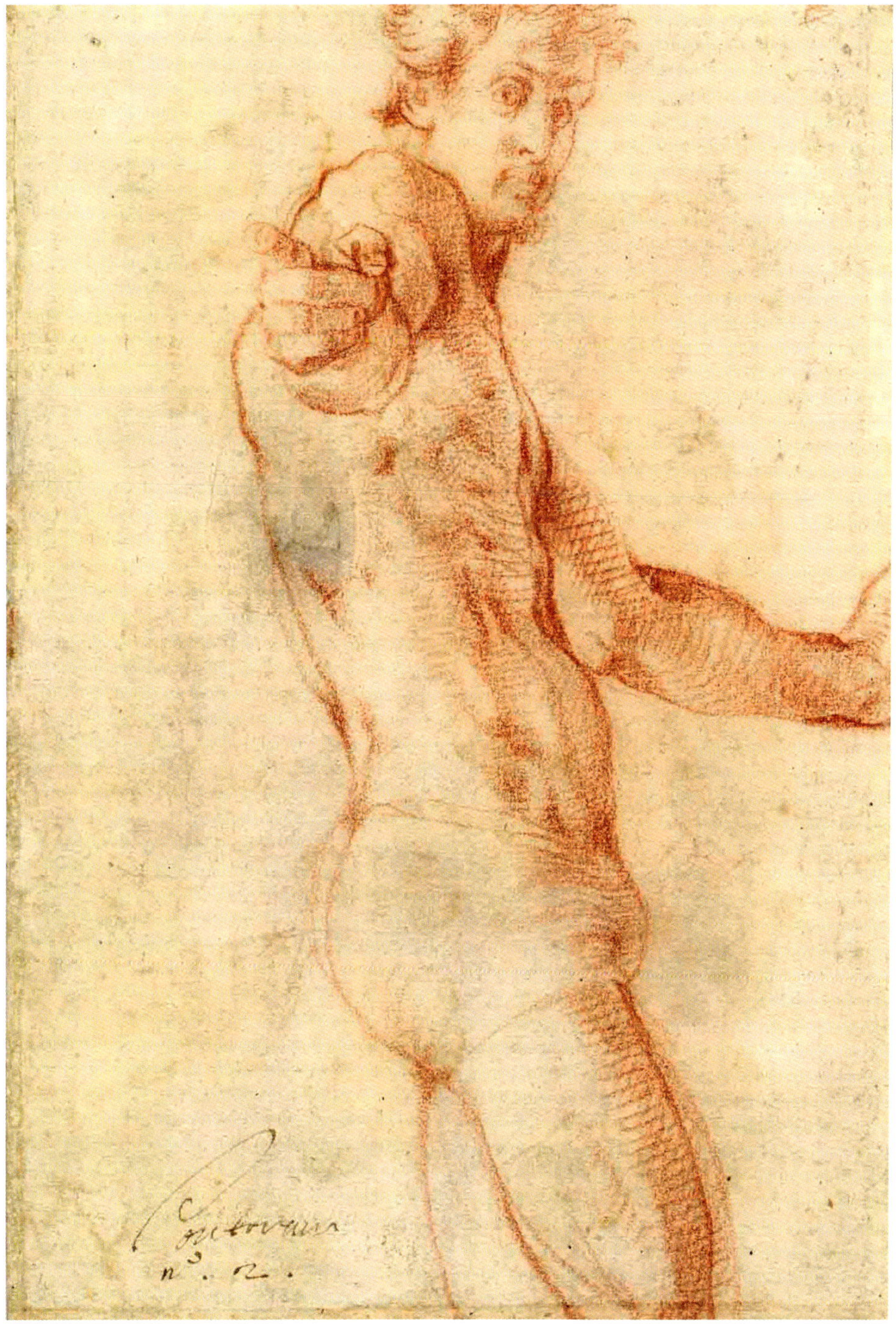

13 Pontormo, *Self-Portrait*, c. 1525. Red chalk on buff paper, 284 × 202 mm. London: British Museum. Photo: © Trustees of the British Museum.

he has just completed, caught between viewing and making in what Michael Fried has identified as a core modality of the reflexive structure of early modern art.[47]

A red chalk self-portrait drawing by Pontormo that dates from the same time as the Parmigianino and the Dürer inscription shows a similar pictorial fusion between the mirror and the drawing, between viewing and making (*plate 13*).[48] Here Pontormo stands in a three-quarter profile pose. With his lower arm he draws himself on a sheet of paper directly in front of him, placed just outside the image to the right. The other hand extends forward towards the surface of the picture plane, seemingly touching it with the tips of his fingers. The lower fingers are loosely curled into the fist, while the index finger and thumb extend forward in a gesture of sizing. His gaze is directed towards the tips of those fingers. Like Dürer's early self-portrait drawing,

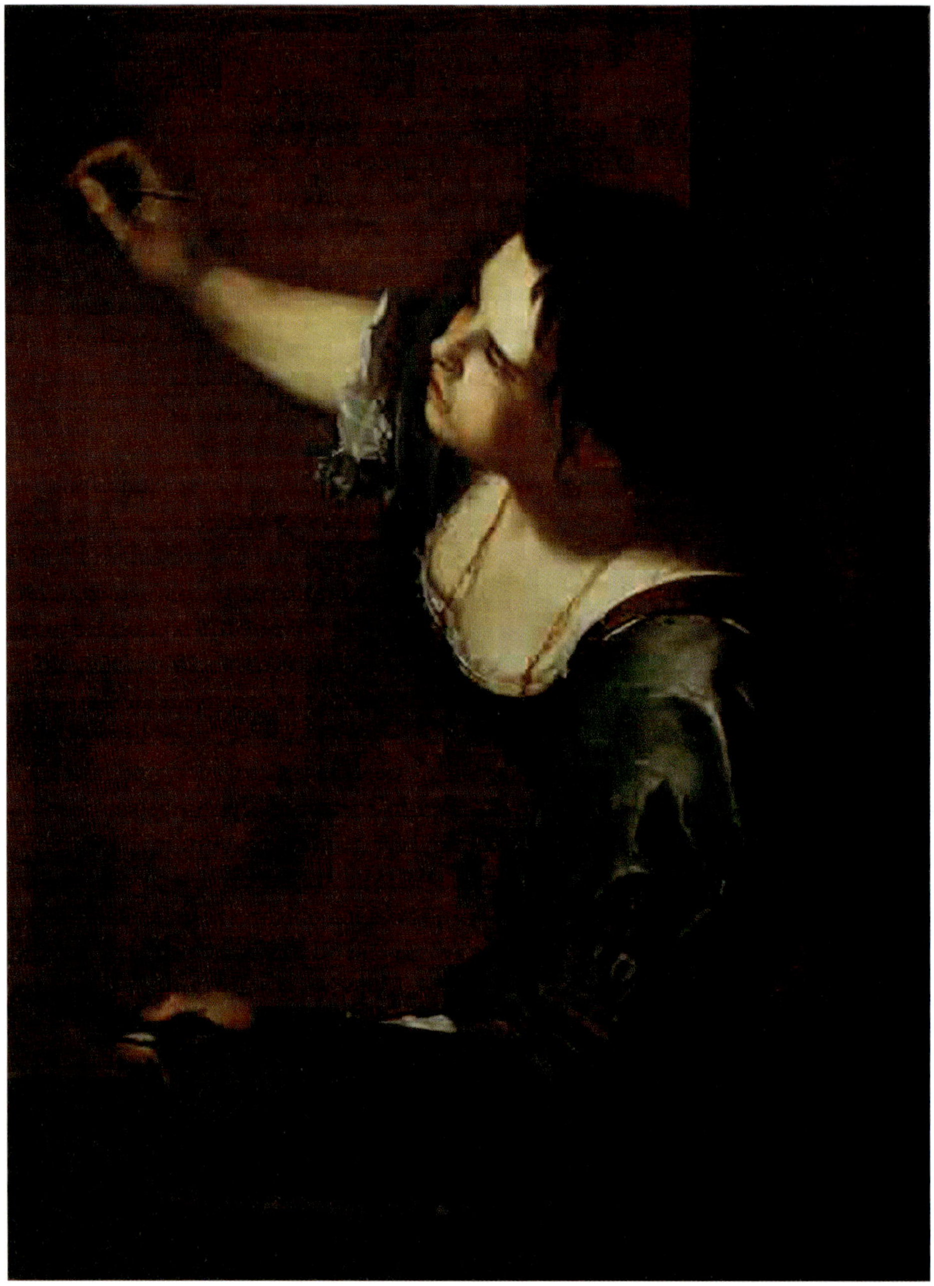

14 Artemisia Gentileschi, *Self-Portrait as an Allegory of Painting*, 1638. Oil on canvas, 96.5 × 73.7 cm. London: Royal Collection. Photo: © Her Majesty Elizabeth II.

Pontormo is apparently measuring what he sees in a mirror, his fingers spanning a detail in a kind of judgment of hand and eye. Pontormo would seem to practise Michelangelo's famed *giudizio dell'occhio*, that the artist's true compass should lie in the judgment of the eye, rather than depending on the devices and grids that Dürer had elaborated for craftsmen in his treatise on measurement, also of 1525.[49] As the study of a reflection in a flat mirror Pontormo's image is reversed, with the consequence that the drawing hand appears to be left but in fact would have been right. Like the Parmigianino, a bifurcation is at play in the pictorial logic of the composition. On the one hand, there is an elision of the drawing with the mirror, in its representation of what Pontormo saw. On the other hand, we understand the drawing as a process occurring to the right of this sheet as we regard it. Thus its temporal structure moves from mirroring to drawing, so pictorializing the means of its own production. We find a similar pictorial strategy in Artemisia Gentileschi's *Self-portrait as an Allegory of*

Painting of the late 1630s, which speaks to the continuing significance of this specular conception of painting, in both representation and process, across the century divide (*plate 14*).[50] The image shows Artemisia in the act of painting. The canvas is before her, her right hand holding a brush raised and about to mark the surface, while in her left hand she holds a palette. But like the Pontormo drawing, her gaze is directed beyond the edge of the canvas to what is surely a mirror in which she studies her reflection before applying her paint. Here too, the absent mirror is both the instrument and the matrix of representation. Artemisia paints the process of her portrait's own production.

The Mirror of Venus

Turning back to an earlier historical moment in the relation between painting and its conceptual resemblance to the mirror, circa 1500 marked a newly intensified approximation between them that extended beyond self-portraiture to other genres. Weaving together the history of mirror-making with the representation of mirrors in art, both saw a distinct increase in production in the early years of the sixteenth century, and particularly in Venice, long the centre for glass-making on which mirror manufacture depended.[51] The expanding production of mirrors as a luxury commodity in Venice had a direct parallel for the theory and practice of Cinquecento painting, where the representation of mirrors also flourished.

Representation of a mirror within painting was most often coupled with a depiction of the female form, and particularly the female nude. Two closely related paintings stand at the inception of the genre, a late work by Giovanni Bellini on a small wood panel, and a somewhat larger canvas by his pupil, the young Titian, that scholars have dated to the same moment. Bellini painted a young woman holding a small looking glass through which she views her hair by means of a further mirror on the wall behind her (*plate 15*).[52] The view of the back of the head reflected in the mirror on the wall is, of course, the view that the subject herself observes in the small glass she

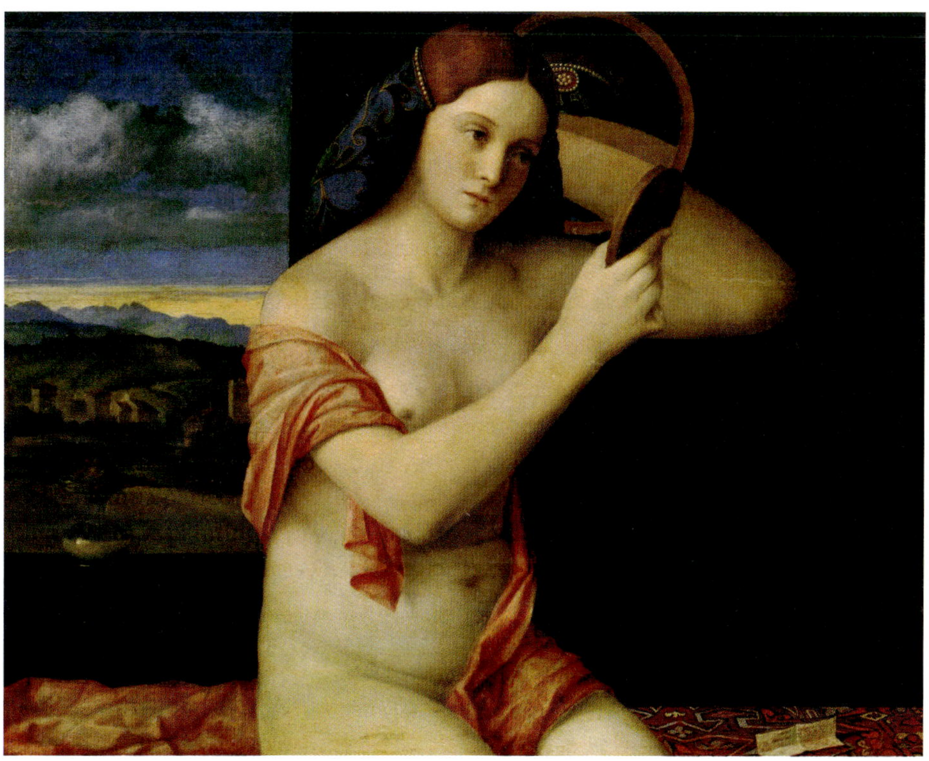

15 Giovanni Bellini, *Woman with Looking Glass*, 1515. Oil on panel, 62 × 79 cm. Vienna: Kusthistorisches Museum. Photo: Scala, Florence.

holds in front of her. Like Brunelleschi's lost panel, the picture calls into play the motif of the mirror as an instrument of observation and as a 'painting within a painting', that is, both a direct reflection on artistic practice and a metaphor of painting's specular art. The force of the picture lies precisely in its doubled structure, signifying both its status as representation and the mirror technology of its own production.

In turn Titian depicted a young woman whose loose hair falls across her shoulder to her waist, which she also tends by means of a doubled mirror reflection. At her side a young man steadies a gilt-framed convex mirror with his left arm while holding up a smaller rectangular mirror in his right hand (*plate 16*).[53] The woman gazes at her face in the small mirror which, as with the Bellini, is seen only from the reverse side. Here she must also see her reflection from the back, like the viewer. In the mirror we can trace the dim outline of her head and shoulders set just under the strong reflection of what appears to be a window, casting its light on her face from the left. To her side in the reflection-image is the profile of the man holding the mirrors, with his arms extended to suggest both an embrace and the act of painting.

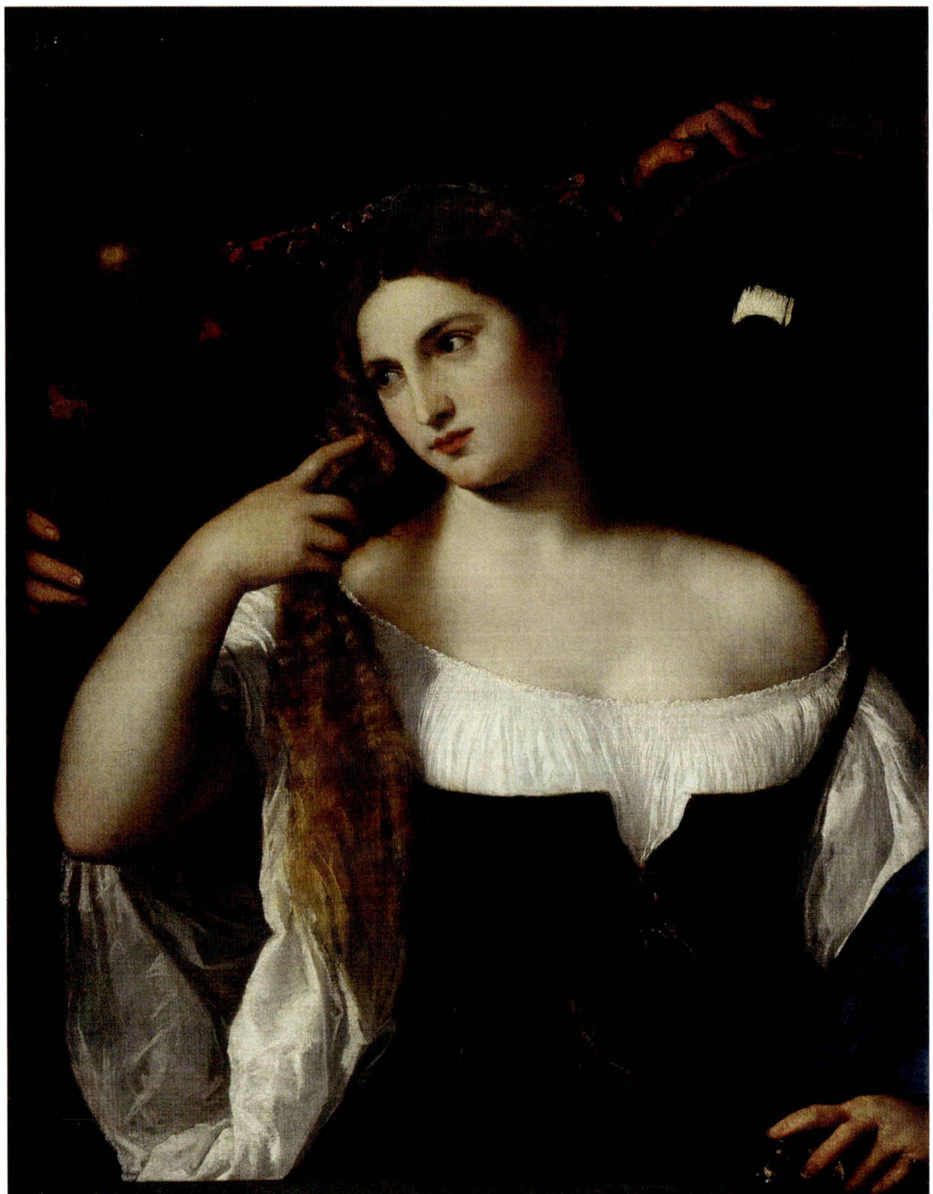

16 Titian, *Woman at Her Toilet*, c. 1515. Oil on canvas, 99 × 76 cm. Paris: Musée du Louvre. Photo: Scala, Florence.

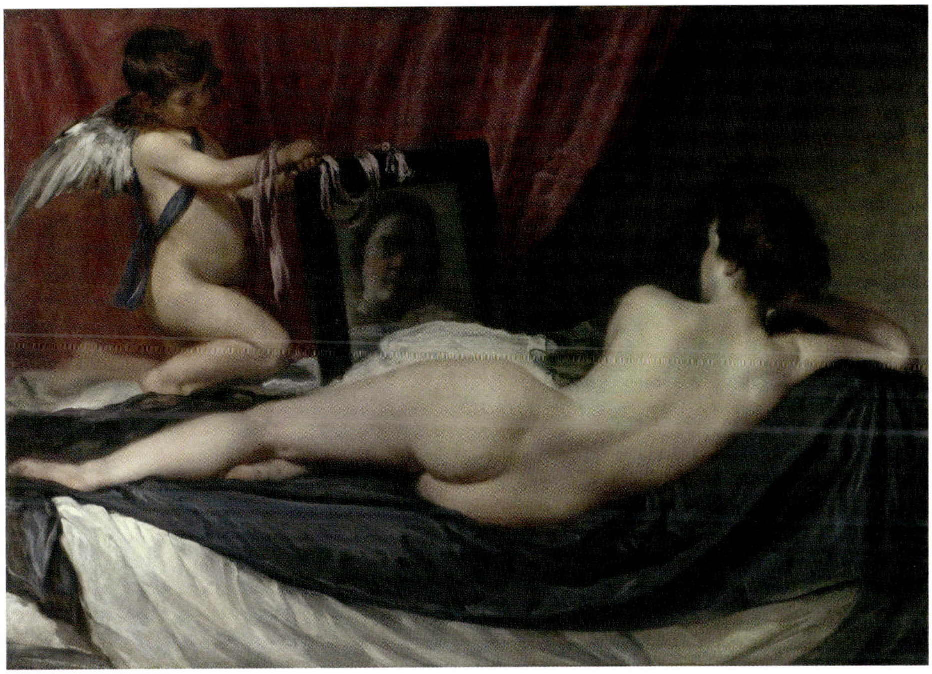

17 Velazquez, *Rokeby Venus*, c. 1650. Oil on canvas, 122 × 177 cm. London: National Gallery. Photo: Scala, Florence.

Sometimes simply termed paintings of *belle donne*, these works of the Venetian Cinquecento of female beauties enhanced by their mirror reflection approach the typology of a Toilet of Venus. This subject, also new to Cinquecento art, similarly connected the mirror reflection and the female nude as conjoined allegories of the beauty of art.[54] The rich issue of this pictorial tradition also lies behind Lomazzo's re-attribution of Giorgione's lost work of reflections as a female nude, and the allegory of *pittura*, with which this essay began.

Such play with the viewer's attention by means of the painted mirror reflection was something that Velazquez, some 150 years after Giorgione, also observed in paint. His study of Venetian painting, through his travels in Italy and through his position as keeper of the Spanish Royal collections, included a now-lost Toilet of Venus and her mirror reflection by Titian.[55] Velazquez's knowledge of this pictorial tradition, and of Titian in particular, informed his own great painting of Venus at her mirror c. 1650 (*plate 17*).[56] A reclining female nude seen from behind, Velazquez's rendition is unusual as a Venus without jewels, her identity instead manifest in the beauty of her form and colour alone. At the heart of the painting's composition lies the reflecting surface of a mirror, its frame echoing that of the painting itself. The mirror captures the face of Venus centrally within its shimmering surface, depicting this female body from the front which the spectator would not otherwise see. Yet the mirror reflection remains the least-resolved aspect of this celebrated painting, presenting the viewer with a series of optical conundrums. The reflection is apparently misaligned, for the angle of the mirror should properly reflect us rather than Venus. As this is so, this image within its courtly viewing context suggested a flattery of the viewer as either a Venus herself, or as the goddess's lover. The ambiguity was surely intended, with the intimation that the circuit of gazes include Venus regarding the viewer as well as the viewer regarding Venus. Notable also are the differences in paint surfaces. While a rich creamy impasto is consistently applied across the majority of the canvas, in comparison the mirror is rendered with a thinner, more fleeting application of paint through which the mid-tone of the ground and the canvas weave remain visible.[57] Its sketchy surface finish is that of a painting in progress. Thus the mirror reflection becomes a further variation of the artist's subject – his art. But the

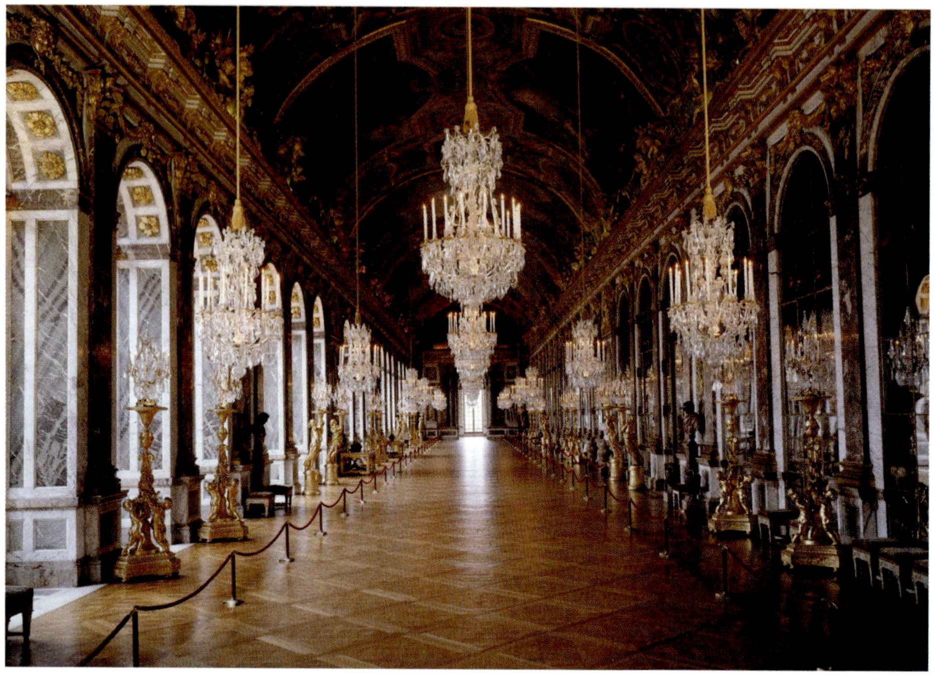

18 Galerie des glaces, Palais de Versailles, 1678. Photo: Scala, Florence.

mirror is also the painter's viewing instrument, as this painting so brilliantly declares. If this is so then Velazquez's contrast between the smoothly modelled surface of the body and the sketchiness of the mirror suggests a considered critical engagement with a theory of practice. The polished body figures painting as representation, while the *alla macchia* reflected image of the face stages the mode of its production. Like the reflection of the studio in Caravaggio's carafe, this reflective image represents both the acme of early modern pictorial illusion and a view into its making. As court art, the painting's conceit is its artfulness, and it is in this, the play of its pictorial illusionism, that its status as a meta-image resides. Nowhere might this be more so than in a rendition of Venus and her mirror as conjoined signs of the beauty of art and the technology of its making, and thus a commentary on what a painting is.

The early modern preoccupation with a mirroring art, in which paintings acted as if reflections of themselves, was to become a leitmotif of the seventeenth-century court. Within the early modern palace interior, display of art in princely galleries would, over the century, include ever-larger mirrors alongside painting and sculpture, as equivalent pictorial fields of vision. In the mirror the surrounding artworks would be seen again in reflection. Improved mirror-making technology now made possible the manufacture of extensive flat mirrors that could occupy the place of painted wall panels. This new decorative order of the early modern court would culminate in Versailles' most celebrated *Galerie des glaces (plate 18)*.[58] Here, facing mirrors produced a cascading play of reflection that Foucault famously recognized as the cultural paradigm of early modernity. New to the scale and grandeur of the palace interior, the unfolding reflections of facing mirrors was an effect that Leonardo had of course already noted in his study of optics some 150 years before:

> Now the first [mirror] being reflected in the second carries to it its own image together with all the images reflected in it, among these being the image of the second mirror, and so it continues from image to image on to infinity, in such a way that each mirror has an infinite number of mirrors within it, each smaller than the last, and one inside another.[59]

If Versailles was the magnificent culmination of this genre of display, decoration with mirrors of ever-greater dimensions was already diffuse in early modern palace decoration. The Alcàzar then had a Hall of Mirrors (destroyed by fire in 1734) whose decoration with paintings Velazquez had orchestrated in the 1630s, as visible in a portrait by Juan Carreño de Miranda of Charles II of Spain depicted in front of a mirror in this hall (*plate 19*).[60] Presumably the portrait intentionally mirrored the actual décor of the hall with its depiction of richly ornamented mirrors framed by gilt lions and eagles as the symbols of kingship. Within the hall the representational force of the portrait would have worked like a further mirror, seemingly reflecting back the room in which it hung and so multiplying the effects of its surroundings. In this

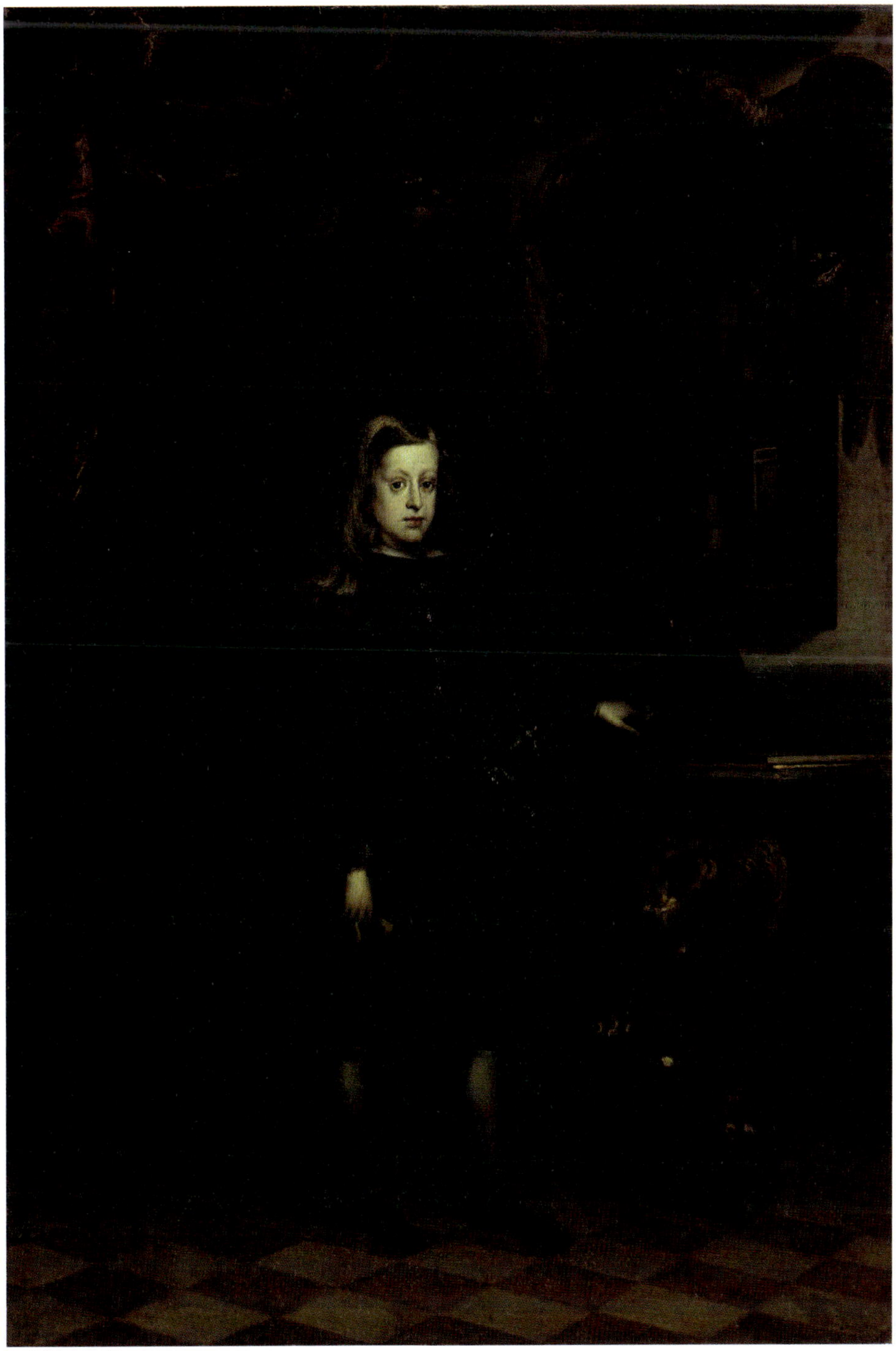

19 Juan Carreño de Miranda, *Charles II in the Hall of Mirrors at the Alcàzar Palace*, c. 1673-75. Oil on canvas, 204 × 142 cm. Madrid: Museo del Prado. Photo: Scala, Florence.

regard it manifests a further type of portrait, in addition to the example of the self-portrait, that is bound to the motif of the mirror. It extends the genre to encompass the study of the mirror reflection within painting more broadly, in a specular doubling of both painted and architectural space, and of the production of painting within its own representational field.

Some five years after the Rokeby Venus Velazquez painted his great court portrait and culminating mirror image work, *Las Meninas* (*plate* 20). At a first level of signification it represents the Spanish Infanta or princess at play with her maids of honour while Velazquez paints. The room depicted in fact served as the artist's studio at the Alcàzar palace.[61] Much of the painting's renown, however, rests on its conceptualization as a mirror image that multiplies the artifice of its reflexive illusion serially, the study of which has dominated the extensive scholarly literature on this work.[62] At the same time it reflects on its own production, and on the central place of mirrors within that process. An inventory of Velazquez's possessions housed in his rooms at the Alcàzar at his death includes not only a library with a substantial number of books on optics, perspective and the sciences of projection, but also a number of optical devices used in artistic practice, including glasses, and ten mirrors of various sizes and types.[63] Undoubtedly, *Las Meninas* drew on the Arnolfini double portrait

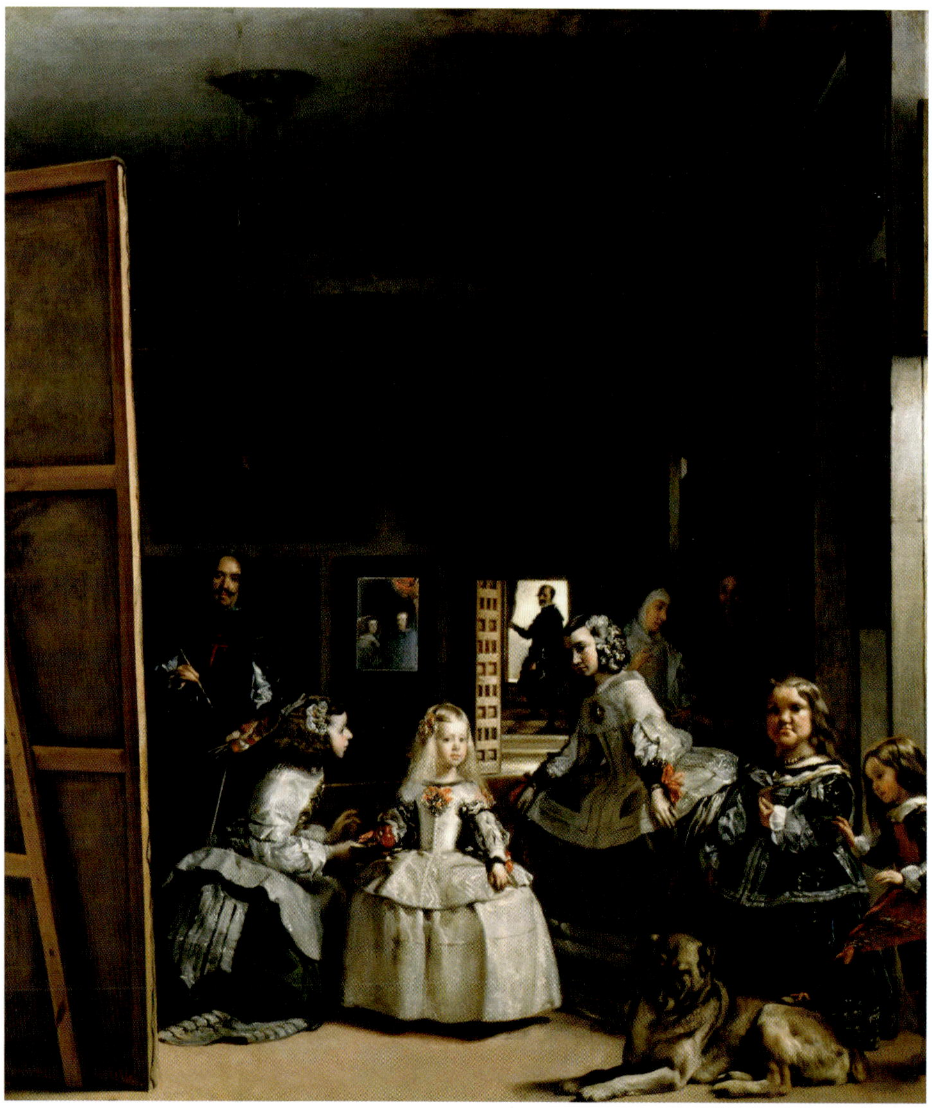

20 Velazquez, *Las Meninas*, 1656. Oil on canvas, 318 × 276 cm. Madrid: Museo del Prado. Photo: Scala, Florence

with its inset mirror, a painting Velazquez would have known well as it was then part of the Hapsburg collection in Madrid and under his aegis as keeper of the royal paintings.[64]

On the back wall of *Las Meninas*, displayed alongside a group of paintings, hangs a mirror, picked out by the shimmer of light across its surface and most intensely at its bevelled edge. In the reflecting surface of this mirror we see what Velazquez's fellow court artist Antonio Palomino identified for us as the King and the Queen.[65] The reflection implies the presence of the royal couple before the canvas, in the space of the viewer, but also of Velazquez as he painted this work. They are surely the subject of Velazquez's large easel painting within the painting. As viewers, however, we see this canvas from the back. Through the mirror Velazquez reveals to the viewer what he also withholds – his art. The complexity of mirror images at play pulls the viewer into the labyrinth of reflections that this painting seems to contain within it. As well as the mirror depicted within the painting, there is also the implication of a further mirror within this work – the mirror that Velazquez used to paint his self-portrait. By means of a mirror, Velazquez paints a mirror-as-painting within the painting, which reflects the painting's subjects along with the easel on which he works.[66] As a portrait of the artist at work in the company of the Infanta, the mirror reflection that *Las Meninas* holds within it is a portrait-within-a-portrait, as the canvas-within-the-canvas on which Velazquez paints must also be. The painted mirror that is at the same time a royal portrait hangs on a wall of further paintings-within-the-painting, which makes of it a picture gallery-within-a-gallery of the royal collection in which, then as now, it surely hung. If the paint surface of this mirror reflection double portrait is sketchy in comparison with the rest of the canvas, it is fitting, for it reflects a painting in the process of its making. Here reflecting what is apparently represented on the artist's canvas, the metaphor of the inset mirror within the painting is direct. Distinct from the miniaturized details of Caravaggio's carafe or Vermeer's glass globe, this reflexive vision of art as a mirror is central to the work. As a summation of the pictorial trajectory of the Renaissance '*miroire-peinture*', *Las Meninas* declares itself the mirror of art. Depicting a painted mirror, by means of a mirror, *Las Meninas* recalls the long history of the painted mirror reflection to argue that a painting may indeed be a perfect double with the mirror technology of its own production.

Notes

The research for this essay was funded by the AHRC. My thanks to Sam Bibby, Sam Cohn, Richard Taws and Jack Warwick for their thoughts and comments.

1 Giorgio Vasari, *The Lives of the Artists* (1568), trans. George Bull, London, 1965, 276. On Giorgione's lost painting see especially Rona Goffen, *Renaissance Rivals: Michelangelo, Leonardo, Raphael, Titian*, London and New Haven, CT, 2002, 60–4 and notes.

2 Paolo Pino, *Dialogo di Pittura* (1548), in Paola Barocchi, ed., *Trattati d'arte del Cinquecento, fra Manierismo e Controriforma*, Bari, 1960, I: 131; G. P. Lomazzo, *Idea del Tempio della Pittura*, Milan, 1590, ed. Robert Klein, Florence, 1974, I, 151.

3 On mirrors in painting, Jonathan Miller, *On Reflection*, exh. cat. National Gallery London, London and New Haven, CT, 1998; and with specific reference to the Renaissance the seminal essay by Daniel Arasse, 'Les miroirs de la peinture', *L'Imitation: Aliénation ou source de liberté? Rencontres de l'Ecole du Louvre*, September 1984, 63–94 ; Rossella Vodret, 'Riflessioni sugli specchi nell'arte', in Valeria Merlini and Daniela Storti, eds, *Donna allo specchio: Tiziano a Milano*, Milan, 2010, 127–38.

4 André Chastel, 'Le tableau dans le tableau', *Stil und Überlieferung in der Kunst des Abendlandes: Akten des 21. Internationalen Kongresses für Kunstgeschichte in Bonn, 1964*, vol. I, *Epochen Europäischer Kunst*, Berlin, 1967, 19.

5 Michel Foucault, *Les mots et les choses: une archéologie des sciences humaines*, Paris, 1966, 7; Victor Stoichita, *L'instauration du tableau: Metapeinture à l'aube des temps modernes*, Paris, 1993. The implication of the mirror reflection is at the centre of Michael Fried's *The Moment of Caravaggio*, A. W. Mellon Lectures in the Fine Arts, Princeton, NJ, 2010.

6 Lucien Dällenbach's much-cited *Le récit spéculaire: essai sur la mise en abyme*, Paris, 1977, opens with André Gide's journal reflection of 1893 on the mirror in early modern painting alongside Shakespeare, 15.

7 Chastel, 'Le tableau'.

8 Mirrors would also come to play a key role in the development of early modern scientific instruments of vision – the telescope, the microscope, binoculars – where small mirrors were used to cast reflected light and to magnify the object under scrutiny. These instruments were, of course, also deployed by artists as, for example, with Lodovico Cigoli in his collaboration with Galileo. On the interface between early modern instruments of vision, art, and science see especially Horst Bredekamp, *Antikensehnsucht und Maschinenglauben: Die Geschichte der Kunstkammer und die Zukunftder Kunstgeschichte*, Berlin, 1993; and Bredekamp, ed., with Vera Dünkel and Birgit Schneider, *Das technische Bild: Kompendium für eine Stilgeschichte wissenschaftlicher Bilder*, Berlin, 2008; Barbara Maria Stafford and Frances

Terpak, object list by Isotta Poggi, *Devices of Wonder: From the World in a Box to Images on a Screen*, exh. cat., Los Angeles, 2001; and Alina Payne, ed., *Vision and its Instruments: Art, Science, and Technology in Early Modern Europe*, University Park, PA, 2015.

9 Patricia Falguières, 'Poètique de la machine', in Philippe Morel, ed., *L'art de la Renaissance entre science et magie*, Collection d'histoire de l'art de l'Académie de France à Rome, 5, 2006, 401–49; and for a different view on this development, Pamela H. Smith, *The Body of the Artisan: Art and Experience in the Scientific Revolution*, Chicago, IL, 2004. David Hockney, *Secret Knowledge: Rediscovering the Lost Techniques of the Old Masters*, London, 2006, posited the use of the mirror as a copying device, on which see the corrective A. M. Smith, 'Reflections on the Hockney-Falco thesis: Optical theory and artistic practice in the fifteenth and sixteenth centuries', *Early Science and Medicine*, 10: 2, 2005, 163–85, and the larger special issue edited by Sven Dupré on the question of optics in artistic practice. See Leon Battista Alberti, *On Painting*, trans. John R. Spencer ed., rev., New Haven, CT, 1966, 68–9: 'Nor will I hear what some may say, that the painter should not use these things, because even though they are great aids in painting well, [they] may perhaps be so made that he will soon be able to do nothing without them. I do not believe that infinite pains should be demanded of the painter, but paintings which appear in good relief and a good likeness of the subject should be expected. This I do not believe can ever be done without the use of the veil. Therefore, let us use this intersection, that is the veil, as we have said.' Filarete also defended the artist's uses of visual technologies, Antonio di Piero Averlino, Filarete's *Treatise on Architecture*, 2 vols, trans. and ed. John R Spencer, New Haven, CT, 1965, I, 305–6: 'If you should desire to portray something in an easier way, take a mirror and hold it in front of the thing you want to do. Look in it and you will see the outlines of the thing more easily, whatever is closer to you, and that which is farther away will appear to diminish. Truly I think it was by this method that Pippo di Ser Brunellesco discovered this perspective.'

10 As described c. 1480 in Howard Saalman, ed., *The Life of Brunelleschi* by Antonio di Tuccio Manetti, trans. Catherine Enggass, University Park, PA, 1970, 42–4. See Samuel Y. Edgerton Jr, *The Renaissance Rediscovery of Linear Perspective*, New York, 1975; and his *The Mirror, the Window, and the Telescope: How Renaissance Linear Perspective Changed Our Vision of the Universe*, Ithaca, NY and London, 2009, 44–53; and Martin Kemp's corrective 'Science, non-science and nonsense: The interpretation of Brunelleschi's perspective', *Art History*, 1, 1978, 134–61.

11 Jurgis Baltrušaitis, *Le Miroir: Essai sur une légende scientifique: Révélations, science-fiction, et fallacies*, Paris; Sabine Melchoir-Bonnet, *The Mirror: A History*, trans. Katharine H. Jewett, prefaced by Jean Delumeau, London and New York, 1994; *Miroirs: jeux et reflets depuis l'antiquité*, Paris, 2000; M. Anderson, *The Book of the Mirror: An Interdisciplinary Collection Exploring the Cultural Story of the Mirror*, Cambridge, 2007.

12 W. Patrick McCray, *Glassmaking in Renaissance Venice: The Fragile Craft*, Aldershot, 1999.

13 Vasari, at the opening of his life of Antonello da Messina (eds Luciano Bellosi and Aldo Rossi, Turin, 1986, 360). See the summary of the problem by Ashok Roy and Raymond White, 'Van Eyck's technique: The myth and the reality', I and II, in Susan Foister, Sue Jones and Delphine Cool, eds, *Investigating Jan Van Eyck*, Turnhout, Belgium, 2000, 97–106; and the useful discussion by Francis Ames-Lewis, 'Sources and documents for the use of the oil medium in fifteenth-century Italian painting', in Ingrid Alexander-Skipnes, ed., *Cultural Exchange between the Low Countries and Italy (1400–1600)*, Turnhout, 2007, 47–62.

14 On the Arnolfini double portrait and its mirror see, from an extensive literature, Lorne Campbell, *The Fifteenth Century Netherlandish Paintings, National Gallery Catalogues*, London, 1998, 174–211.

15 See D. G. Carter, 'Reflections in armor in the *Canon van der Paele Madonna*', *The Art Bulletin*, 36: 1, 1954, 60–2; D. Farmer, 'Further reflections on a van Eyck self-portrait', *Oud Holland*, 83, 1968, 107–9.

16 The connection between the two panels is disputed by Campbell, *Fifteenth Century Netherlandish Paintings*, 174–211, who notes that the portrait had shutters by 1516. See Margaret L. Koster, 'The Arnolfini double portrait: A simple solution', *Apollo*, 158: 499, 2003, 3–14, for an alternative interpretation of the evidence.

17 See Paola Barocchi's edition of Varchi and Vincenzo Borghini on the *paragone, Pittura e scultura nel Cinquecento*, Leghorn, 1998; Claire Farago, *Leonardo da Vinci's Paragone: A Critical Interpretation with a New Edition of the Text in the Codex Urbinas*, Leiden, 1992.

18 Alberti, *On Painting*, 56.

19 Alberti, *On Painting*, 68–9. See Edgerton's discussion of Alberti's window or velum, *The Mirror, the Window, and the Telescope*, 126–32.

20 Alberti, *On Painting*, 64. See further Ulrich Pfisterer, 'Künstlerliebe. Der Narcissus-Mythos bei Leon Battista Alberti und die Aristoteles-Lektüre der Frührenaissance', *Zeitschrift für Kunstgeschichte*, 64, 2001, 308, 324–8.

21 Rossella Vodret, 'Il Narciso di Caravaggio: ritratto di artista allo specchio', *Art e dossier*, 11: 109, 1996, 9–13, for a discussion of the painting's composition by means of mirrors. On Caravaggio's use of mirrors more generally see also Roberta Lapucci, ed., *Caravaggio and Optics*, Florence, 2005.

22 As related by G. P. Bellori, *Vite de pittori, scultori et architetti moderni*, Rome, 1672, 212, with reference to Caravaggio's depiction of a gypsy. On the larger question of Caravaggio's art of visible realism see especially Ferdinando Bologna, *L'incredulità del Caravaggio e l'esperienza delle 'cose naturali'*, Turin, 2006.

23 Published in S. Corradini and M. Marini, 'Inventarium omnium et singulorum bonorum mobilium di Michelangelo da Caravaggio "pittore"', *Artibus et historiae*, 14: 28, 1993, 161–76.

24 Giovanni Baglione, *Le vite de' pittori, scultori, et architetti*... Rome, 1642, 136–9. This has been taken to mean that Caravaggio painted his self-portrait within these early mythological works, on which see the helpful corrective by Mina Gregori, *The Age of Caravaggio* (exh. Metropolitan Museum, New York and Capodimonte, Naples), Milan, 1985, 241–2: 'What Baglione records is the widespread practice of employing a mirror as an aid in achieving a realistic representation – a method that, as a painter, he was certainly familiar with.'

25 Fried, *Caravaggio*, 60. See also the useful summary by Gregori, *The Age of Caravaggio*, 236–41.

26 Daniel Arasse, *Vermeer: Faith in Painting*, trans. Terry Grabar, Princeton, NJ, 1994, 22–39. On the contested question of Vermeer's use of a camera obscura see Philip Steadman, *Vermeer's Camera: Uncovering the Truth behind the Masterpieces*, Oxford, 2001; and the corrective H. Perry Chapman, 'The imagined studios of Rembrandt and Vermeer', in Michael Cole and Mary Prado, eds, *Inventions of the Studio, Renaissance to Romanticism*, Chapel Hill, NC and London, 2005, 108–46.

27 On the various critical editions of Leonardo's notes see Martin Kemp, ed., *Leonardo on Painting*, New Haven, CT and London, 1989.

28 Kemp, *Leonardo on Painting*, 202; and the brilliant discussion by Michael Baxandall in *Shadows and Enlightenment*, London and New Haven, CT, 1995, 'Three notes on Leonardo and early Renaissance shadow', 146–55. For a recent discussion of the intellectual and practical foundations of Leonardo's working methods issuing from conservation research see Larry Keith, 'In pursuit of perfection: Leonardo's painting technique', in Luke Syson with Larry Keith *et al.*, *Leonardo da Vinci: Painter at the Court of Milan*, London, 2012, 54–79.

29 Kemp, *Leonardo on Painting*, 205: 'And above all he must liken his mind to nature, giving it a surface like a mirror.'

30 Vasari, *Vite*, 1986 edn., 554, 559.

31 Kemp, *Leonardo on Painting*, 206.

32 Martin Kemp, *The Science of Art: Optical Themes in Western Art from Brunelleschi to Seurat*, London and New Haven, CT, 1990, especially 170–1. See also the compendium of Leonardo's notes on painters' aids in Kemp, *Leonardo on Painting*, 191–217. See also the introduction to this volume.

33 For example, trans. Irma A. Richter, *The Notebooks of Leonardo da Vinci*, Oxford, reprint 1986, 122. Leonardo noted that: 'if you look into the eye of anyone you will see your image there .... Your own image mirrored in the eye.'

34 Albrecht Dürer, *The Painter's Manual of Measurement*, Facsimile ed. and trans., Walter L. Strauss, New York, 1977.

35 Erwin Panofsky, *The Life and Art of Albrecht Dürer*, Princeton, NJ, 1943, 257.

36 Ludovico Cardi detto Il Cigoli, *Trattato pratico di prospettiva di Ludovico Cardi detto Il Cigoli: Manoscritto MS2660A del Gabinetto dei disegni e delle stampe degli Uffizi*, ed. Rodolfo Profumo, Rome, 1992; and Filippo Camerota, *Linear Perspective in the Age of Galileo: Ludovico Cigoli's Prospettiva pratica*, Florence, 2010; and the useful summary in Kemp, *The Science of Art*, 167–88.

37 Dürer, *The Painter's Manual*, 435.

38 Edgerton, *The Mirror, the Window, and the Telescope*, 126–32; and Kemp, *Science of Art*, 167–88.
39 Erwin Panofsky, *Dürers Kunsttheorie: Vornehmlich in ihrem Verhältnis zur Kunsttheorie der Italianer*, Berlin, 1915, 41 references also Leonardo showing his use of both the grid and the glass pane method.
40 Alberti, *On Painting*, 83
41 Falguières, 'Poètique de la machine', 401–49.
42 From a vast literature see Kemp, *Leonardo on Painting*, on the ordering of Leonardo's notes around themes of perspective and the relation between linear, colour and atmospheric perspective, 47–116.
43 Joseph Koerner, *The Moment of Self-Portraiture in German Renaissance Art*, Chicago, IL, 1993.
44 The translation is taken from Koerner, *Self-Portraiture*,, 43. Vasari, *Vite*, 1986 ed., 356, relates that Alberti had made a self-portrait from a round mirror.
45 Koerner, *Self-Portraiture*, 43–4. The portrait of Dürer the Elder, also in the Albertina collection, has been identified both as a self-portrait and as a portrait by the young Dürer of his father; in both cases it is seen as linked to Dürer's 1484 self-portrait drawing.
46 See David Ekserdjian's discussion of Vasari's slightly differing accounts of the painting from the 1550 to the 1568 edition of his *Vite*... in *Parmigianino*, New Haven, CT and London, 2006, 130. See also Sylvia Ferino Pagden, 'L'autoritratto del Parmigianino: La consistenza (im)material dell'autoritratto di Vienna', in Lucia Fornari Schianchi, ed., *Parmigianino e il manierismo europeo* (Atti del convegno internazionale di studi, Parma, 2002), Milan, 2002, 67–82.
47 Fried, *Caravaggio*, 96.
48 Carol Plazzotta in *Renaissance Faces: Van Eyck to Titian*, exh. cat., London, 2008, 256.
49 On Michelangelo and the *giudizio dell' occhio* see especially the discussion by David Summers, *Michelangelo and the Language of Art*, Princeton, NJ, 1981, 332–79.
50 Judith Mann, 'The myth of Artemisia as chameleon: A new look at the London "Allegory of Painting"', in Judith Mann, ed., *Artemisia Gentileschi*, exh. cat., St Louis, 2005, 51–77.
51 McCray, *Glassmaking*.
52 See especially Sylvia Ferino Pagden's summary of a complex literature on the 'belle donne': 'Pictures of women, pictures of love', *Bellini, Giorgione, Titian and the Renaissance of Venetian Painting*, David Alan Brown and Sylvia Ferino Pagden, National Gallery of Art, Washington and Kunsthistorisches Museum Vienna, London and New Haven, CT, 2006, 190–9 and 219–23. With regard to Bellini there is also his earlier, small panel usually termed an *Allegory of Prudence* (Galleria dell'Accademia Venice) which depicts a standing female nude on a pedestal, as if a sculpture, holding a convex mirror up while gesturing to its reflection of a torso, which is by implication the viewer. This has been the subject of much varying interpretation on which see *Tiziano: Amor Sacro e Amor Profano*, exh. cat., Rome, 1995, 241–2.
53 On this painting see the recent *Donna allo specchio: Tiziano a Milano*, eds Valeria Merlini and Daniela Storti, exh. cat., Paris, 2010.
54 Hubert Damisch, *Le Jugement de Paris: Iconologie analytique*, Paris, 1992, on the theme and its Renaissance expansion as an allegory of art.
55 Fernando Checa Cremades, *Tiziano/Rubens: Venus ante el espejo*. Contextos de la Colección in permanente, no. 13 (exh. Museo Thyssen-Bornemisza), Madrid, 2002. Titian's lost work is known through another autograph version now in the National Gallery of Art, Washington, and through Rubens' imitation of it on his visit to Madrid, now in the Thyssen Bornemisza Museum, which Velazquez would also have known.
56 Andreas Prater, *Venus at her Mirror: Velazquez and the Art of Nude Painting*, Munich, London and New York, 2002.
57 On Velazquez's paint handling more generally see especially Gridley McKim-Smith, Greta Anderson-Bergdoll and Richard Newman, eds, *Examining Velazquez*, London and New Haven, CT, 1988.
58 See Nicolas Milovanovic and Alexandre Maral, *La Galerie des Glaces: Charles Le Brun, maître d'oeuvre*, exh. Versailles, Paris, 2007; Nathalie Volle and Nicolas Milovanovic, *Galerie des glaces après sa restauration*, Rencontres de l'Ecole du Louvre, 26, 2013.
59 Trans. Irma A. Richter, *The Notebooks of Leonardo da Vinci*, Oxford, reprint 1986, 111–12.
60 See Steven N. Orso, *Philip IV and the Decoration of the Alcàzar of Madrid*, Princeton, NJ, 1986; and Fernando Checa, *El Real Alcàzar de Madrid: Dos Siglos de Arquitectura y Coleccionismo en la corte de los Reyes de España*, Madrid, 1994.
61 See J. M. Serrera, 'El palacio como taller y el taller como palacio. Una reflexion más sobre la meninas', in F. Marìas, ed., *Otras meninas*, Madrid, 1995, 231–46.
62 From an extensive and celebrated bibliography, Michel Foucault's introduction to *Les mots et les choses*; John R. Searle, 'Las Meninas and the paradoxes of pictorial representation', *Critical Inquiry*, 6, 1980; Leo Steinberg, 'Velazquez's Las Meninas', *October*, 19, 1981, 45–54.
63 Francisco Javier Sanchez Canton, 'Como vivá Velázquez: inventario discubierto por d.f. Rodriquez Marin', *Archivio español de arte*, XL, 1942, 69–91.
64 For discussion of the sources regarding Velazquez's interest in the Arnolfini double portrait see especially Madlyn Millner Kahr, 'Velazquez and *Las Meninas*', *Art Bulletin*, 52: 2, 1975, 225–46.
65 Antonio Palomino, *El museo pictõrico*, 3 vols, Madrid, 1796, III, 508–10.
66 Whether the mirror reflects the King and Queen as they sit for the painting, or whether it reflects what is painted on Velazquez's canvas, has been greatly debated, with a general though by no means universal consensus for the former, but see also Joel Snyder, 'Las Meninas and the Mirror of Princes', *Critical Inquiry*, 11: 4, 1985, 539–72.

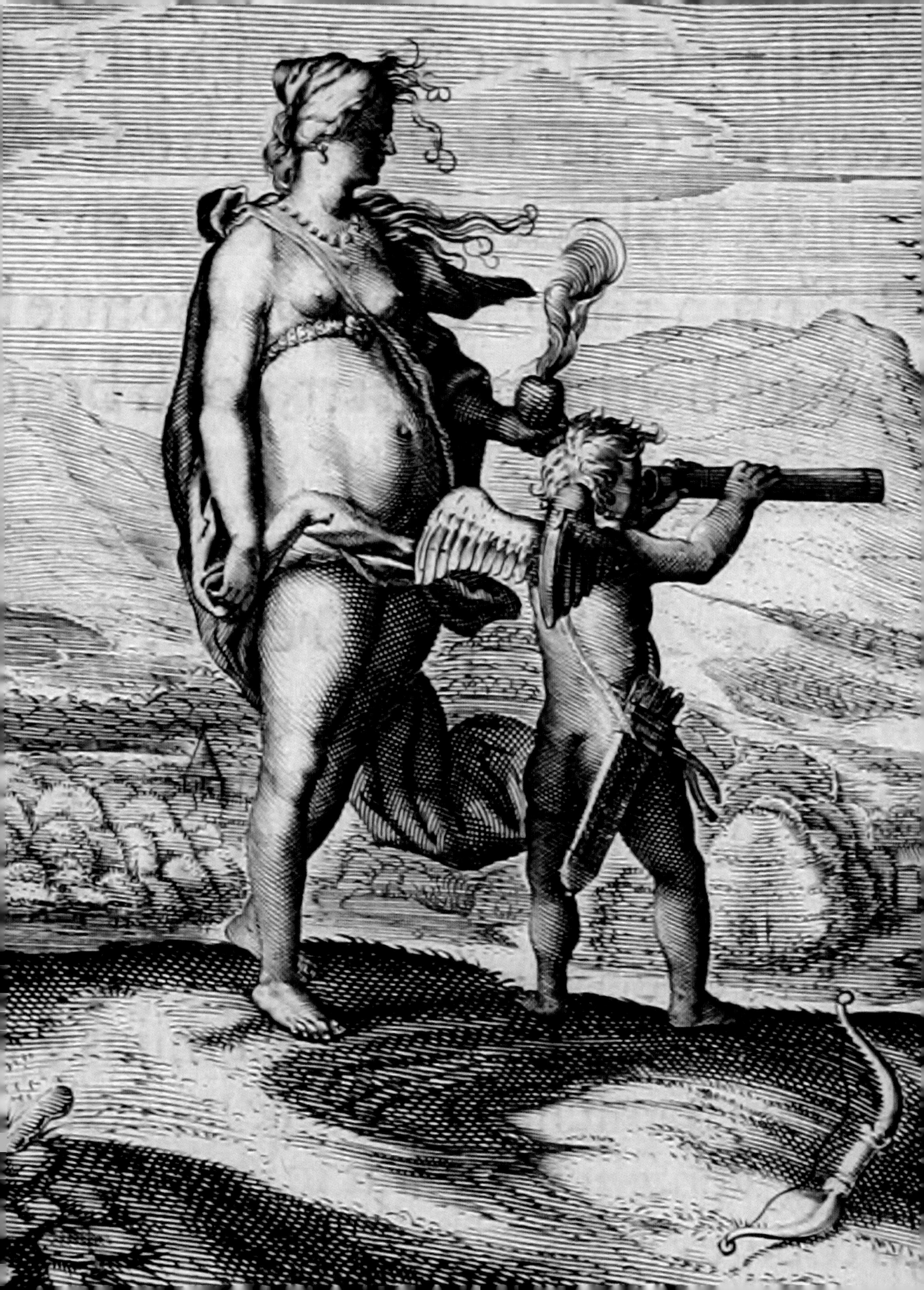

Chapter 5
Squaring the Circle: The Telescopic View in Early Modern Landscapes

Amy Knight Powell

In his *Natural Magic* of 1589, Giambattista della Porta describes the image that a camera obscura projects through its lens as a roundel: 'In a small circle of paper . . . you shall see as it were an Epitomy of the whole world.'[1]

In fact, relatively few early modern images have a circular format. Even those acquired with the aid of a lens were generally translated into the standard rectilinear format of picture or page: for example, Galileo Galilei's wash drawings of the phases of the moon, which were models for the copperplate etchings included in *The Starry Messenger* of 1610 (*plate 1*). Despite the circularity of the lenses in the telescope through which he observed the moon, and despite the sphericalness of the moon itself, Galileo embedded several of his moons in dark squares that represent the night sky. Lighter versions of these squares reappear in the etchings that illustrate the book.[2] Pictures, it seems, tended toward the rectilinear, even as optical instruments with round lenses became commonplace in the seventeenth century.[3]

Given this tendency, it is noteworthy that a good number of landscapes were painted in a round format in the Netherlands in the second half of the sixteenth and the first half of the seventeenth century. (The circular image that della Porta discovers projected through his lens is also a landscape: 'and so shall the beholder see more fitly Birds flying, the cloudy skies, or clear and blew, Mountains that are afar off.'[4]) Roundel landscapes, especially those made in the Dutch context, have been somewhat marginalized in the art-historical literature.[5] Some have even been made to square literally with the canon. In the 1980s, conservators at the National Gallery in London discovered that a landscape by Hendrick Avercamp, which had long masqueraded as a square, was in fact a circle with later extensions (*plate 2*).[6] A good portion of the surviving early modern landscape roundels circulate in the imprecisely documented world of auction houses and private collections, since they do not look quite right in museum galleries filled with rectilinear paintings and do not conform to the rectangular pages of books. When they do appear in either of these contexts, their format is rarely discussed at any length.

I will argue in this essay that the roundel became a fitting format for landscape in the second half of the sixteenth century because the genre was intimately bound up with fantasies of telescopic vision, which is to say that telescopic landscape painting predates the invention of the device itself.[7] This somewhat counterintuitive argument is based in part on a belief, now widely

Detail from illustration for Jacob Cats, 'Cupido Brilleman', after Adriaen van de Venne, in *The Zeeland Nightingale*, Middelburg, 1623, page 20 (*plate 5*).

Art and Technology in Early Modern Europe *Edited by Richard Taws and Genevieve Warwick*.

held by historians of science, that telescopic technologies themselves predate the appearance of the telescope proper, in late 1608. If proto-telescopic and early telescopic landscape views were shaped by a fantasy of seeing everything, however, the emergence of real telescopes in the first decade of the seventeenth century and their popularization during the ensuing decades made this fantasy difficult to sustain.[8] In these decades, what constituted a telescopic view changed from a vision of the whole world to a conspicuously limited take on things.

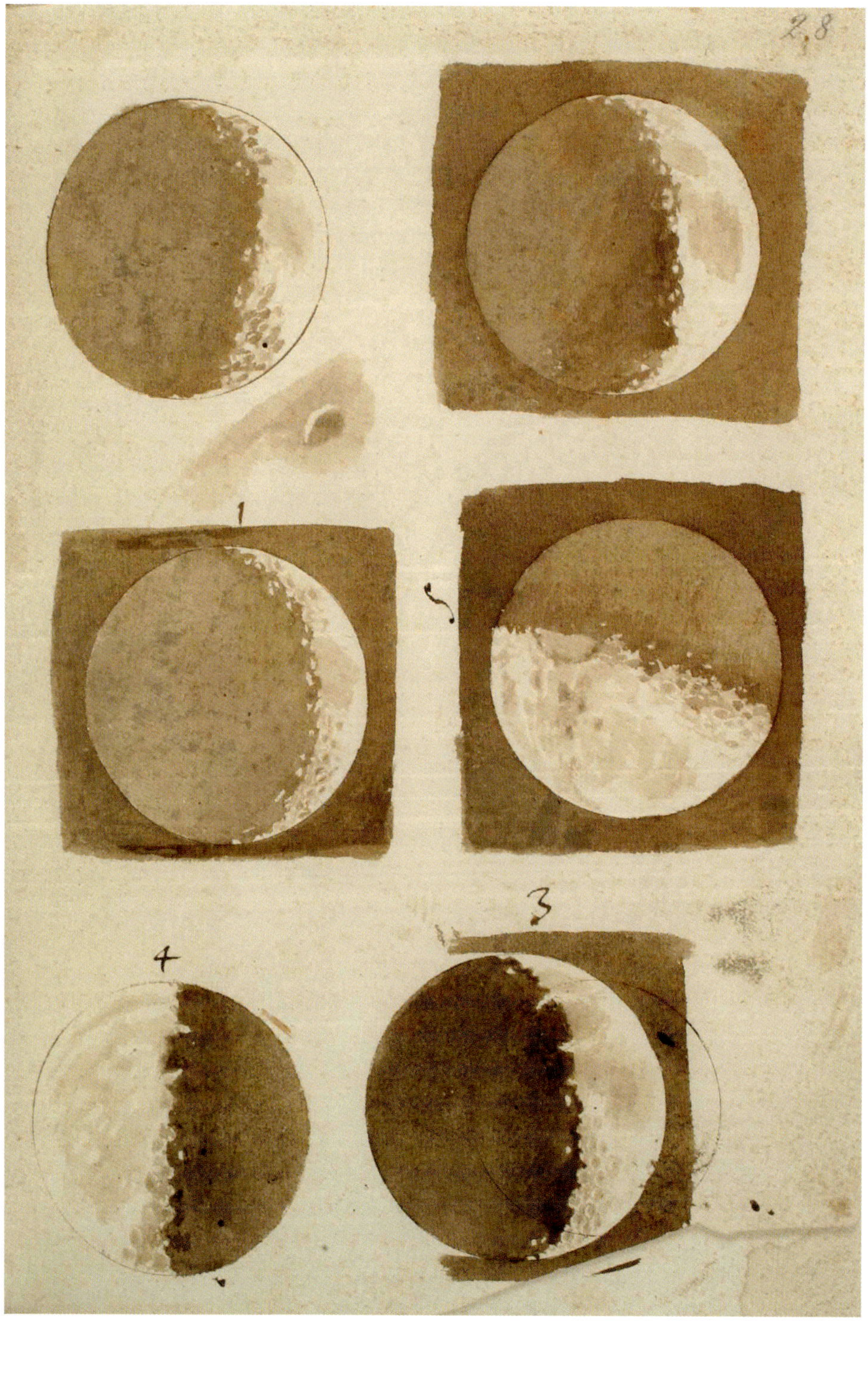

1 Galileo Galilei, *Phases of the Moon*, 1610. Ink drawing, 25.3 × 17.7 cm. Florence: Biblioteca Nazionale Centrale di Firenze (MS Gal. 48, fol. 28r). Photo: GAP Digital Imaging.

2 Hendrick Avercamp, *A Winter Scene with Skaters Near a Castle*, c. 1608–09. Oil on panel, diameter 40.7 cm. London: National Gallery. Photo: National Gallery.

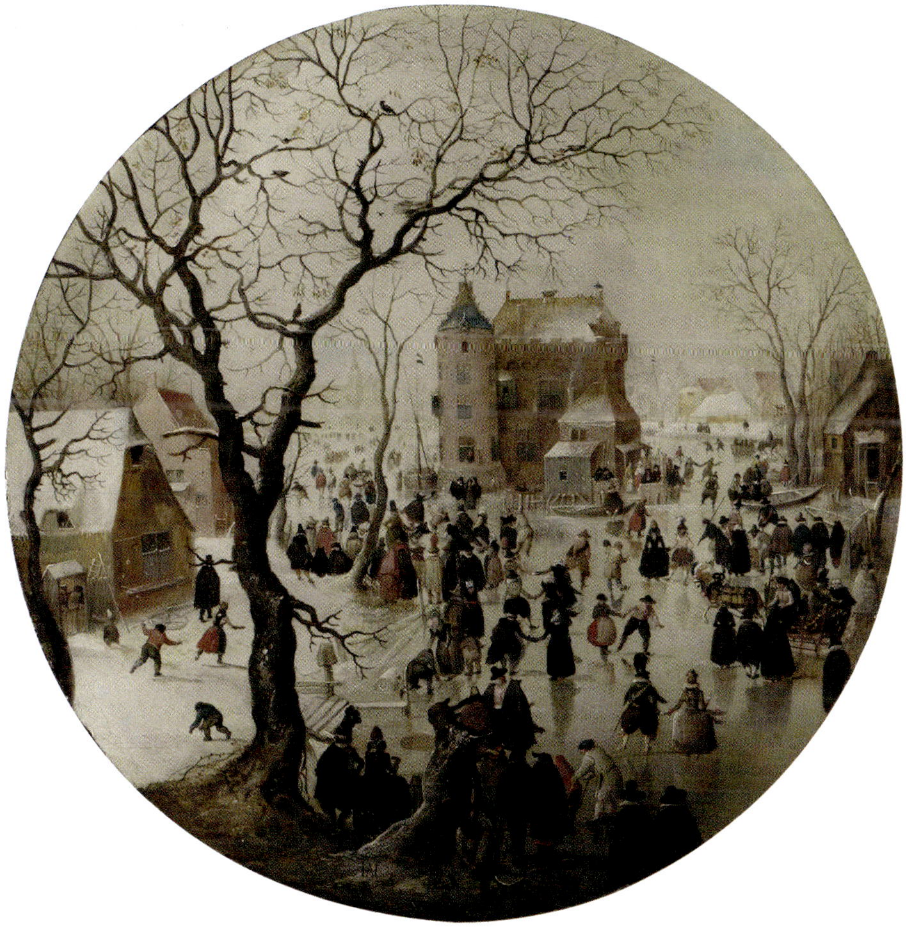

Prospects and Perspectives

Though it was Galileo's celestial observations described in *The Starry Messenger* that made the telescope an international sensation, the instrument was more popularly used during the seventeenth century for terrestrial viewing, both on land and at sea. One of the earliest depictions of a telescope is to be found in *Landscape with a View of Mariemont Castle* by Jan Brueghel the Elder, court painter to Archduke Albert of Austria and Archduchess Isabella Clara Eugenia, the Spanish Governors of the southern Netherlands (*plate* 3).[9] Situated close to Brussels, Mariemont was a favourite residence and hunting lodge of the two. In the lower left corner of Brueghel's painting, the Archduke and a small entourage are hunting herons. Albert trains his telescope – perpendicular to the picture plane – on a bird in the trees on the right side of the painting. Based on letters written by Brueghel and on the state of the castle, whose various stages of renovation are well documented in his paintings, *Landscape with a View of Mariemont Castle* has been dated to the fall of 1609, when the telescope was still a brand new invention.[10]

On 25 September 1608, the provincial government of Zeeland sent word to The Hague informing the States-General – the governing body of the newly formed Dutch Republic – that a spectacle maker from the city of Middelburg had 'a certain device by means of which all things at a very great distance can be seen as if they are nearby'.[11] A few days later, the States-General considered the Middelburg spectacle maker Hans Lipperhey's patent application. Shortly after that, two other men, Jacob Metius of Alkmaar and Zacharias Janssen of

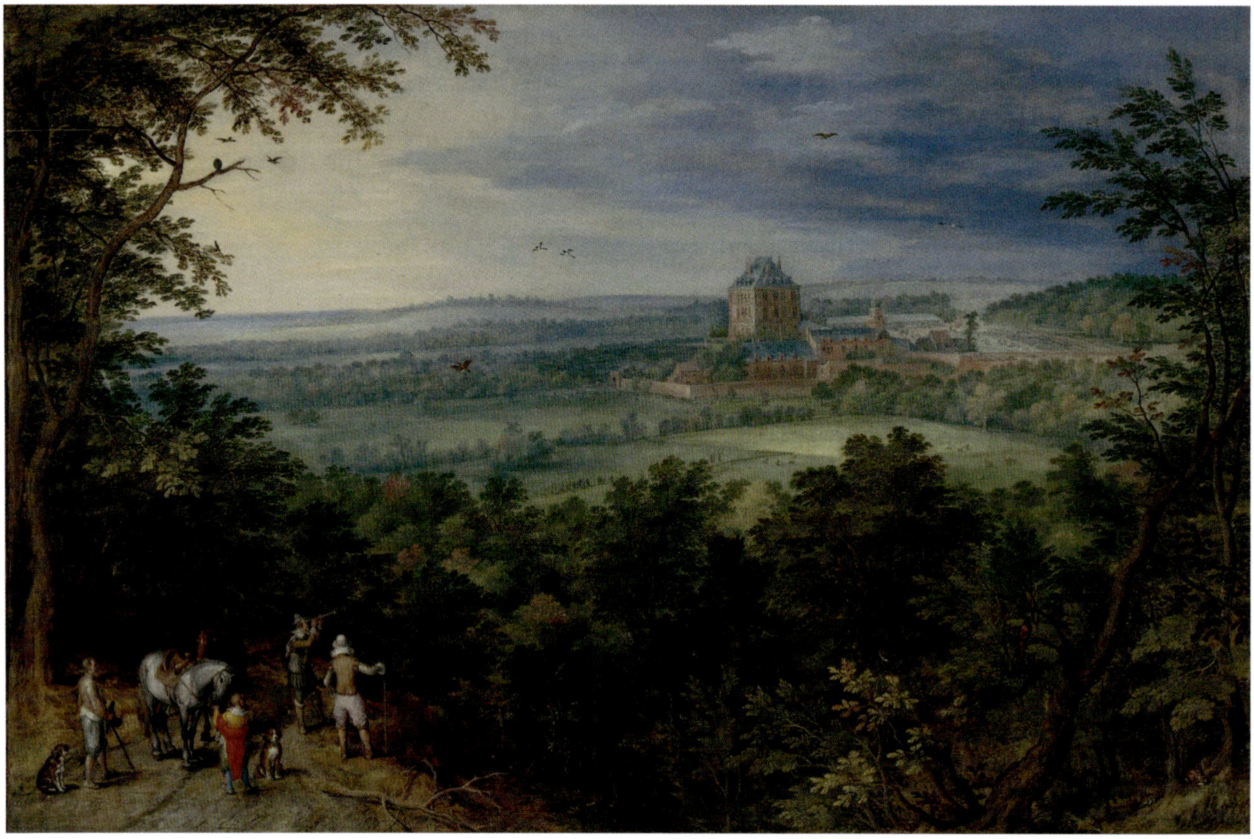

3 Jan Brueghel the Elder, *Landscape with a View of Mariemont Castle*, c. 1609. Oil on canvas, 84.7 × 130.8 cm. Richmond: Virginia Museum of Fine Arts (Adolph D. and Wilkins C. Williams Fund). Photo: Katherine Wetzel/ Virginia Museum of Fine Arts.

Middelburg, came forward with competing claims to the invention. This sudden appearance of not just one but several telescopes in September and October of 1608 happened to coincide with the end of the negotiation of the Twelve-Year Truce between the Dutch Republic and Spain.

Because of the negotiations, diplomats from Spain and elsewhere were in The Hague and quickly sent home news and, when they could, examples of the telescope. Among these diplomats was Ambrogio Spinola, chief commander of the Spanish army in the southern Netherlands, who returned to Brussels from The Hague in the first days of October, news of the invention in hand. Moved by Spinola's report, Albert was soon able to acquire a telescope. In a letter dated 2 April 1609, Guido Bentivoglio, the papal nuncio in Brussels, writes: 'recently I had occasion to test the instrument belonging to his Highness [the Archduke], being present with him and the Most Serene Infanta at a heron hunt here outside the gates of Brussels.'[12] It is precisely such a hunt, absent the Infanta Isabella, that Brueghel pictures in *Landscape with a View of Mariemont Castle*.

Spyglasses, as they were also called, quickly became available beyond the courtly milieu. Already in 1609, they were being sold in the shops of spectacle makers in France.[13] And by the 1620s, images of them were cropping up in emblem books and other printed materials.[14] The eclectic painter and poet Adriaen van de Venne was clearly familiar with them. From about 1608 to 1625, he lived and worked in Middelburg, not far from Lipperhey's workshop, alongside his brother Jan van de Venne, who ran a publishing house. In one of his contributions to *The Zeeland Nightingale* (1623) – a celebration of Zeeland culture published by his brother – Adriaen enumerates the things available in the shop associated with Jan's press:

> Would anyone like Paintings or Prints, Books, Spyglasses [*Buysen*],
> Or pretty lacquer-ware, for decorating houses.
> Compass and Time-piece, or Maps small and large,
> Or whatever serves the needs of a Sea-pilot,
> That finds one for sale here.[15]

Adriaen groups the spyglass with the images and other decorative objects rather than with the tools for navigation, as if he saw the view through the spyglass as a picture, to be appreciated primarily on aesthetic grounds.[16]

The role that the telescope came to play in emblem books and allegories suggests that it called to mind not just any pictures but landscapes in particular. Adriaen's illustration for the emblem 'Envy enjoys the misfortune of others' in Johan de Brune's *Emblemata* of 1624 shows a well-heeled man pointing his spyglass at someone we cannot see beyond the frame (*plate* 4).[17] Though his spyglass is ostensibly focused on the 'others' of the emblem's title, the setting compels us to imagine the view through the telescope as a landscape. As our gaze is drawn along the path beside the river into the distance, so the man's gaze is drawn through the tube of his spyglass into a distance that runs perpendicular to the one that captivates our own. Adriaen's illustration for Jacob Cats's poem 'Cupido Brilleman', also in *The Zeeland Nightingale*, has a similar structure. It pictures Cupid peering out at a vast panorama through a telescope, scouting out targets for his love-tainted arrows (*plate* 5). As in 'Envy enjoys the misfortune of others', our gaze into the distance crosses at right angles his telescopic view, which we cannot see but imagine – thanks especially to the roundel format of the emblem – to be a landscape like our own.[18]

4 After Adriaen van de Venne, *Envy Enjoys the Misfortune of Others*, emblem XLVII in Johan de Brune, *Emblemata of Zinne-werck*, Middelburg, 1624. Photo: J. Paul Getty Trust.

5 After Adriaen van de Venne, illustration for Jacob Cats, 'Cupido Brilleman', in *The Zeeland Nightingale*, Middelburg, 1623, page 20. Photo: J. Paul Getty Trust.

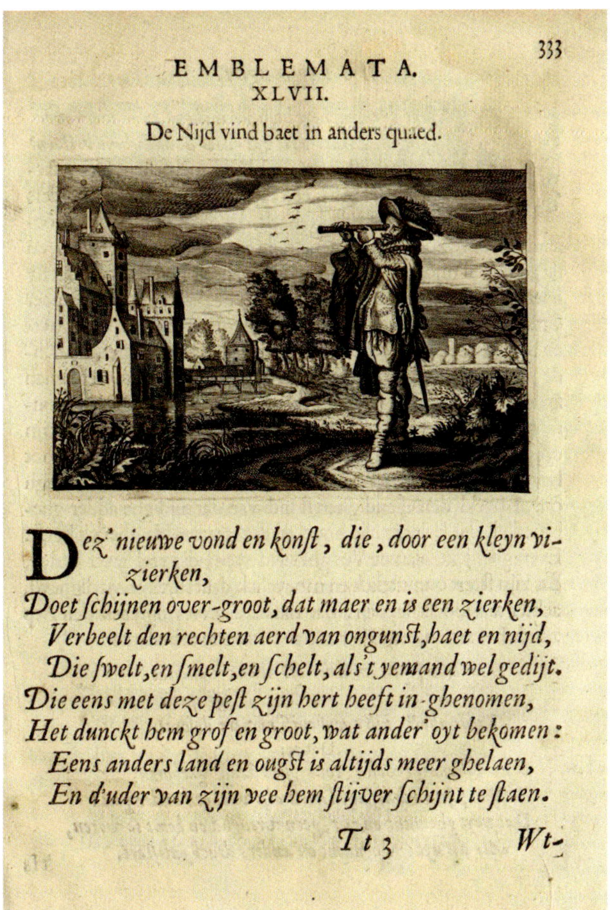

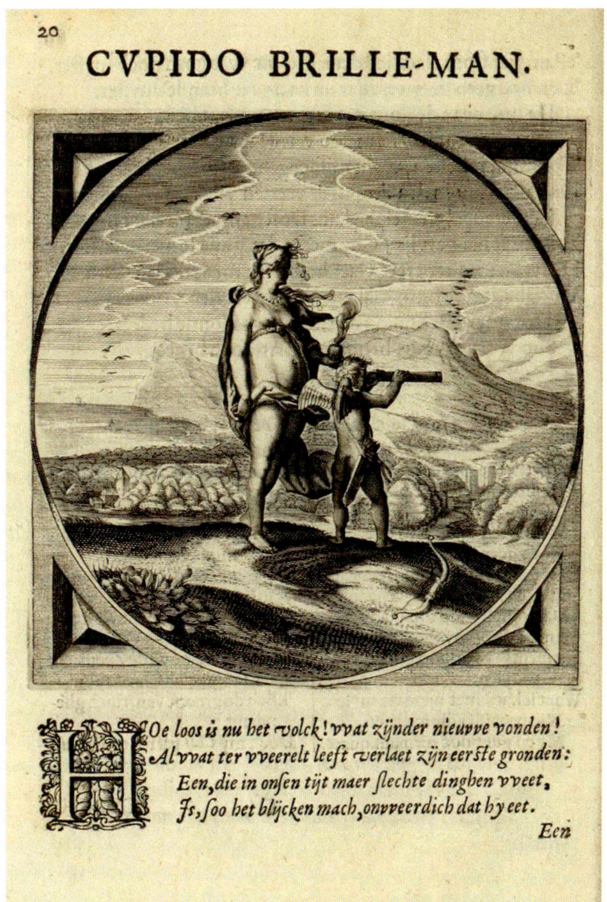

In the beginning, the genre of landscape was nearly synonymous with distant prospects.[19] An anonymous, probably late sixteenth-century, English manuscript on perspective defines landscape – and this is one of the earliest English definitions – as that 'which expresseth places of larger prospecte, as whole contries where the eye seemeth not to be hindered by any objectes ... but to passe as farre as the force therofe can pierce'.[20] Because of their shared association with seeing far, the genre of landscape and the telescope had several names in common. In Dutch, a landscape was sometimes called a *perspectief*, a telescope a *perspectiefglas*. In English, the words 'perspective' and 'prospect' were used to designate a landscape painting, while 'perspective tube' and 'prospective tube' named the optical device.[21] All of this is to say that the genre and the device were strongly associated with one another.

World-Landscapes

Brueghel's *Landscape with a View of Mariemont Castle* is an example of what has come to be called a Flemish world-landscape. This style of landscape painting seemed to have run its course already by the 1580s, when more intimate views came into vogue, but Brueghel along with several other Flemish painters revived it briefly around the turn of the century. World-landscapes are miniature paintings. While Brueghel's painting of Mariemont measures 84 × 130 centimetres, most world-landscapes are less than sixty centimetres wide, many are smaller than that, and regardless of their size, their scale relative to the worlds they depict is always very small.[22] The poet Maurice Scève probably had in mind a Flemish painting when he wrote, 'in a single glance, at a rather small painting, one saw a large landscape, the sky, the earth, and the water'.[23] Federico Borromeo, an important collector of Flemish landscapes, marvels over some of the paintings in his collection: '[they] enclose in narrow places, the space of the earth and the heavens; [and this allows us to] go wandering, and making long journeys standing still in our room'.[24]

World-landscapes can be traced back to views glimpsed through windows and arches in the paintings of Jan van Eyck and Rogier van der Weyden. In the last decades of the fifteenth century, Hans Memling incorporated landscape backgrounds into his portraits. And around 1500, Hieronymus Bosch allowed these partial views to fill the frame. Building on this tradition, Joachim Patinir established an enduring formula for the illusion of vastness in the first decades of the sixteenth century: a high horizon that accommodates a superabundance of anecdotal detail, and a colour scheme that moves from brown in the foreground, to green in the middle distance, to blue at the horizon.

World-landscapes tend to defy the laws of atmospheric perspective, showing very distant things in minute detail almost all the way to the horizon, where the atmosphere suddenly sets in with a vengeance, sucking the visible into the invisible and thereby signalling that, though we have reached the limits of our own field of vision, there is still more to see. Van Mander urges painters to show that this fading into obscurity at the horizon is a matter of faltering sight: 'You cannot show the distance too vaguely. . . . Take into account the thickness of the air, which contains a blue substance, obstructing sight, and forming a haze through which it is impossible to see really clearly.'[25] Working, around midcentury, with a version of Patinir's formula and a keen sense of the pathos of the horizon, Pieter Bruegel the Elder (Jan Brueghel the Elder's father) composed worlds of breathtaking dimensions, which earned him Van Mander's praise.[26]

Because horizons like the one at the end of the valley in Bruegel's *Rustic Solicitude* indicate that the world continues beyond them, his landscapes have sometimes

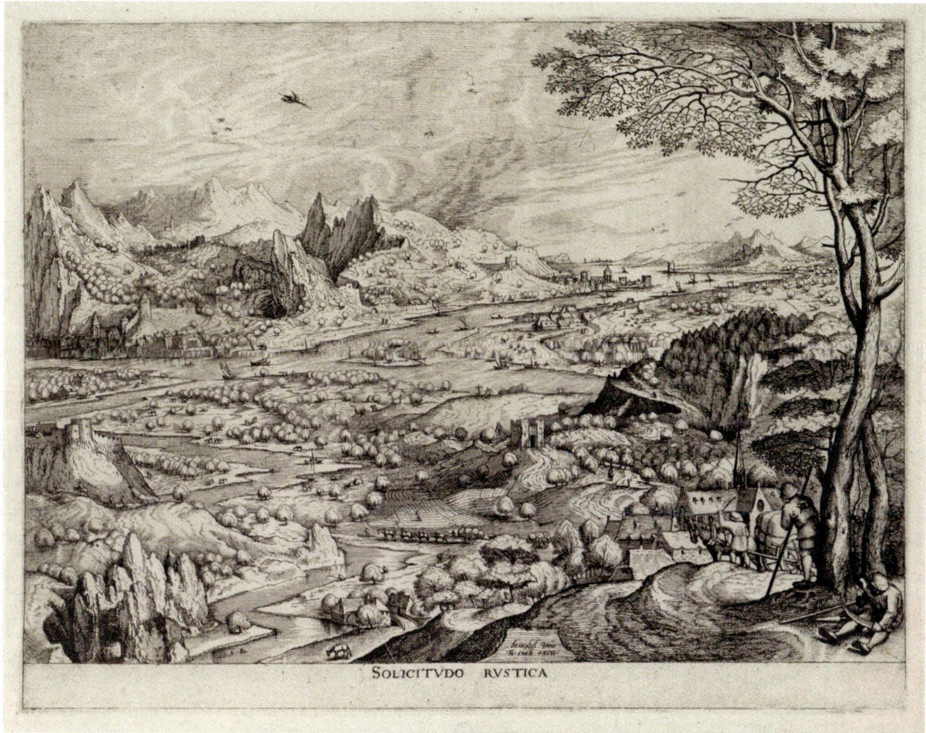

6 Johannes or Lucas van Doetechum after Pieter Bruegel the Elder, *Rustic Solicitude*, c. 1553–58. Etching and engraving, 32.4 × 42.9 cm. Amsterdam: Rijksmusem. Photo: Rijksmusem.

been interpreted as evidence of a new, modern sense of geographic as well as cultural contingency, that is, a new sense that the world is vaster than any one person can know (*plate* 6).[27] But Bruegel's landscapes are really no more modern than those of the previous generation or two. Copernicus and the discovery of the New World notwithstanding, his sense of space is deeply traditional. The horizon is still high, and our point of view is not, in fact, delimited by it. While the men in the lower right corner of *Rustic Solicitude* find themselves embedded in the landscape and their view determined by their contingent location, we are lifted up into a quasi-omniscient position. We have, in fact, no point of view, properly speaking. From our elevated perch, we are privy to not just a lot but rather the entirety of the world; we see some of it directly, and we are given enough information to feel that we can infer the rest.

In the sixteenth century, the still new concept of 'landscape' implied this kind of totality. By definition, a landscape was not just large but all-encompassing. Bringing together metaphorical usage in English literature and Van Mander's technical usage, James Turner explains that, '[landscape] comes to mean a definitive model or microcosm: within every creature is "an adumbration or landskip of the whole Univers", London is "our England of England, and our Landskip and Representation of the whole island" . . . Landscape painting itself, according to Van Mander should be designed as a small universe. Summary plenitude and aesthetic wholeness combine in "a perfect peece of Lantskip".'[28]

Landscape paintings could be made to seem even more perfectly whole through the use of a circular format. This format was used occasionally in early Netherlandish paintings, before which it was found primarily in the decorative arts.[29] In his early celebration of Van Eyck, the humanist Bartolomeo Facio describes 'a circular representation of the world', which is now lost.[30] The roundel, or tondo, then found its first true exponent in Bosch, who used it on the exterior of *The Garden of Earthly Delights* and elsewhere.[31]

Around midcentury, the prolific landscape painter Herri met de Bles made a circular painting of the earthly paradise that epitomizes the Flemish world-landscape tradition. It is dominated by the Garden of Eden in the foreground, amid whose dark trees are scattered scenes from Genesis: the creation of Eve, the forbidding of the fruit, temptation, and expulsion (*plate 7*). Behind the wooded garden, the lands into which the angel drives the first couple extend out toward a misty horizon. All of this is set within a circular band of sky studded with birds, sun, moon, and stars, which is circled, in turn, by a narrower band of primordial waters.[32]

When it comes to the genre of landscape especially, the tondo format suggests the entirety of the orb of the world as well as the all-seeing gaze that is capable of taking it in. Nicholas of Cusa describes God's sight as spherical: 'when our eye turneth itself to an object, 'tis because our sight seeth but through a finite angle. But the angle of thine Eye, O God, is not limited, but is infinite, being the angle of a circle, nay, of a sphere.'[33] Met de Bles's *Paradise* opens like the pupil of an eye onto a landscape that implies not only the full extent of the world east of Eden but also the full duration of human history. It thereby lends the viewer God's 'eye of sphericity and of infinite perfection', as Nicholas of Cusa calls it, 'wherefore it seeth at one and the same time all things around, above and below'.[34] Later, after the invention of the telescope, Joseph Glanvill would describe Adam, in paradise, as not yet in need of what he calls 'Galileo's tube' in order to see the total expanse of creation. Only after the fall, would optical prosthetics become necessary: '[Adam] could command distant

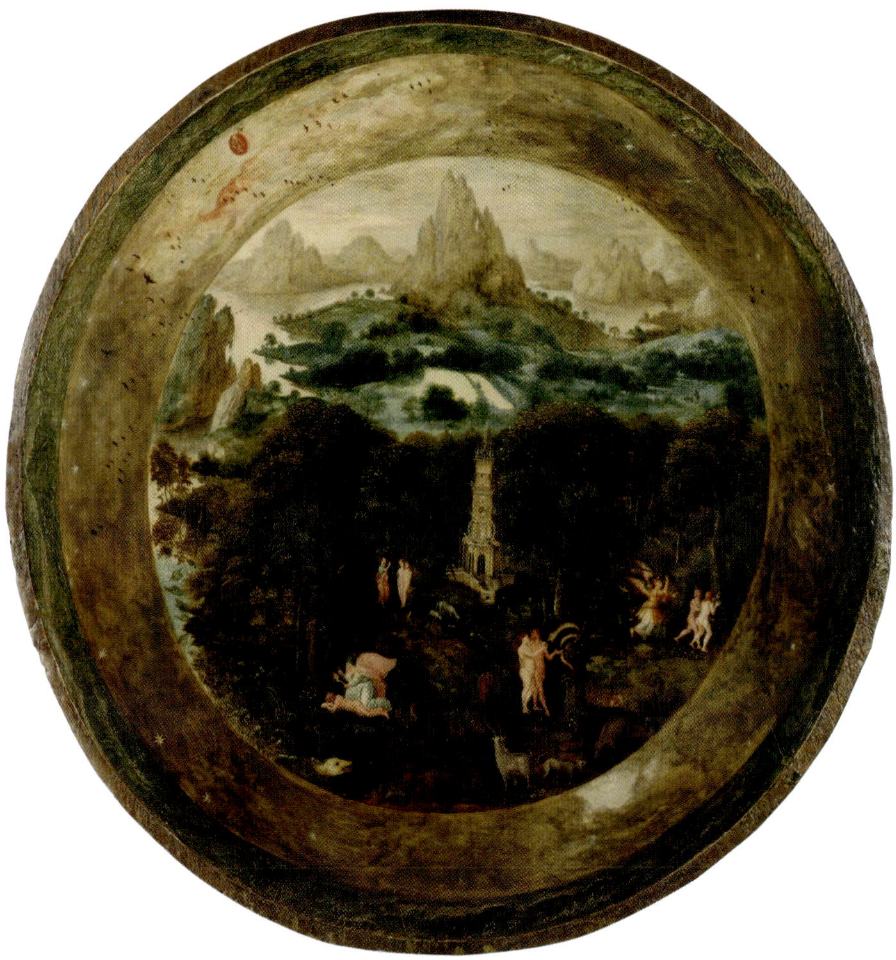

7 Herri met de Bles, *Paradise*, c. 1541–50. Oil on panel, 46.6 × 45.5 cm. Amsterdam: Rijksmuseum. Photo: Rijksmusem.

objects by natural, as we do by artificial advantages.'[35] In his poem 'On the Telescope', Constantijn Huygens likewise celebrates the telescope as the expansion of the human gaze to godlike proportions: 'At last mortals may, so to speak, be like gods, / If they can see far and near, here and everywhere.'[36] Like optical devices, world-landscapes, including Met de Bles's *Paradise*, give their (postlapsarian) viewers enhanced powers of vision.

Legendary Mirrors

Though there were no full-fledged telescopes at the time that Met de Bles painted his *Paradise*, there were real and imaginary proto telescopic contrivances. Popular legends told of mirrors and combinations of mirrors and lenses that allowed people to see across great distances. In *The Letter of Prestor John*, a tale of a mythical ruler of a lost Christian nation in Asia that first appeared in the mid-twelfth century, John boasts of mirrors powerful enough that, 'there is no land so distant / Where someone desires to make war on us / Nor treason by any people / That we do not see immediately. / ... Our enemies and their preparations, / Nothing can be hidden from us.'[37] Various European romances of the thirteenth and fourteenth centuries likewise mention a large concave mirror mounted high on a tower that permitted Roman rulers to anticipate the approach of enemies.[38]

By the later sixteenth century, rumours had emerged in European sources that the Lighthouse of Alexandria, then ruined by several earthquakes, had once been surmounted by a concave mirror used to survey and ignite threatening ships as they approached the Egyptian port.[39] Along with allowing one to see across great distances, these legendary glasses allowed one to gaze into the future. When Hans Christoph Teufel travelled to Egypt around 1587, he described a 'tower with enchanted mirrors..., mirrors in which one could see, fifty days in advance, the arrival of all armies that approached with the intention of harming the kingdom of Egypt'.[40] These legends tell of telescopic contrivances in faraway times and places, but there were also reports of contemporary European devices. Already in the thirteenth century, Roger Bacon had described an optical instrument with the capacity to magnify.[41] In the sixteenth century, speculation and experimentation with telescopic machines became more widespread. Della Porta would always maintain that all the information necessary to build a telescope was already available in the book on refraction that he published in 1593. And others cited della Porta's *Natural Magic* (1589) as a source for the Dutch telescope.[42]

So it is not surprising to learn that the Dutch telescope that Brueghel includes in *Landscape with a View of Mariemont Castle* was not the first telescopic device that Albert and Isabella owned. Before they obtained a telescope proper, they were already in possession of at least one instrument for seeing far. According to an inventory record, they reimbursed Count Charles of Arembourg for 'perspective glasses [*voires perspectifs*]' in 1599.[43] A letter written by the Italian Francesco Angelo Coccio in 1557 suggests that proto-telescopic glasses of this kind were associated with Flemish world-landscape paintings: 'I am taking up my glasses for seeing far [*occhiali da veder lontano*]', he writes to his friend the Venetian physician Ettore Ausonio, 'and without getting up from the table, I see Venice in the way that one sees certain distant cities represented in the landscapes of Flemish painters'.[44]

After centuries of debate about who should be given the title of inventor of the telescope, scholars have come to the conclusion that the invention was an attenuated process. Technically speaking everything was in place long before 1608, and many people had conceived or constructed optical devices intended to make faraway things

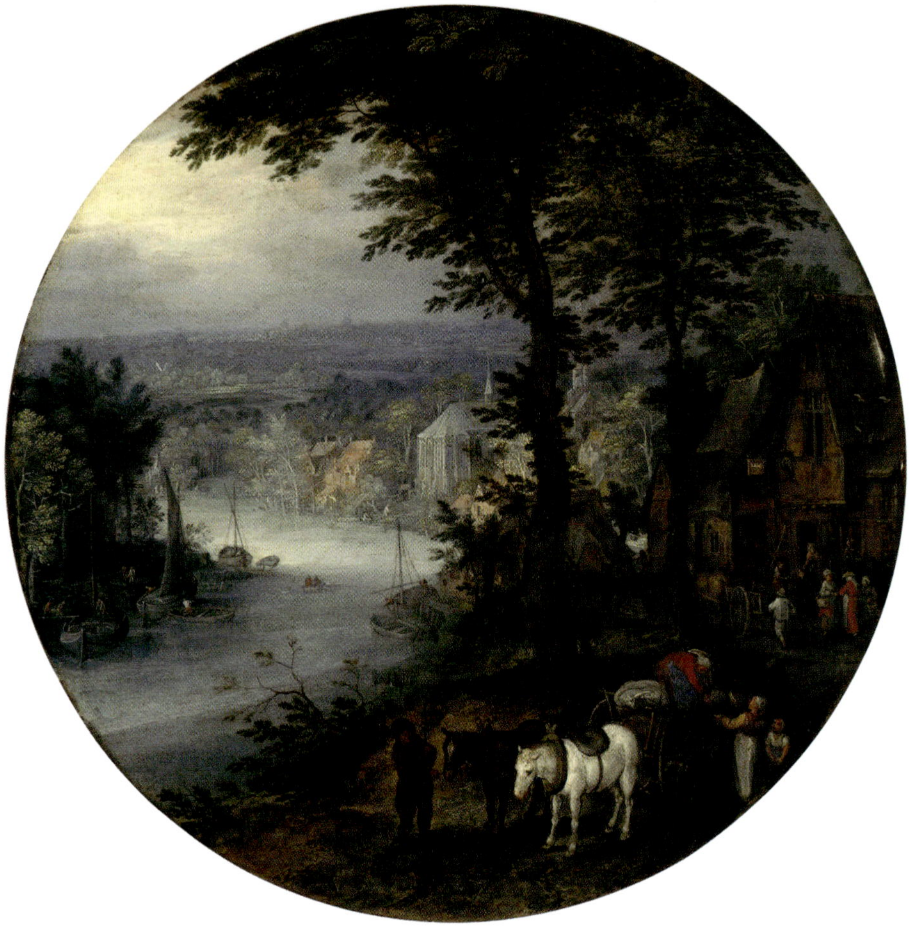

8 Jan Brueghel the Elder, *Tavern on a Riverbank*, c. 1600. Oil on copper, diameter 21 cm. Munich: Alte Pinakothek, Bayerische Staatsgemäldesammlungen. Photo: bpk, Berlin/Bayerische Staatsgemäldesammlungen/ Art Resource, NY.

appear close well before that date.[45] Moreover, even after its invention, the telescope would remain the stuff of legend, as Eileen Reeves has brilliantly shown. Galileo initially reacted to news of the novel instrument as if it were simply another spin on the old tales, and a rather disappointing one at that. The legends told of machines much more powerful than the earliest telescopes, which magnified objects only about three times.[46]

Taking this longer view of the invention of the telescope allows us to see the extent to which sixteenth-century world-landscape paintings, especially those with a circular format, are telescopic views *avant la lettre*. In the half-century preceding the emergence of the telescope in Middelburg, a relatively large number of tondo landscapes were made in Flanders. Many of these – including works by Jacob Grimmer (*c.* 1526–90) and Abel Grimmer (*c.* 1570–1620) – belonged to series of the four seasons or twelve months or summer–winter pendant pairs, which added a temporal dimension to the project of picturing the totality of space.[47] Many of these combine aspects of panoramic worldviews with more intimate scenes of everyday life.

Brueghel chose a circular format for a number of the belated worldviews that he painted around 1600, including *Tavern on a Riverbank* (*plate* 8).[48] On the right, the foreground, with its magnificent trees pushing up against the upper limits of the circle, forms a frame within the frame, through which one views a blue river snaking toward an equally blue horizon. Peering at this painting – a copper roundel only about 21 centimetres in diameter – with its river flowing toward the futurity of the horizon, is like looking into a crystal ball.

Paintings like this one show that Brueghel had already marshalled all he needed to paint a persuasive telescopic view by the time an actual telescope arrived at the Brussels court in 1609. When he painted his *Landscape with a View of Mariemont Castle* (see plate 3), where the telescope looks perfectly at home, he reverted, in fact, to what one scholar has called a 'hopelessly outmoded' type of composition that includes a frame within its frame, as does *Tavern on a Riverbank*.[49] The figures, including the Archduke with his eye glued to his telescope, stand in a small clearing in the lower left corner of the painting, beneath a tree that frames the left side of our view. They are pictured in Mariemont Castle's famous, wooded deer park, which extends across the width of the canvas, ending on the right in a coulisse of trees that complements the one on the left. These trees form an oval through which we view a vast, verdant panorama that fades into the river Haine toward the horizon. With its oval frame, our (outmoded) view through the opening in the trees is like the Archduke's view through his (brand new) perspective tube.

Telescopic Views

If landscape roundels were being made in the southern Netherlands already by the mid-sixteenth century, they only began to be made in the northern Netherlands, that is, in the Dutch Republic, at about the time that Brueghel painted *Landscape with a View of Mariemont Castle*. This, as I have said, was just shortly after several nearly simultaneous patent claims were filed for the telescope in The Hague. Among the earliest of these Dutch roundels is Hendrick Avercamp's *A Winter Scene with Skaters Near a Castle* (see plate 2).[50] This is the panel from the National Gallery in London that was long disguised as a square. A good deal was temporarily lost in the squaring of this circle, whose composition has a carefully calibrated relationship to its frame.

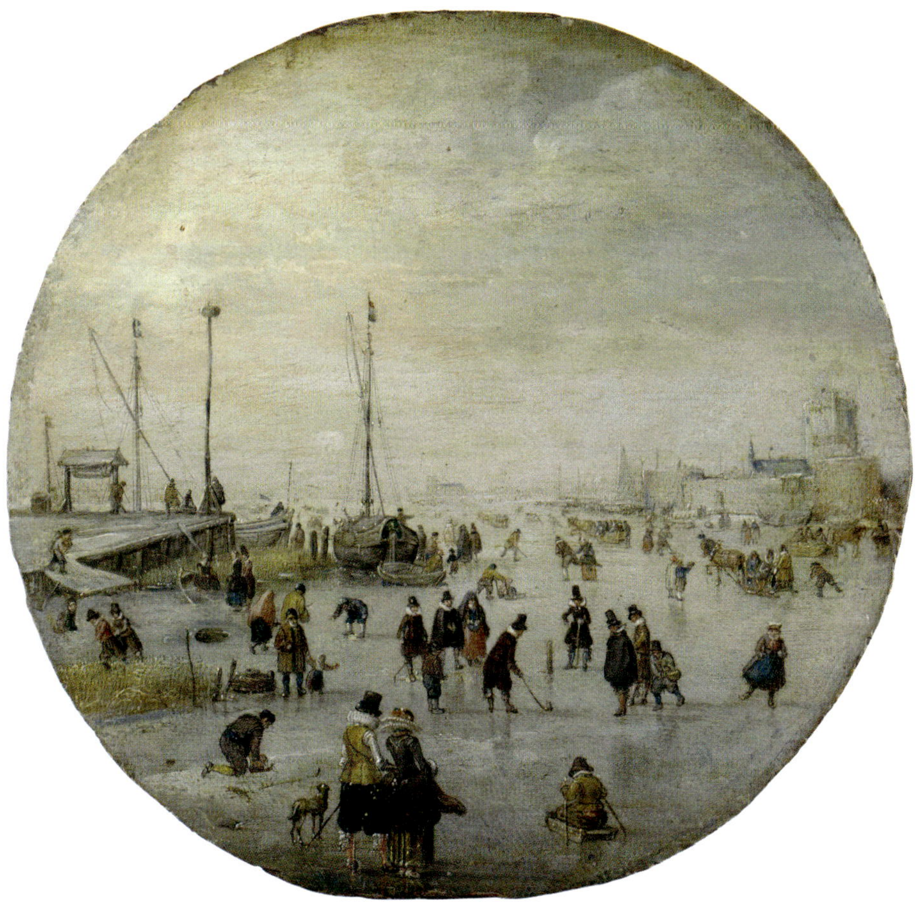

9 Hendrick Avercamp, *Winter Landscape*, c. 1610. Oil on panel, diameter 18.5 cm. Hamburg: Hamburger Kunsthalle. Photo: bpk, Berlin/Hamburger Kunsthalle/Art Resource, NY.

The mound of earth supporting the tree that dominates the foreground nestles into the curve of the panel's lower edge. Where that tree branches off to the right, it bends in accordance with the circle, while its smallest branches form a chorus of arcs. We view all this as if we were perched high above the scene, perhaps in another denuded tree, telescope in hand, spying on the skaters, most of whom are turned away.

Though it is difficult to date Avercamp's works and though they did not develop in a purely progressive manner, there is an overall movement in his oeuvre away from the all-encompassing structure of the world-landscape. His earliest paintings, including the tondo in the National Gallery, have a relatively high horizon, which derives from Flemish models. But he soon began to place the horizon close to the middle of the frame, as in his *Winter Landscape* in Hamburg (*plate* 9).[51] Only 18.2 centimetres in diameter, this diminutive roundel nonetheless captures a great deal of space – this it owes to the world-landscape tradition – but now the vast stretch of ice is mirrored by an equally expansive stretch of sky.

Avercamp grew up in the relatively isolated city of Kampen in the province of Overijssel. Though he returned to Kampen for good around 1613, he spent the preceding decade or so in Amsterdam, where he would have become aware of the telescope soon after its invention. It is also possible that he had access to telescopes even in Kampen through his relatively well-educated family. (His father was an apothecary and his brother a doctor.) While we can only speculate about whether Avercamp ever looked through a real telescope, we can be certain that Adriaen van de Venne, whom I discussed earlier, did. Along with writing about the telescopes in his brother's shop and picturing them in paintings and emblems (*plate* 4 and *plate* 5), the Middelburg poet and artist painted a pair of roundel landscapes with fantastically deep prospects that imply a familiarity with the new apparatus.[52]

Though old fashioned in their dependence on Flemish models, Avercamp's roundels made an impression on one of the most progressive painters of his generation, Esaias van den Velde, an important figure in the development of Dutch realism. Esaias, who was born in Amsterdam in 1587 and remained there until his move to Haarlem in 1610, would have known Avercamp's work well. During his Haarlem period, Esaias

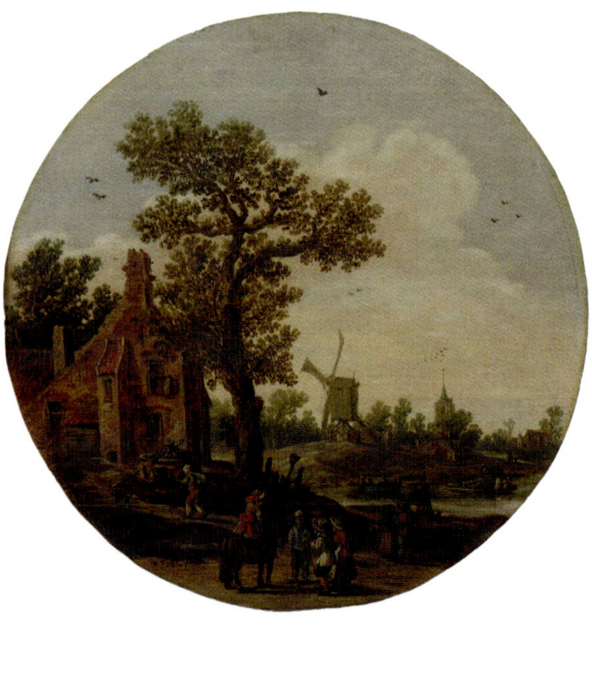

10 Jan van Goyen, *Summer*, 1625. Oil on panel, diameter approximately 44 cm. Amsterdam: Rijksmuseum. Photo: Rijksmusem.

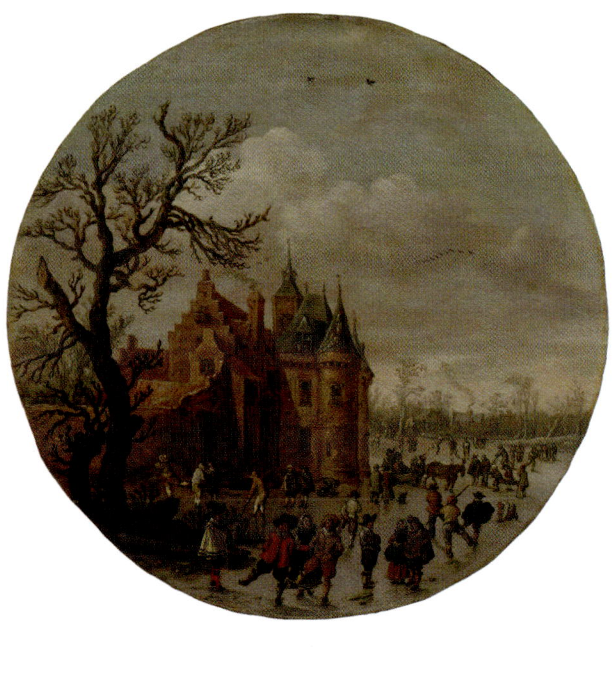

11 Jan van Goyen, *Winter*, 1625. Oil on panel, diameter approximately 44 cm. Amsterdam: Rijksmuseum. Photo: Rijksmusem.

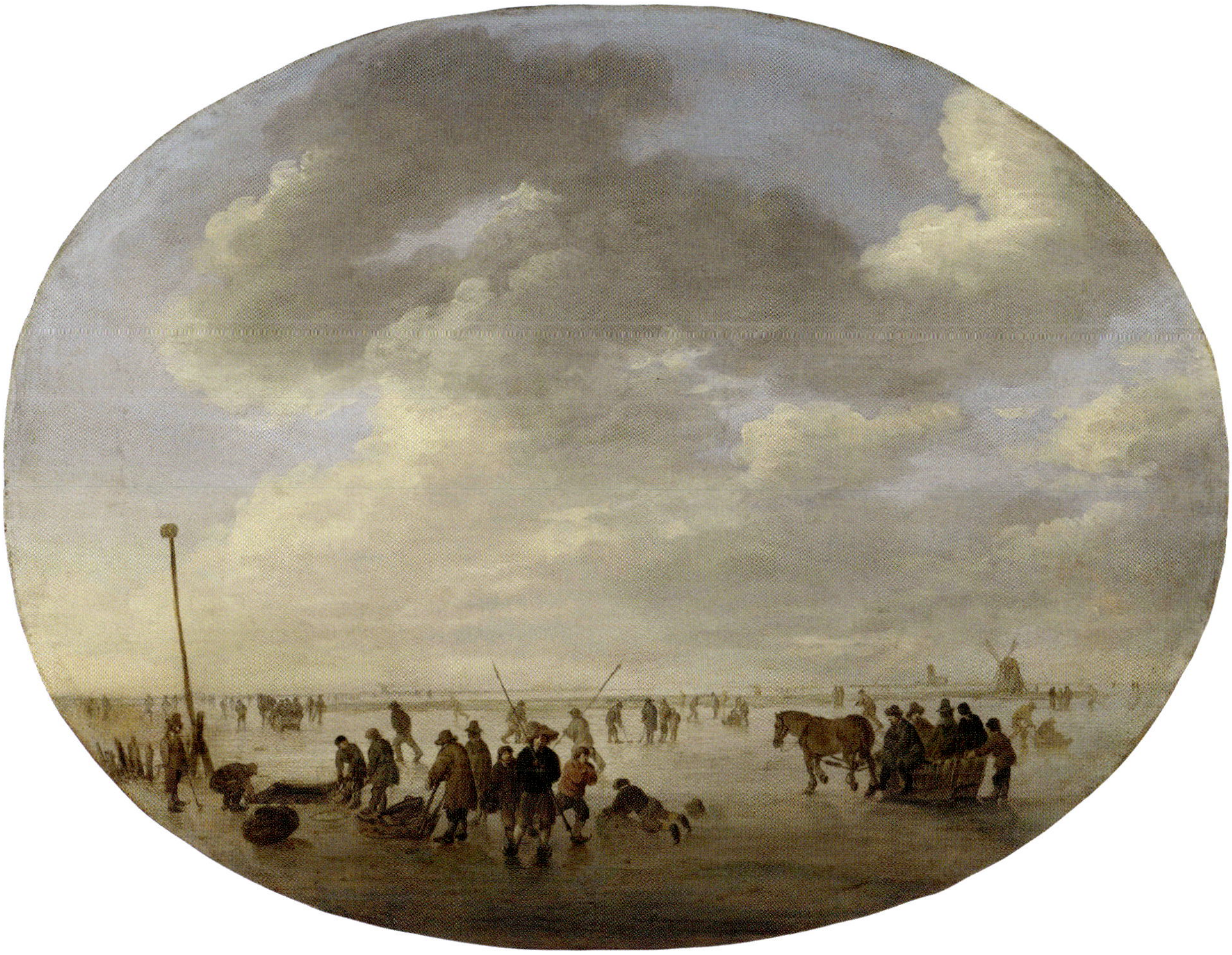

12 Jan van Goyen, *Skaters on a Wide River*, 1641. Oil on panel, 40.5 × 54.5 cm. Karlsruhe: Staatliche Kunsthalle. Photo: bpk, Berlin/Staatliche Kunsthalle/Annette Fischer/Heike Kohler/Art Resource, NY.

began to paint summer–winter pendant roundels inspired by Avercamp's. These must have sold well, for he continued to produce them throughout his career.[53]

Esaias's roundels were influential in turn. Among the earliest extant works by Jan van Goyen, who apprenticed with Esaias in Haarlem, are a number of circular summer–winter pendants from the first half of the 1620s. Very close in conception to those of his teacher, the pair in the Rijksmuseum in Amsterdam is dotted with bright colours and anecdotal details (*plate 10* and *plate 11*).[54] Details of both seasons agree with the round format, but especially those of winter, in which skaters in round hats circle the castle with its cylindrical towers. Van Goyen would go on to paint about sixty roundels over the course of his life. Most of those from the 1620s are summer–winter pairs like the one in the Rijksmuseum. Closer to midcentury, he turned to making individual circular or oval compositions with a variety of landscape and seascape subjects.[55] In these later roundels, a new sense of telescopy takes shape.

Already in Avercamp's icy worlds, we begin to feel that what we are experiencing, as our gaze moves across the distance, and well before it reaches the horizon, is the failure of our eyesight. In his later roundels, for example, in his oval *Skaters on a Wide River* of 1641, now in Karlsruhe, Van Goyen deepens this experience (*plate 12*). Here, only the figures closest to us are legible. The distance immediately swallows the rest. If world-landscapes are dominated by an extended middle ground, Van Goyen's oval has almost none to speak of. Everything beyond the immediate foreground is already far away.

Metaphors of travel and movement across space dominate early writing about the telescope. In his *New Organon* of 1620, Francis Bacon explains that the new device allows one to 'open up and practice a closer approach to the stars, as if by ferries or dinghies'.[56] The telescope was routinely described as an 'invention to stretch out sight', and one French journalist explained that, 'as one looks at the object which one wishes to inspect, it appears to come near'.[57] These metaphors of overcoming distance correspond well to the world-landscape, which, as Borromeo says, allows us to 'go wandering'. Looking through the oval frame of Van Goyen's Karlsruhe painting, by contrast, is less an experience of conquering distance than of submitting to it. The familiarity with real telescopes gained by the time Van Goyen made this painting did not have the expected effect of bringing the world close. On the contrary, it pushed it further away.

The Starry Messenger

I have emphasized that the telescope was commonly used for terrestrial viewing, and for viewing the landscape in particular. But no one was more closely associated with the instrument than Galileo, who has often been mistaken for its inventor. Galileo had higher ambitions than just seeing over the next hill: 'It would be entirely superfluous', he writes, 'to enumerate how many and how great the advantages of this instrument are on land and at sea. But having dismissed earthly things, I applied myself to explorations of the heavens.'[58] For Galileo's purposes, the early telescopes sold in the shops of spectacle makers would have been useless. As soon as he understood the Dutch invention, he set about improving it, quickly devising telescopes with magnifying powers of eight and later twenty times, and in March of 1610, he published *The Starry Messenger*, which describes his telescopic observations of the heavens. Along with finding new stars, Galileo discovered that the cloudy swathe of the Milky Way is made up of many separate points of light, and that the moon's surface is rough and irregular, contrary to what had been believed about the perfection of the heavenly bodies. The book was an instant success, and the fame of the telescope grew accordingly.[59]

For the average reader or for the person who knew *The Starry Messenger* only through hearsay, what exactly Galileo saw in the sky was probably of less importance than his discovery that there were discoveries there to be made at all. This contradicted received ideas about the universe, including what Hans Blumenberg calls the 'postulate of visibility':[60] 'That the world could contain things withdrawn and inaccessible to man's natural optical capacity, not only at times and provisionally but definitively, was an idea unknown to the ancient world and the Middle Ages and also, by their philosophical assumptions, an impossible one.'[61] Galileo's own expectations were initially shaped by precisely these assumptions, insofar as he hoped that the telescope might arm him with visual evidence in support of Copernicus's world system. Only later, according to Blumenberg, did he come to see that the human eye is not capable of perceiving systems of this kind, even when it is aided.[62]

One can trace a similar movement away from the postulate of visibility in the history of telescopic landscape painting. Proto-telescopic and early telescopic worldviews like Met de Bles's paradise tondo attempt to show us the full extent (and duration) of creation. They thereby accord us a godlike omniscience predicated on the belief that everything is, finally, visible. In Van Goyen's later telescopic views, by contrast, the world withdraws from our gaze. That the Book of Nature is God made manifest and visible was a Christian commonplace with considerable currency during the seventeenth century. In Calvin's words, '[God] clothes himself, so to speak, with the image of the world in which he would present himself to our contemplation . . . magnificently arrayed in the incomparable vesture of the heavens and the earth.'[63] But, when seen through real rather than imaginary telescopes, the world pulls away, each new discovery a sign that there are still more to be made.[64]

It is this effect that Blaise Pascal has in mind when he writes: 'the whole visible world is only an imperceptible atom in the ample bosom of nature. No idea approaches it. We may enlarge our conceptions beyond an imaginable space; we only produce atoms in comparison with the reality of things.'[65] Before real telescopes were constructed, telescopy held out the promise of correcting the flaws of human perception. The production and dissemination of real telescopes in the several decades following the invention, however, seem to have instilled a belief that human vision is essentially incommensurate with what there is to see and no amount of correction or augmentation can change this fact.

Once the telescope had shown the world to be beyond human comprehension, painting could no longer hope to mirror it, at least, not in its entirety. In this context, the roundel acquired new connotations. No longer suggestive of the gaze of an infinite god – Nicholas of Cusa's 'eye of sphericity and of infinite perfection' – it now marked the tightly circumscribed field of human vision and the sphere of the human eye. The Hermitage owns a group of very small tondos painted by Van Goyen in 1641, including *Boats on a Riverbank* (plate 13).

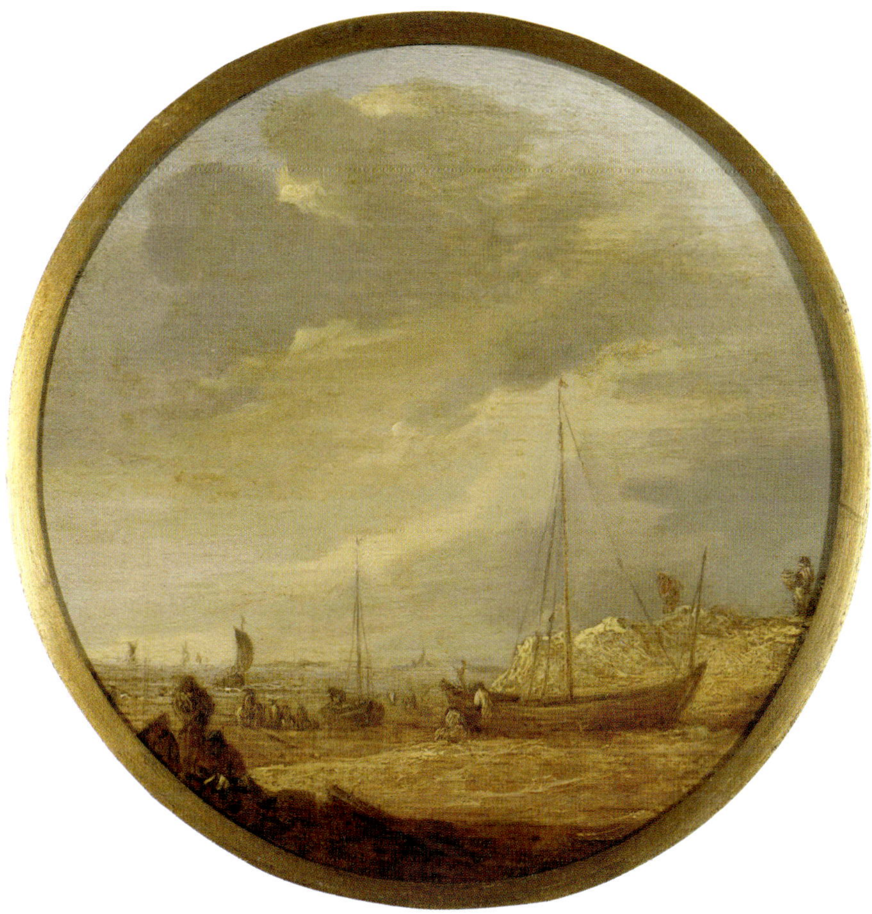

13 Jan van Goyen, *Boats on a Riverbank*, c. 1648. Oil on panel, diameter 15.5 cm. Saint Petersburg: The State Hermitage Museum. Photo: The State Hermitage Museum/Vladimir Terebenin.

Though these form a series of sorts – produced in rapid succession during an especially productive year – they have none of the totalizing aspirations of those earlier series of months or seasons.

At a mere 15.5 centimetres in diameter, these views are like Pascal's atoms, each picturing some corner of the world where land, water, and sky meet, a bit of shore dotted with boats, buildings, and people in the foreground, set off against a low and distant horizon, crowned by golden clouds against a pale sky. Painted in a restricted palette of mostly browns and yellows, they look like views of the world seen through a tinted lens.[66] Della Porta describes this phenomenon: 'when glass is melted with heat in the furnace, with any little color it will be tainted; if you cast in yellow, the face of him that looks into it, will seem to have the yellow Jaundies.'[67] In Van Goyen's golden tondos, which show us a world that is coloured as well as shaped by the gaze and its optical prosthetics, the seeing subject, with all his or her ineluctable limitations, clearly comes to define the seen.

The First Disk, 1913 – A Coda

The history of Renaissance and post-Renaissance painting is a history dominated by rectangles. While it ostensibly freed painting from the obligation to make pictures, twentieth-century abstraction did not free painting from the rule of the rectangular, at least not during its early phases. Painters would only begin to make 'shaped' canvases – that is canvases that deviate from the standard rectangle – in any significant numbers around midcentury. In fact, one of the

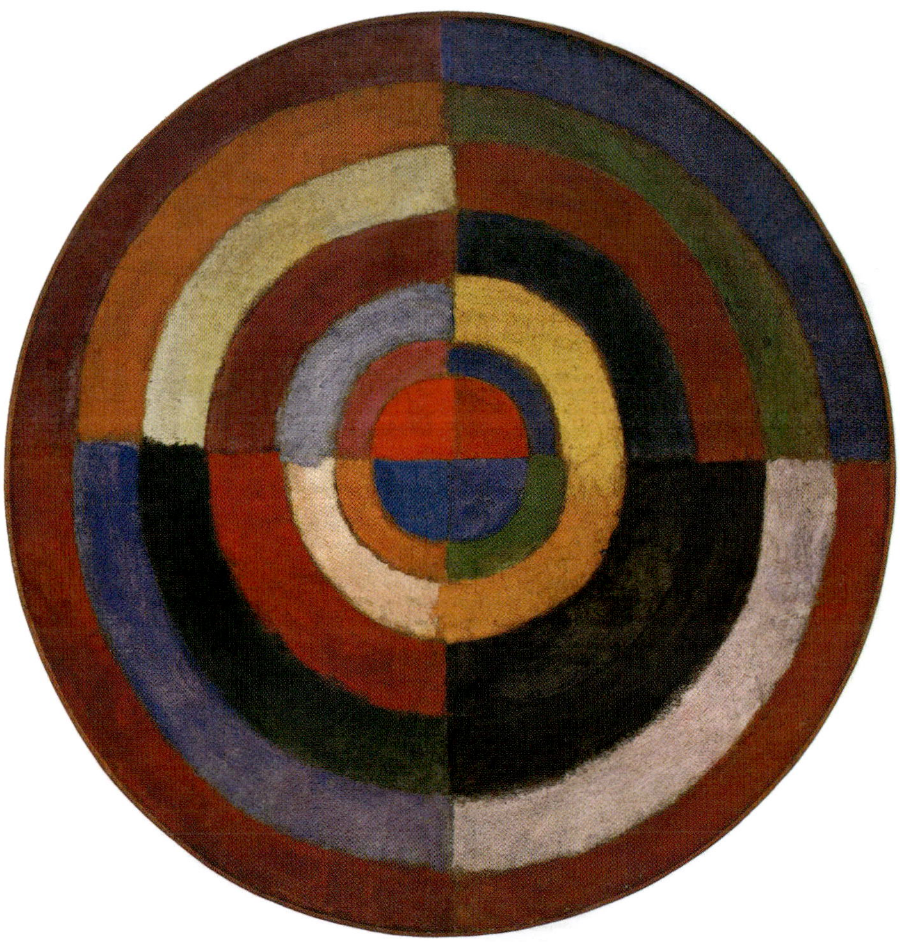

14 Robert Delaunay, *The First Disk*, 1913. Oil on canvas, diameter 134 cm. Private collection.

most important things that twentieth-century abstraction did was bring this rectangle, long taken for granted, into focus, through the use of deductive compositional structures.

Robert Delaunay's *The First Disk* (1913) is a remarkable departure from the norm of the rectangle – to which he himself otherwise conformed (*plate 14*). Sonia Delaunay-Terk describes her husband's process of making a series of paintings titled *Circular Forms, Sun* during the period just before he painted *The First Disk*: 'He forced himself to stare into [the sun] to the point of being nearly blinded. Upon returning to the house, he sought to throw onto the canvas what he saw both with his eyes open and with his eyes closed: all the contrasts registered on his retina ... He discovered spots in the form of disks.'[68] Though circles proliferate in Delaunay-Terk's account – the sun, the retina, spots, and disks – the paintings in the series *Circular Forms, Sun* are rectangular. In this sense, they are representational paintings: pictures of circular things.

The First Disk, by contrast, is a manifesto of abstraction. With its concentric bands of colour that conform to the structure of the whole, it is insistently self-referential. Though it is not the first abstract painting, its title's claim to primacy is not without merit insofar as it is among the first paintings in the Western canon to bring the shape of its support into such perfect alignment with the shapes it depicts. This alignment is a heady mix of eye and sun – vision itself as the merging of the two: 'Our eyes go up to the sun', writes Delaunay.[69] But it is more than that. In *The First Disk*, the shape of the eye and the shape of the sun merge not only with one another but also with the support of the painting. More than just a picture, the disk is an embodiment of the eye caught up in the process of seeing.

That Delaunay had to depart from one of the most entrenched conventions of Western painting, namely, the rectilinear format, in order to merge the structure of the gaze, the object, and the picture is a good indication of how significant the tension among these structures has been to the history of Western art. Delaunay never re-created the kind of homology between vision, object, and picture that governs *The First Disk*. He returned immediately to making pictures of things in the world, framed as if seen through rectangular windows.[70] Van Goyen's telescopic views, which register the defamiliarizing effects of a new technology, were likewise quickly forgotten, relegated to the margins of the history of Western painting.

Notes

I would like to thank Genevieve Warwick and Richard Taws as well as *Art History*'s anonymous reader.

1 Giambattista della Porta, *Natural Magick*, trans. anonymous, London, 1658, 364.

2 In the pirated edition printed in Frankfurt in 1610, the woodcuts of the moon have no framing edges, but in the first authorized edition printed in Venice in 1610, the copper engravings, financed and perhaps in part made by Galileo himself, have square-shaped backgrounds. On these drawings and their translation into print, see Horst Bredekamp, 'Gazing hands and blind spots: Galileo as draftsman', in Jürgen Renn, ed., *Galileo in Context: An Engineer-Scientist, Artist, and Courtier at the Origins of Classical Science*, Cambridge, 2000, 170–84; Irene Brückle and Oliver Hahn, eds, *Galileo's Sidereus Nuncius: A Comparison of the Proof Copy (New York) with Other Paradigmatic Copies*, Berlin, 2011, 105–14; Horst Bredekamp, *Galilei der Künstler: der Mond, die Sonne, die Hand*, Berlin, 2007, 131–48, 177–216.

3 Many of the illustrations in Robert Hooke's *Micrographia* (1665), the first major publication on the microscope, have circular frames, but these are exceptions to the hegemony of the rectilinear.

4 Della Porta, *Natural Magick*, 364.

5 Three round paintings and several round etchings are reproduced in Wolfgang Stechow, *Dutch Landscape Painting of the Seventeenth Century*, London, 1966. One oval painting and no roundels appear in Christopher Brown, *Dutch Landscape: The Early Years, Haarlem and Amsterdam, 1590–1650*, London, 1986. Though Sutton reproduces only two roundels, he writes: 'viewed in the context of more conventional rectangular paintings from the seventeenth century, these small circular paintings appear to us as exceptional deviations. However, this probably was not the case in Van Goyen's day, when such charming roundels, with their small-scale figures and vibrant accents of color, enjoyed considerable popularity.' Peter C. Sutton, *Masters of 17th-Century Dutch Landscape Painting*, Boston, MA, 1987, 319.

6 http://www.nationalgallery.org.uk/paintings/hendrick–avercamp–a–winter–scene–with–skaters–near–a–castle. Accessed 8 July 2015.

7 Not a great deal has been written about the telescope and the genre of landscape, though Larsen speculates in a number of publications on the connection between the device, the genre, and Cartesian philosophy. See, for example, Erik Larsen, 'Descartes and the rise of naturalistic landscape painting in 17th century Holland', *Art Journal*, 24, 1, 1964, 12–17; Erik Larsen, 'The proof of the use of the inverted telescope in Dutch 17th century landscape art', *Gazette des Beaux-Arts*, 89, May–June 1977, 172–4. Some of the problems with Larsen's argument are indicated in Peter J. P. Whitehead, 'Frans Post and the reversed Galilean telescope', *The Burlington Magazine*, 128, 1986, 805–7.

8 On the popularization of the telescope, see Huib J. Zuidervaart, 'The "invisible technician" made visible: Telescope making in the seventeenth and early eighteenth-century Dutch Republic', in Alison D. Morrison-Low, ed., *From Earth-Bound to Satellite: Telescopes, Skills and Networks*, Leiden, 2012, 92–8.

9 Paolo Molaro and Pierluigi Selvelli, 'On the telescopes in the paintings of Jan Brueghel the Elder', *The Role of Astronomy in Society and Culture, Proceedings of the International Astronomical Union*, 260, 2009, 327–32.

10 Brueghel painted a number of other paintings closely related to this one, none of which have telescopes. Klaus Ertz, *Jan Brueghel der Ältere (1568–1625): kritischer Katalog der Gemälde*, Luca, 2008, vol. 3, 1206–17. See also Brod Gallery, *Jan Brueghel the Elder: A Loan Exhibition of Paintings*, London, 1979, cat. nos 18 and 34; Anne T. Woollett, *Rubens and Brueghel: A Working Friendship*, Los Angeles, CA, 2006, 104–5.

11 Albert Van Helden, 'The invention of the telescope', *Transactions of the American Philosophical Society*, 67: 4, 1977, 36. Around the same time, a Dutch merchant was trying to sell a telescope with a cracked lens at a fair in Frankfurt. These events as well as the ones I will recount below are narrated in Van Helden, 'The invention', 20–8; Engel Sluiter, 'The telescope before Galileo', *Journal for the History of Astronomy*, 28, 1997, 223–32.

12 The letter is addressed to Cardinal Scipione Borghese, nephew of Paul V and papal secretary of state, the letter ends: 'Therefore I am sending the instrument by the next post to my brother Henry, in order that in my name he may deliver it to your Eminence', as trans. in Sluiter, 'The telescope', 226.

13 Van Helden, 'The invention', 44.

14 William S. Heckscher, 'Renaissance emblems: Observations suggested by some emblem-books in the Princeton University Library', *The Princeton University Library Chronicle*, 15: 2, 1954, 60–1; Christine Göttler, *Last Things: Art and the Religious Imagination in the Age of Reform*, Turnhout, 2010, 1–30.

15 Adriaen van de Venne, 'Zeeusche Mey-clacht', in *De Zeeuwsche Nachtegael*, Middelburg, 1623, 59, as trans. in Laurens J. Bol, *Adriaen Pietersz. van de Venne, Painter and Draughtsman*, Doornspijk, 1989, 29. The term used here is *buysen* or 'tubes'; *verrekijkersbuysen* was also sometimes used. See the entry for *verrekijker* in *Woordenboek der Nederlandsche taal*, 's-Gravenhage, 1942.

16 This recalls Alpers's description of Constantijn Huygens's reaction to the microscope: 'Huygens looks into a lens and calls for a picture.' Svetlana Alpers, *The Art of Describing: Dutch Art in the Seventeenth Century*, Chicago, IL, 1983, 7.

17 'De Nijd vind baet in anders quaed' is emblem XLVII in Johan de Brune, *Emblemata of Zinne-werck*, Middelburg, 1624. For discussion, see Heckscher, 'Renaissance emblems', 60–1. Telescopes also appear in the artist's several versions of *Garden Party (Allegory of the Five Senses)*, also from the 1620s, which are discussed and reproduced in Edwin Buijsen, 'Roodkrijttekeningen naar schilderijen van Adriaen van de Venne en hun mogelijke functie', *Oud Holland*, 118: 3/4, 2005, 132, 178–9. Landscape plays an important role in these garden party scenes, though not in Adriaen's illustration of the idiom 'Besiet u selven' (c. 1640–52), in which a telescope also appears. Early images of telescopes in landscape settings by other artists include: a garden party scene by David Vinckboons from around 1615, reproduced in Otto Schmitt, ed., *Reallexikon zur deutschen Kunstgeschichte*, Stuttgart, 1937, vol. 8, 263–4; the engraving *Young Man with a Spyglass*, which decorates emblem III (*In iactorest, vel qui sua nimis amant*) in the *Emblemata* of *Paolo Maccio*, Bologna, 1628.

18 *De Zeeuwsche Nachtegael*, 20. Though the tondo format became common for emblems around 1600, it takes on a special significance here in light of the subject. For discussion of the telescope in Van de Venne's work, see Bol, *Adriaen Pietersz*, 33n.10.

19 James Turner, 'Landscape and the "art prospective" in England, 1584–1660', *Journal of the Warburg and Courtauld Institutes*, 42, 1979, 290; Walter S. Melion, *Shaping the Netherlandish Canon: Karel van Mander's Schilder-Boeck*, Chicago, IL, 1991, 7–8, 197n.18.

20 As quoted in Henry V. S. Ogden and Margaret S. Odgen, *English Taste in Landscape in the Seventeenth Century*, Ann Arbor, MI, 1955, 1.

21 See the entries for *perspectief*, *prospect*, and *prospectief* in *Woordenboek der Nederlandsche taal*; Guido Guillén, *Literature as System: Essays Toward the Theory of Literary History*, Princeton, NJ, 1971, 301–3; Van Helden, 'The invention', 13–15; Turner, 'Landscape', 291–2; Eileen Reeves, *Galileo's Glassworks: The Telescope and the Mirror*, Cambridge, 2008, 59–60; Reeves, 'Complete inventions: The mirror and the telescope,' in Albert Van Helden et al., *The Origins of the Telescope*, Amsterdam, 2010, 173–4.

22 Walter S. Gibson, *Mirror of the Earth: The World Landscape in Sixteenth-Century Flemish Painting*, Princeton, NJ, 1989, xxi, 76.

23 Maurice Scève, *Microcosme* (1562), ed. Enzo Giudici, Paris, 1976, 236–7, as trans. in Gibson, *Mirror of the Earth*, 57.

24 Federico Borromeo, *De suis studiis commentaries*, Milan, 1628, fol. 252r–253r, as trans. in Pamela Margaret Jones, *Federico Borromeo's Pinacoteca Ambrosiana*, Ann Arbor, MI, 1985, 94–5. See also Gibson, *Mirror of the Earth*, 57, where other relevant examples can also be found. Angel Day, *The English Secretary or Methode of Writing of Epistles and Letters*, London, 1599, 23–4.

25 Karel van Mander, *Het schilder-boeck*, Haarlem, 1604, chapter 8, no. 10, as trans. in Brown, *Dutch Landscape*, 37. For discussion of Van Mander's interest in the representation of distance, see Melion, *Shaping The Netherlandish Canon*, 12.

26 Van Mander, *Het schilder-boeck*, chapter 8, nos 19, 25, as trans. in Brown, *Dutch Landscape*, 38–9.

27 Koerner argues this with characteristic brilliance: 'moving with us as we travel, the horizon tells us that the visible is only a small segment, or eccentric foreground, of reality. Part cause and part consequence of a new cosmology that removed man from the world's center . . . the experience of the horizon as at once inescapable and infinitely displaceable is characteristic of the modern world view.' Joseph Leo Koerner, 'The printed world', in Barbara Butts, ed., *The Printed World of Pieter Bruegel the Elder*, St Louis, MO, 1995, 25. My argument here only shifts the advent of this modernity to a slightly later moment. On the pace of the spread of Copernicanism, see Thomas S. Kuhn, *The Copernican Revolution: Planetary Astronomy in the Development of Western Thought*, Cambridge, 1957, 185–227.

28 Turner, 'Landscape', 290.

29 The format was particularly popular in glass, see William Cole, *A Catalogue of Netherlandish and North European Roundels in Britian*, Oxford, 1993, xx; Timothy B. Husband, *The Luminous Image: Painted Glass Roundels in the Lowlands, 1480–1560*, New York, 1995, 11.

30 Michael Baxandall, 'Bartholomaeus Facius on painting: A fifteenth-century manuscript of the *De Viris Illustribus*', *Journal of the Warburg and Courtauld Institutes*, 27, 1964, 102; Gibson, *Mirror of the Earth*, 39.

31 Lynn F. Jacobs, 'The triptychs of Hieronymus Bosch', *The Sixteenth Century Journal*, 31: 4, 2000, 1028–39.

32 This tondo is sometimes said to be unique in Bles's oeuvre, but he did make others. See Norman E. Muller, Betsy J. Rosasco and James H. Marrow, eds, *Herri met de Bles: Studies and Explorations of the World Landscape Tradition*, Princeton, NJ, 1998, 33–4. The painting's iconography probably derives from a German woodcut printed in a Bible of 1534, but the scale and complexity of the painting far exceed those of the print. Jacques Toussaint, ed., *Autour de Henri Bles*, Namur, 2000, cat. no. 1; Hanna Benesz, ed., *Die Flämische Landscaft, 1520–1700*, Luca, 2003, cat. no. 103.

33 Nicholas of Cusa, *On the Vision of God* (1453) in John Patrick Dolan, *Unity and Reform: Selected Writings of Nicholas of Cusa*, Notre Dame, IN, 1962, 147. Quoted in Walter S. Gibson, 'Hieronymus Bosch and the mirror of man: The authorship and iconography of the "Tabletop of the Seven Deadly Sins"', *Oud Holland*, 87: 4, 1973, 217.

34 Dolan, *Unity and Reform*, 147.

35 Joseph Glanvill, *The Vanity of Dogmatizing or Confidence in Opinions*, London, 1661, 11. Cited in Hans Blumenberg, *The Genesis of the Copernican World*, trans. Robert M. Wallace, Cambridge, 1987, 636–7.

36 Constantijn Huygens, *De gedichten*, Groningen, 1892–99, vol. 2, 236, as trans. in Alpers, *The Art of Describing*, 17.

37 This is an Old French version, as trans. in Reeves, *Galileo's Glassworks*, 18–19.
38 For a list of these, see Reeves, *Galileo's Glassworks*, 172–3n.2.
39 Reeves, *Galileo's Glassworks*, 21–6.
40 Serge Sauneron, ed., *Voyages en Egypte pendant les années 1587–1588*, trans. Ursula Castel and Nadine Sauneron, Paris, 1977, 147; Reeves, *Galileo's Glassworks*, 40.
41 This description is to be found in his *Letter on the Secret Works of Art and Nature*, which was not published until 1542. Bacon's glass for seeing far would be mentioned in many later works on optics as well as in many of his biographies. Reeves, *Galileo's Glassworks*, 27–9.
42 Van Helden, 'The invention', 44–6; Reeves, *Galileo's Glassworks*, 8–9. For other examples of sixteenth-century experiments with magnification, see Reeves, *Galileo's Glassworks*, 47–80.
43 Alexandre Desplanque, Jules Finot and Chrétien César Auguste Deshaisnes, eds, *Inventaire sommaire des Archives départementales antérieures à 1790, Nord: Archives civiles, Série B*, Lille, 1885, vol. 5, 371; Reeves, 'Complete inventions', 173.
44 Paolo Procaccioli, 'Note e testi per Francesco Angelo Coccio', *La Cultura*, 27: 2, 1989, 416–17, as trans. in Reeves, *Galileo's Glassworks*, 53.
45 Van Helden, 'The invention', 24; Vincent Ilardi, *Renaissance Vision from Spectacles to Telescopes*, Philadelphia, PA, 2006, 219; Rolf Willach, 'The long route to the invention of the telescope', *Transactions of the American Philosophical Society*, 98: 5, 2008, 93–7; Huib J. Zuidervaart, 'The "true inventor" of the telescope: A survey of 400 years of debate', in Albert Van Helden *et al.*, *The Origins of the Telescope*, 9–44.
46 Reeves, *Galileo's Glassworks*, 10–11.
47 Reine de Bertier de Sauvigny, *Jacob et Abel Grimmer: catalogue raisonné*, Brussels, 1991, 16.
48 Mirjam Neumeister, ed., *Brueghel: Gemälde von Jan Brueghel d.Ä.*, Munich, 2013, 228.
49 Ertz calls this '*Guckkasten-Komposition*' '*hoffnungslos altmodish*'. Ertz, *Jan Brueghel der Ältere*, vol. 3, 1206.
50 On Avercamp's roundels, see Clara J. Welcker, *Hendrick Avercamp, 1585–1634*, Doornspijk, 1979, cat. nos 47–54; Albert Blankert, *Hendrick Avercamp, 1585–1634, Barent Avercamp, 1612–1679: Frozen Silence, Paintings from Museums and Private Collections*, Amsterdam, 1982, 45; Sutton, *Masters*, 320.
51 On the dating and development of Avercamp's oeuvre, see Sutton, *Masters*, 321; Ger Luijten, ed., *Dawn of the Golden Age: Northern Netherlandish Art, 1580–1620*, Amsterdam, 1993, 634–5, Pieter Roelofs, *Hendrick Avercamp: Master of the Ice Scene*, Amsterdam, 2009, 42–54.
52 Bol, *Adriaen Pietersz*, 20, figs 5A, 5B. They are undated, but they are similar to a rectangular pair painted around 1615. On these, see Stechow, *Dutch Landscape*, 85–6. A series of circular etchings of the seasons made in Amsterdam by the Flemish artist Hans Bol around 1580 were a source for both Van de Venne and Avercamp. Stechow, *Dutch Landscape*, 85; Roelofs, *Hendrick Avercamp*, 41. Bol, who emigrated from Antwerp to Amsterdam in 1584, was an important figure in the translation of the Flemish style to the north. Christiaan Vogelaar, *Jan van Goyen*, Zwolle, 1996, 84.
53 George Keyes, *Esaias van den Velde, 1587–1630*, Doornspijk, 1984, passim; Vogelaar, *Jan van Goyen*, 84–5. In 1618, Esaias relocated permanently to The Hague, where he worked for both Prince Maurits and Frederick Hendrick, who would have been able to provide him with access to telescopes.
54 Sutton, *Masters*, 321; Vogelaar, *Jan van Goyen*, 89.
55 These range in size from about 10 to 68 centimetres in diameter. Hans-Ulrich Beck, *Jan van Goyen, 1596–1656: ein Oeuvreverzeichnis*, Amsterdam, 1972, vol. 2, 2–14, 53–76. On his production of roundels, see also Stechow, *Dutch Landscape*, 88; Sutton, *Masters*, 319–20; Vogelaar, *Jan van Goyen*, 84–5.
56 Francis Bacon, *The New Organon*, ed. Lisa Jardine and Michael Silverthorne, Cambridge, 2000, 171.
57 The first phrase appears, for example, in a letter to Jacob Metius from the States-General dated 17 October 1608, in Van Helden, 'The invention', 40. The second appears in the report of a French journalist dated 30 April 1609, in Van Helden, 'The invention', 44.
58 Galileo Galilei, *Sidereus Nuncius or The Sidereal messenger*, trans. Albert Van Helden, Chicago, IL, 1989, 38.
59 See, for example, Elizabeth L. Eisenstein, *The Printing Press as an Agent of Change*, Cambridge, 1979, 652–3.
60 Blumenberg, *The Genesis*, 629.
61 Hans Blumenberg, *The Legitimacy of the Modern Age*, trans. Robert M. Wallace, Cambridge, 1983, 364–5. See also Alexandre Koyré, *From the Closed World to the Infinite Universe*, New York, 1958; David Freedberg, *The Eye of the Lynx: Galileo, His Friends, and the Beginnings of Modern Natural History*, Chicago, Il, 2002, 108.
62 Hans Blumenberg, 'Das Fernrohr und die Ohnmacht der Wahrheit', in Hans Blumenberg, ed., *Sidereus Nuncius: Nachricht von neuen Sternen*, Frankfurt am Main, 1965, 5–73.
63 John Calvin, *Commentaries on the Book of Moses, Called Genesis*, trans. John King, Grand Rapids, MI, 1948, vol. 1, 60.
64 Karsten Harries, *Infinity and Perspective*, Cambridge, 2001, 267.
65 Blaise Pascal, 'Man's disproportion', in *Pensées*, trans. W. F. Trotter, New York, 1958, 16.
66 Toward the end of the 1620s, Van Goyen and other Dutch painters began to paint with a very reduced palette. This is referred to as the tonal or monochrome phase of Dutch painting. From this point on in his career, Van Goyen would use a reduced palette selectively, returning to a more colourful palette whenever it suited him.
67 Della Porta, *Natural Magick*, 355.
68 Sonia Delaunay, *Nous irons jusqu'au soleil*, Paris, 1978, 44, as trans. in Gordon Hughes, 'Abstraction chez Delaunay' in Leah Dickerman, ed., *Inventing Abstraction, 1910–1925: How a Radical Idea Changed Modern Art*, New York, 2012, 75.
69 Robert Delaunay and Sonia Delaunay, *The New Art of Color: The Writings of Robert and Sonia Delaunay*, New York, 1978, 47, see also 53–4.
70 On *The First Disk*, see the excellent recent work of Gordon Hughes: 'Envisioning abstraction: The simultaneity of Robert Delaunay's "First Disk"', *The Art Bulletin*, 89: 2, 2007, 306–32; *Resisting Abstraction: Robert Delaunay and Vision in the Face of Modernism*, Chicago, Il, 2014, 96–137.

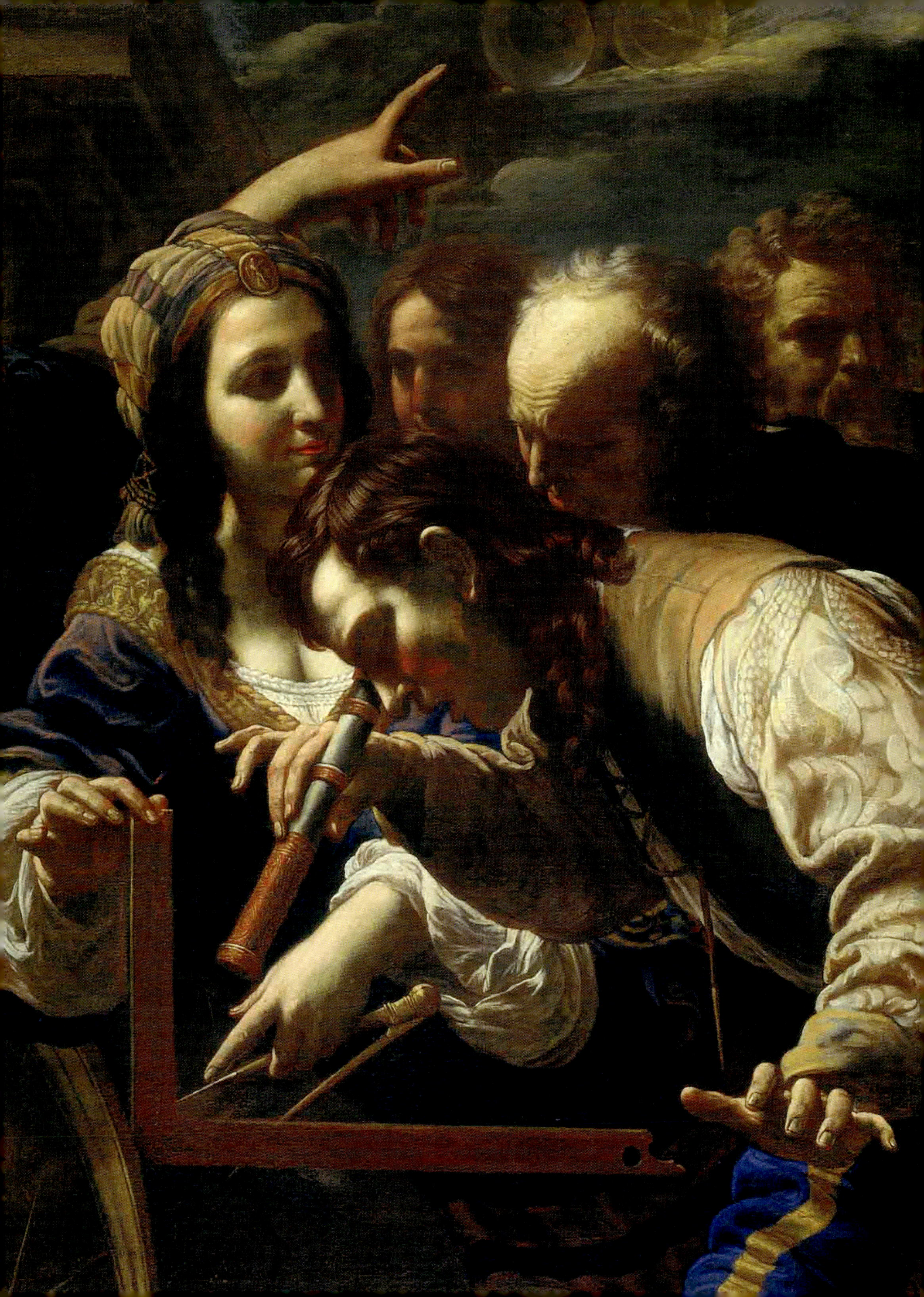

Chapter 6
After Galileo:
The Image of Science in Niccolò Tornioli's *Astronomers*

Giulia Martina Weston

The leading work in the artistic production of the Sienese painter Niccolò Tornioli (1606–51), the *Astronomers* (*plate* 1) represents a dispute on opposed astronomical theories.[1] Conceived in the aftermath of Galileo's death in Rome in the 1640s, the canvas portrays an animated dialogue among the supporters of geocentricism and heliocentricism, drawing on the dichotomy between the Ptolemaic and Copernican systems introduced by Galileo Galilei's *Dialogue Concerning the Two Chief World Systems* of 1632. By exploring Tornioli's iconographic and compositional choices, this contribution sets out to unveil the painter's response to the technological and epistemological novelties in the history of scientific observation introduced by Galileo and carried on by his legacy. Whilst the accurate depiction of a telescope in Tornioli's *Astronomers* stands out as an overt reference to the newly invented precision tool, a subtler deploy of painterly strategies, such as the simultaneous representation of different phases of the moon, may be read as a pictorial disquisition on the post-Galilean science of vision. Touching upon the interweaving of artistic, scientific and technological achievements of early modernity, the present research seeks to reveal Tornioli's particular response to Galileo's controversial use of images as both truthful and deceptive entities. More specifically, the impact of the Galilean revolution on the credibility of the visual will be taken into special account. Thus the essay sheds light on Tornioli's engagement in the debate centred on the role of the visual arts in the process of knowledge-making, and at clarifying the pictorial devices employed to reaffirm the epistemological possibilities of painting.

Further, this contribution reconsiders the genesis of the *Astronomers* in the context of Tornioli's patronage by the Spada family in Rome, with specific attention to the chief role played by both Father Virgilio Spada (1596–1662) and Cardinal Bernardino Spada (1594–1661) in shaping the artist's imagery. As will be argued, the scientific content of the canvas should be understood as a pictorial manifesto of Virgilio's support of Galileo's theories, which were anathema to the Church. The depiction of technological tools was surely also predicated on Tornioli's familiarity with Virgilio's museum of curiosities, which contained such technological instruments of vision. In addition, an enquiry into the rhetorical content of the verses hosted in the sundial built by Cardinal Bernardino in the early 1640s in the Spada palace corroborates Tornioli's ability to make his patrons' programmatic *concetti* in scientific matters his own. Finally, analysis of Tornioli's process of bringing the *Astronomers* into being demonstrates the artist's significant contribution to the Spada cultural agenda and his fulfilment of his patrons' reputation.

Detail from Niccolò Tornioli, *Astronomers*, before 1645 (*plate 1*).

Art and Technology in Early Modern Europe *Edited by Richard Taws and Genevieve Warwick*.

Tornioli's *Astronomers*

As will be discussed shortly, the complexity of the subject of this painting raises a series of unexplored or unresolved questions, including those of identifying the depicted figures. The young man at the heart of the composition has been interpreted as Nicolaus Copernicus, for his gesture of pointing at the phases of the moon clearly epitomizes his heliocentric theory. The older man on the left is identifiable as Aristotle; an explicit allusion to Aristotelian geocentric theories is made by the illustration of concentric spheres in his bulky volume. The figure characterized by an extravagant Oriental hat has been so far interpreted as Ptolemy on the basis of his physical (and therefore ideological) proximity to Aristotle. Supporters of heliocentricism, instead, are grouped on the right-hand side of the composition. Here, a young astronomer is represented in the act of using a telescope under the guidance of the personification of Astronomy and an unrolled graph records the concentric motion of the planets. These references to contemporary scientific and technological achievements culminate in Galileo's portrait in the upper right corner.

The demanding theme of the canvas, centred on the lively dialogue among historical figures and their interplay with a variety of tools of knowledge, poses a series of questions. Scholarly research has particularly focused on the inclusion of Galileo's portrait, arguing that the astronomical dispute painted by Tornioli may have a source in Stefano Della Bella's etched frontispiece of Galileo's edition of the *Dialogo sopra i due massimi sistemi del mondo* of 1632 (*plate* 2).[2] Portraying Aristotle, Ptolemy and Copernicus, the image represents Copernicus in the likeness of the elderly Galileo. However, the statuesque and motionless characters of Della Bella's invention are replaced, in Tornioli's *Astronomers*, by a more dynamic composition, where the eloquence of gestures is highlighted by vibrant chiaroscuro effects.

Curiously, no attempt has been made to identify the other figure in the right background. It seems reasonable to put forward the hypothesis that this is a portrait of Niccolò Tornioli himself, who was at the time in his late thirties. Unfortunately, the only alleged record of Tornioli's appearance, included in an anomymous caricature drawing held at the Gabinetto Disegni e Stampe of the Uffizi, does not offer useful ground for comparison.[3] In the latter, Tornioli's facial features are in fact quite sketchy and even grotesque, possibly immortalized with sharp irony. Nevertheless, the young man depicted in the *Astronomers* is featured with those visual clues that may signal an artist's self-portrait. In particular, the figure uniquely engages the bystander in a dialogue, thus facilitating the latter's inclusion in the painting. The suggestion advanced here that this is Tornioli's self-portrayal will be further related to his aspirations at this point in his career, in light of the impact of the Galilean revolution on the reliability of paintings as tools of knowledge.

The two figures represented on either side of Aristotle present other unresolved iconographic questions. The man on the far left has been identified with Ptolemy on the basis of the Oriental hat and the astronomer's physical, and therefore intellectual, proximity to Aristotle. However, a comparison between this figure and other contemporary representations of Ptolemy does not allow corroboration of this hypothesis. Conversely, a closer look into the character's hat reveals affinities with Alexander the Great's carved helmet, crowned by long white feathers.[4] Tornioli's 'antiquarian' accuracy in the depiction of the metallic helmet and the soft plumes seems to betray the painter's deliberate attempt to reproduce Alexander's hat and finds notable parallels in contemporary images of the emperor, such as Pietro da Cortona's *Battle of Alexander versus Darius*, frescoed in 1644 (Rome, Musei Capitolini), or Ilario Casolani's *Alexander ceding Campaspe to Apelles* (Siena, Chigi Saracini collection, the

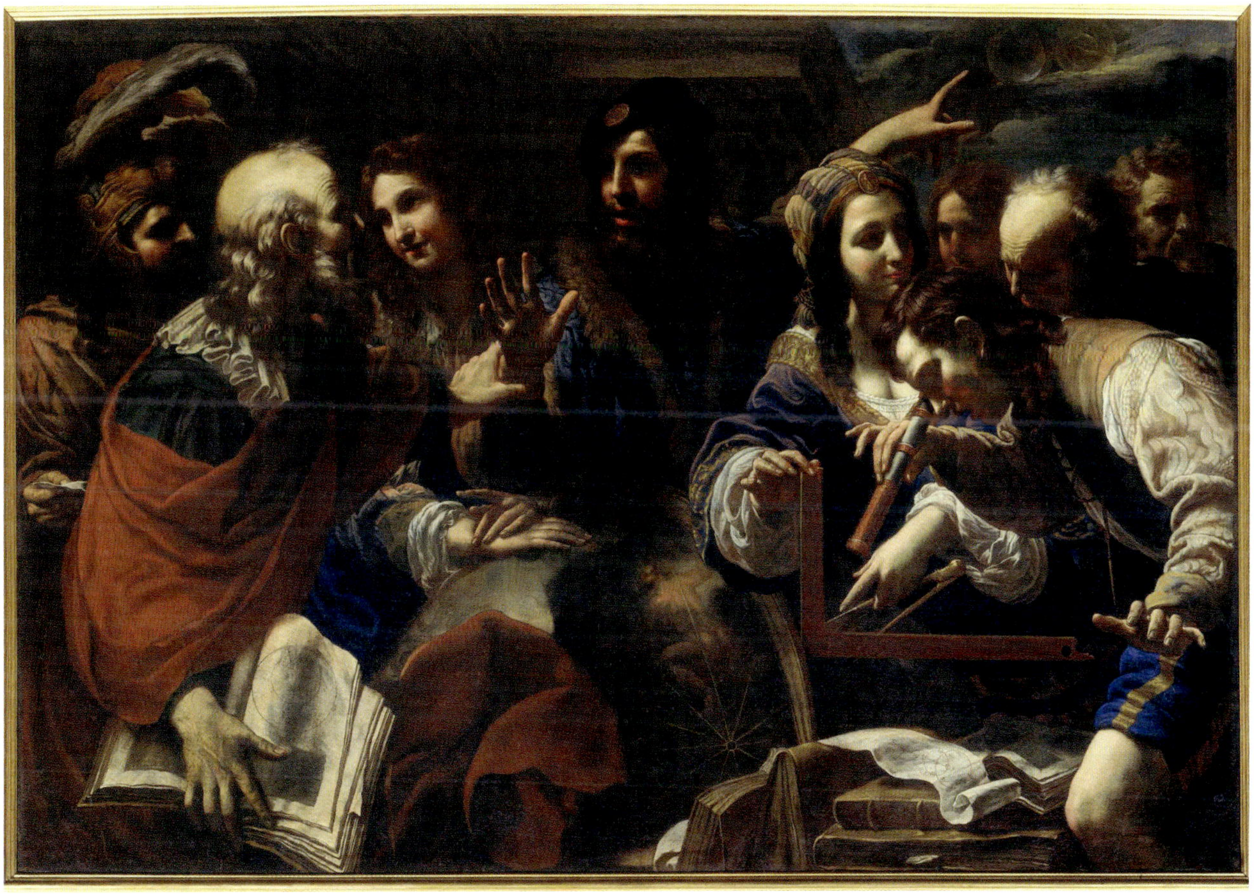

1 Niccolò Tornioli, *Astronomers*, before 1645. Oil on canvas, 148 × 218.5 cm. Rome: Galleria Spada. Photo Scala, Florence/Courtesy of the Ministero Beni e Attività Culturali.

property of Monte dei Paschi di Siena).[5] In the light of this alternative iconographical reading, the relationship between this vigorous figure and Aristotle could thus be explained in philosophical and historical, rather than in scientific terms, as the Greek polymath was Alexander's influential mentor. Accordingly, the female figure of the trio could be interpreted as a personification of Geography, alluding to Alexander's conquest of most of the known world. Paralleled by the allegorical embodiment of Astronomy on the opposite side, the presence of Geography would reinforce the visual and conceptual balance between the two halves of the canvas, enhancing the dialectical tension between earth and sky, between terrestrial and astronomical geographies.[6]

For their prominence and scale, the scientific instruments represented in the foreground appear co-protagonists of the scene. Tornioli could use both Virgilio's and Bernardino's collections of precision tools of scientific enquiry kept in the former's museum of curiosities or displayed by the latter in the rooms of the family palace. It has been wisely suggested, for instance, that the celestial globe depicted in the *Astronomers*, showing the constellation of Erichtonius, should be identified with Willem Blaeu's globe, purchased by Bernardino in 1640 and located in the Studio Grande.[7] Furthermore, contemporary diplomatic and court gifts comprised paintings which potrayed scientific and technological tools. An example is given by Guercino's *Atlas*, showing a massive globe with the constellations of Libra, Scorpio and Sagittarius, purchased by don Lorenzo de' Medici in 1646 and presented as a gift to a member of the Capponi family.[8]

Curiously, the telescope represented in Tornioli's *Astronomers* is not pointed at the moon above the right side group, but is oddly pointed downwards to spot a star on the celestial globe. The inclusion of this tool in the painting celebrates its inventor Galileo, who had introduced it to the scientific community with the publication of *Sidereus Nuncius* in 1610. The faithful depiction of this instrument has been related to Guercino's detailed telescope painted in the *Endymion*, a canvas commissioned by don Lorenzo de' Medici and offered as a gift to a member of the Pamphili family.[9] According to Greek mythology, the young shepherd Endymion was beloved by Diana, the goddess of chastity usually symbolized by the moon. As the first man interested in lunar motion in Pliny the Elder's account, Endymion is celebrated by Nonnus' *Dionysiaca* as an astronomer *ante litteram*.[10] Providing the shepherd with the cutting-edge technological device of his times, Guercino presented a renewed version of Endymion, visualizing the symbolic overlap between the first astronomer and Galileo already codified by Giovan Battista Marino's *Adone* in 1627.[11] In Marino's verses, the scientist is celebrated as a novel Endymion and the modern lover of the Moon, which, thanks to the newly invented telescope, he is allowed to spy on and lay bare.

These intersections between collectible precision tools and scientific-related paintings, or between technology and its representation, corroborate Paula Findlen's findings on the cultural implications of the Galilean revolution, epitomized by the existence of strong 'connections between social status and intellectual identity'.[12] In addition, they provide a thread to further unveil the political repercussions and alliances underlying technological achievements in the Galilean age. As recently emphasized throughout an enquiry into the Medici collections of scientific tools, the Florentine patrons' interest in astronomical discoveries was closely linked to new-found systems to calculate latitudes and distances in the sea, thus accruing significant military and nautical advantages for Florence.[13] On a cultural and intellectual level, if the Medici family was closely involved in the Galileo affair, and possibly in the commission of Guercino's work, both the recipients of the *Endymion*, Camillo Pamphili, and of Tornioli's *Astronomers*, Virgilio Spada, were rooted in Rome and related to the most advanced scientific tendencies developed within the academies to carry on Galileo's legacy.[14] Virgilio's involvement in the passing of the torch between Florence and Rome in defence of a culture of scientific experiment, reflected in the patron's museum of curiosities, may have played a significant role in Tornioli's imagery.

The character holding the telescope in the *Astronomers* takes an unusual pose, downward looking and bent forward. Federico Zeri first proposed a source for this pose in Caravaggio's *Narcissus*, where the distinctive damask pattern of the figure's clothing also bears a similarity to Tornioli's young astronomer.[15] Although scholars have highlighted stylistic affinities with contemporary works by Luca Giordano, Giovanni Lanfranco and Pietro da Cortona for the selected chromatic values and dramatic chiaroscuro effects, it is especially in the 'Caravaggesque manner of crowding figures together tight against the picture plane in intense interaction', as Alina Payne has argued, that Tornioli betrays his knowledge of Merisi's art. Thus, by reference to Caravaggio, Tornioli found 'the ideal visual vehicle for the topic of a *disputa*', strengthening the bond between the form and content of the painted subject.[16]

If this hypothesis addresses the broad issue of Tornioli's selective 'revival' of Caravaggio's art to fulfil the rhetorical needs of Virgilio's demanding commission, the relationship between Tornioli's young astronomer and Caravaggio's *Narcissus* stands out as an explicit one. Tornioli faithfully quotes the hairstyle, embroidered fabrics and prominent bent knee of this painting's mythological self-lover, from a work that is documented as being in Rome until 1645 and was very likely known by Spada and his milieu.[17] This suggests that the visual bond between these two figures goes beyond stylistic affinities. In particular, it seems relevant that the astronomer's downward use of the telescope, making the latter impractical, recalls Narcissus' act of self-contemplation. One wonders if this iconographic choice is to be related to the growing use of telescopes as expensive and emblematic court gifts, in the ineluctable shift from scientific knowledge to court discourse that Galileo had experienced firsthand.[18] On this ground, the reference to Narcissus would unveil Virgilio's and Tornioli's acknowledgement of the crisis of Galileo's scientific legacy and the progressive weakening of the scientist's ideas within the framework of court society and culture. This suggestion leads us back to the main strand of our analysis: the investigation of both the patron's and the painter's ambitions in the execution of the *Astronomers*.

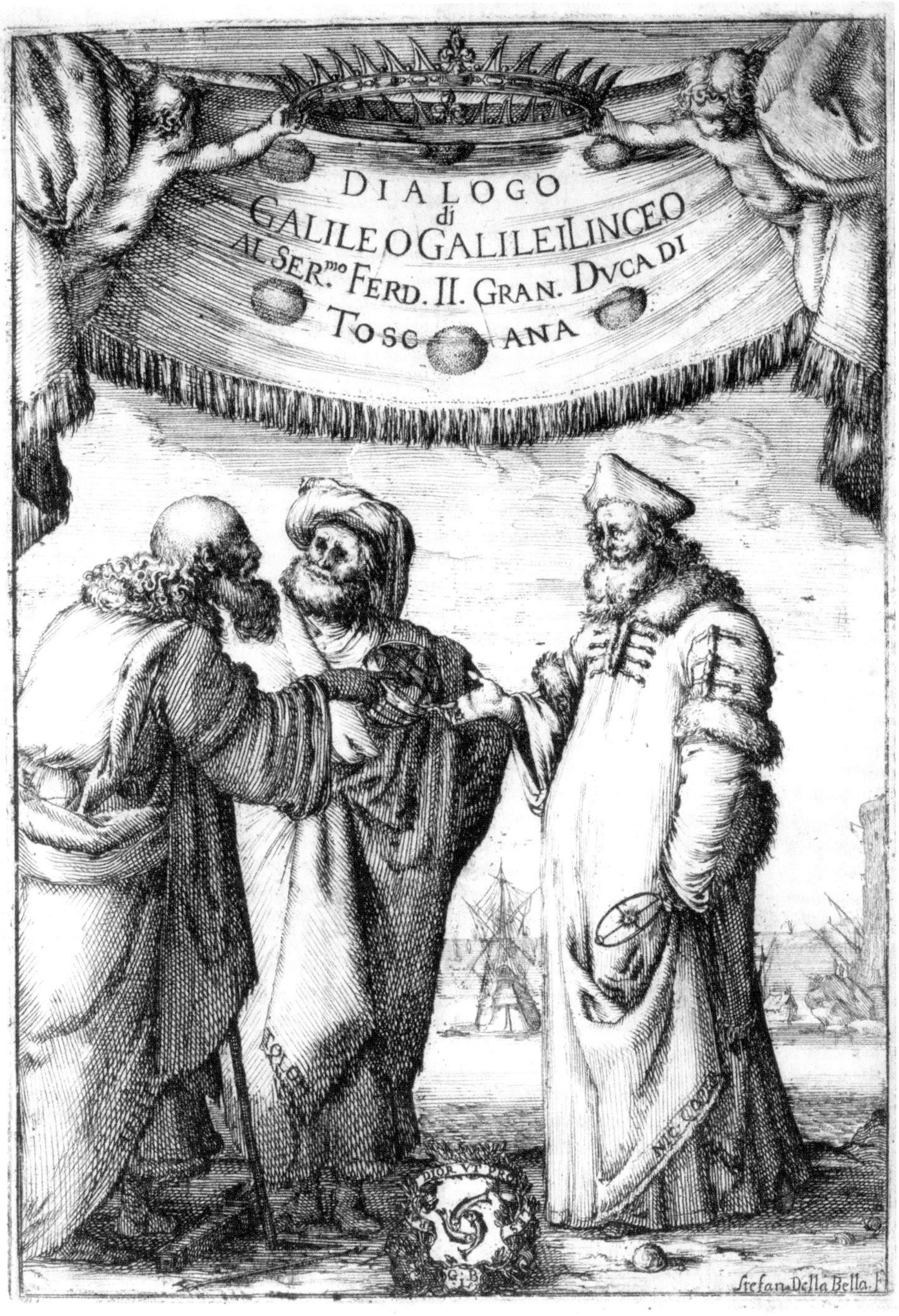

2 Stefano della Bella, *Aristotle, Ptolemy and Copernicus*, etched frontispiece from Galileo Galilei, *Dialogo sopra I due massimi sistemi del mondo*, Florence: Giovanni Battista Landini, 1632. Photo: Museo Galileo, Firenze.

The New Science and the Collection of Curiosities

Tornioli's canvas played a peculiar role within the cultural and collecting programme of the Spada milieu. Looking into the unstudied yet key connections between the painter's output, Virgilio Spada's museum of curiosities and Cardinal Bernardino's verses on the Spada sundial, this essay sheds new light on the particular convergence of scientific objects and concepts, *oggetti* and *concetti*, in Tornioli's *Astronomers*.

Recently the focus of a new scholarly study by Giuseppe Finocchiaro, Virgilio's museum of curiosities had been previously related to the patron's links with the Roman Oratory and interpreted as an outstanding source for Francesco Borromini's architectural solutions and collecting practices.[19] Archival records permit study of the afterlife of Virgilio's cabinet, which formed part of the patron's bequest to the Oratorian Congregation and was recorded by Giovanni Pietro Bellori in 1664.[20] The items were grouped into four categories: *antiquaria* (medals, coins and religious relics), *naturalia* (pieces from the mineral, vegetal and animal realms), *instrumenta* (telescopes, astrolabes, armillary spheres, meridians and terrestrial and celestial globes), and *curiosa* (heterogeneous curiosities wittily ornamented by subsequent craftwork, such as finely decorated ostrich eggs). Reflecting Virgilio's interest in early Christian marvels and historical findings, the Spada museum significantly differs from Cassiano dal Pozzo's or Athanasius Kircher's contemporary collections, but finds a parallel in Borromini's cabinet in Vicolo Orbetelli, which equally displayed a series of *antiquaria*, including a plaster cast of the bust of Seneca and a conspicuous number of silver coins and bronze medals, and *naturalia*, such as gems and seashells.[21]

In Tornioli's *Astronomers* the scope of the Spada cabinet of curiosities materializes in the depicted *instrumenta* of the celestial globe and the telescope, as well as in the *antiquaria* represented by the elaborate bronze medals on Copernicus's hat and the headgear of Astronomy. These medallions have been scarcely considered so far, with the exception of Anna Matteoli's analysis, in which the scholar argued that the figure who personifies Astronomy should rather be interpreted as the muse Urania since the image on the medal portrays Apollo with a bow in the guise of leader of the Muses (Musagète).[22] Even if it has not been possible to identify the coins with any of those recorded among Spada's curiosities, it seems reasonable to suggest that Tornioli faithfully copied them from those held in his patron's cabinet.[23] By virtue of directing and shaping his protégé's work, Virgilio Spada thus combined the fierce defence of Galileo's scientific legacy with a subtle strategy of self-promotion as a virtuous *curioso*. As in a game of mirrors, Tornioli's *Astronomers* reflected and condensed the patron's promotional and self-promotional ambitions. The portrayal of a scientific dispute which included Galileo's effigy and the newly invented telescope was intended as a manifesto of Virgilio's support for the New Science of observation launched by Galileo. The painting thus contained a particular political meaning, also reflected in Virgilio's epistolary and gift exchanges with the members of the Medici and Pamphili families. Remarkably, Leopoldo Medici wrote to Virgilio Spada in 1650 to advocate the cancellation of the anathema against Galileo's *Dialogo sopra i due massimi sistemi del mondo* in order to reprint the work. Moreover, in 1646, Virgilio presented Tornioli's canvas portraying *Cain slaying Abel* (Rome, Galleria Doria Pamphili) as a gift to Pope Innocent X Pamphili.[24]

Simultaneously, the accurate description of precision tools and precious coins in the *Astronomers* created a painted double of Virgilio's museum of curiosities, praising his 'private' activity as a patron and a collector. Giving visibility to this overlap between Virgilio's broad and particular cultural interests, Tornioli's work in a sense confirms Krzysztof Pomian's assumption that Roman cabinets played a distinctively political role, having 'a very real, albeit invisible, influence on urban life'.[25]

Ut natura pictura: The Spada Sundial

Mirroring Virgilio's political culture and private collecting practice, the display of scientific and 'curious' objects in Tornioli's canvas should also be understood more broadly as a reference to the enthusiastic sponsorship of scientific novelties by the Spada family. Thanks to his status of *pittore di casa*, Tornioli could directly profit from Bernardino Spada's scientific interests and patronage. Cardinal Bernardino purchased a stock of seven paintings by Tornioli in 1643, thus becoming one of the painter's major patrons in Rome, yet his contribution to Tornioli's invention has long remained overlooked. An eager collector of scientific tools, as the terrestrial and celestial globes built in Amsterdam by Willem Janszoon Blaeu and the telescope decorated with grotesque motifs by Paolo Maruscelli assembled in the Room of Aeneas make clear, Bernardino also encouraged the construction of sophisticated technological 'marvels' in the Spada palace, such as a catoptric meridian and Borromini's perspective gallery.[26] The programmatic content of the inscriptions included in the Spada sundial should be regarded as a key source that highlights the innovative *concetti* promoted by Cardinal Bernardino and contributes to analysis of the ideological affinities between these verses and Tornioli's *Astronomers*.

Hosted on the first floor of the palace, the Spada meridian was designed by Emmanuel Maignan, a priest of the Order of the Minims of St Francis of Paola, who had already built an analogous sundial in the convent of Trinità dei Monti, and painted in 1644 by Giovan Battista Magni. Unlike traditional meridians, catoptric sundials operate by means of a mirror on the windowsill (instead of the iron gnomon) so sunlight, falling on the mirror, projects a luminous disk over the vault, indicating the time of day and true noon throughout the year. The Spada exemplar was first mentioned in Jean-François Niceron's *Thaumaturgus opticus* (Paris, 1646), and fully described in Maignan's *Perspectiva Horaria sive de horographia gnomonica* (Rome, 1648). Showing time zones and the house of the zodiac and enabling the making of astronomical charts and horoscopes, the sundial is presented as a great achievement of science and as a key example of enlightened patronage. Alongside the description of the figurative apparatus of the sundial, Maignan's book reports all the inscriptions of the gallery, composed by Father Guglielmo du Planté and, in one case, by Bernardino Spada.

These short compositions, edited by Neppi in 1975 but barely considered by later studies, allow us to explore the cardinal's attitude towards scientific novelties. Despite the traditional encomiastic tone, both a poetics of marvel and a new impulse in astronomical researches are to be perceived in a series of rhetorical devices. The technological achievements of the catoptric sundial, in which the rays of the sun, reflected in a set of mirrors, are projected onto the marked surface of the vault, are poetically referred to through the concept of the emulation of nature:

> Aemula naturae manus hic depingere coelos
> Tentavit, Solem pingere non potuit.
> Ne tamen incepto desit Sol ipse labori
> Seque suumque suo lumine pingit iter[27]

If the hand which emulates nature could only depict ('*depingere*') the skies, the sun painted ('*pingit*') itself and its own course out of its own effort ('*ipse labori*') and by means of its own light ('*suo lumine*'). Thanks to the meridian, the sun becomes painter of itself, producing scientifically exact and artistically perfect results. Praising the peculiarity of the catoptric sundial, these verses deal with the poetics of *ut natura pictura* and with the dialectical relationship between art and nature. The alliance of painting and science is also a central theme in Tornioli's *Astronomers*, as indicated by the proximity of the portraits of Galileo and Tornioli and, as will be argued, by the simultaneous presence of a scientific debate and a more hidden artistic dispute in the canvas.

The innovative character of Bernardino's meridian is celebrated by the patron himself, in the last two verses of his inscription:

> Ferreus hinc Gnomon et ferrea pellitur umbra;
> Aurea pro ferro tempora mundus habet.[28]

In this self-celebrating composition, Bernardino succeeds in both praising the sunrays of the sundial, which defeated the shadowy gnomonic system, and alluding to his great contribution to the affirmation of a new golden age, under the sign of Modern Science. The evocative *concetto* of a new era introduced by these verses would gain considerable fortune in the following decades, as witnessed, for instance, in the literary imagery of the phoenix, variously referenced by Gian Lorenzo Bernini and his patrons, which 'symbolizes the new era brought about by an exemplary ruler, a reign of peace engendering the flourishing of science and poetry'.[29] The splendour of Bernardino's clever patronage recurs elsewhere in the inscriptions:

> Hinc igitur Sapiens Princeps, idemque Senatus
> Purpurei lux clara, suas te accersit ad aedes;
> Hinc manus et Magni Sapiens me industria cogit
> Astronomi, tanti votis servire Patroni
> Et nos ergo ΣΟΦΟΙΣ par est parere duobus[30]

Within the rhetorical frame of a dialogue between the Sun and the catoptric mirror, the answer of the latter, from which these verses are extracted, focuses on the figures of Bernardino, 'bright light ("*lux clara*") of the College of Cardinals', and Maignan, the great astronomer gifted with practical and speculative skills. They are both praised as wise and learned personalities, through the repetition of the Latin word '*Sapiens*' and the inclusion of the Greek 'σοφοις', with respect to the deepest and highest possible forms of knowledge.

It seems more than plausible that Tornioli, as Virgilio's trusted *pittore di casa*, profited from Bernardino's meridian in both its physical manifestation and through the encomiastic verses it prompted, to shape the ultimate meaning of his *Astronomers*. Much like the inscriptions, the canvas questions the new boundaries and exchanges between artistic and scientific production in the cultural and moral growth of mankind. At a time – immediately after Galileo's death – when the chill of censorship could still be felt by the Roman *virtuosi*, the fierce defence of the New Science finds a poetic and optimistic dimension in Bernardino's celebration of a new golden age. If in the catoptric sundial 'the new art wins over the old one' ('*Ars vetus in tamtum vincitur arte nova*'), in Tornioli's painting the new Galilean theories win over the old Aristotelian ones, demonstrated by the detailed observation of celestial phenomena enabled by the new scientific tools.[31] Cardinal Spada's verses also reveal a meticulously crafted rhetoric of self-promotion, which projected his scientific patronage against the mythological and super-historical *lieu* of an *aurea aetas*. In this perspective, if Guercino's canvas presented Galileo as a novel Endymion learned in astronomical phenomena, the inscriptions included in the meridian enhanced Bernardino's and Maignan's virtues of novel σοφοι, for they both devote their *ingegni* to a common field of knowledge. By means of celebrating his specific contribution to his historical *hic et nunc*, Bernardino pursued long-lasting fame. These considerations might cast a light on Tornioli's own devices of self-promotion and especially on the inclusion of his self-portrayal among ancient and Modern σοφοι, such as Aristotle and Galileo.

Both Virgilio and Bernardino Spada emerge as luminary yet narcissistic patrons. Shaping the artistic and scientific output produced in the Spada palace, exemplified in Tornioli's canvas or Maignan's catoptric sundial, they combined the promotion of Galilean ideas with their self-promotion as *curiosi* and *virtuosi*. Tornioli's *Astronomers*, at once a result and a celebration of the Spada cultural agenda, fully reflects this two-fold, paradoxical ambition. The subject of a scientific dispute and the representation of contemporary discoveries enable a pictorial defence of Galileo's theories, testifying to Virgilio's (and Bernardino's) official, yet unorthodox support of the recently deceased scientist. At the same time, the explicit or implicit references to *oggetti* in Virgilio's cabinet of curiosities or the *concetti* developed in the inscriptions of Bernardino's sundial, offered a powerful *summa* of the patrons' self-promotional devices, as well as a rhetorical tool for their politics of self-fashioning.

Science and Enquiry after Galileo

In Tornioli's times, a crucial artistic dispute stemmed from Galileo's experimental observations and his ambiguous use of epistemic images, or images 'made with the intent not only of depicting the object of scientific inquiry but also of replacing it'.[32] Whereas Galileo's telescope allowed the acquisition of new forms of visual knowledge, the scientist's use of illustrations generated a debate concerning the scientific status of visual evidence and, more specifically, the role of images in the production of knowledge, thus questioning their epistemological possibilities.

Since the publication of Erwin Panofsky's pioneering study on Galileo as 'a critic of the arts', scholarly attention has been increasingly devoted to synergies between the Tuscan scientist and contemporary artists.[33] Galileo's fame significantly profited from the diffusion of his painted portrait, such as Santi di Tito's lost version or Ottavio Leoni's and Francesco Villamena's engraved effigies, produced after Galileo's death.[34] As Chrysa Damianaki has remarked, Galileo's portrait included in Cecco Bravo's fresco of 1636 in the Florentine house of Michelangelo Buonarroti the Younger, much as a series of busts reproducing that on Galileo's tomb in Santa Croce, played a crucial role in promoting, *per visibilia*, the scientist's legacy.[35] At the same time, a series of Tuscan artists, later known as 'Galilean', cultivated a taste for astronomical observations and, as attested by Ludovico Cigoli's and Domenico Cresti's letters to Galileo, used the telescope to explore the moon's spots.[36] The most remarkable evidence of these experiments is offered by Cigoli's *Immaculate Conception* frescoed in the dome of the Paolina Chapel in Santa Maria Maggiore in Rome between 1610 and 1612, where the Virgin is represented on a spotted and textured moon.[37]

Both a scientist and a philosopher, Galileo not only deployed the visual to validate his theories and enhance his discoveries, but also to promote himself and his Medici patrons. Hence, the scientist's use of images emerges as two-fold. On the one hand, a 'factual' use of naturalistic or astronomical illustrations was adopted in pursuit of probative or didactic aims, as witnessed, for instance, by the representations of the moon spots included in Galileo's *Sidereus Nuncius* of 1610, specifically addressed to his 'image-friendly' patrons.[38] In other respects, Galileo shaped a 'rhetorical' and court use of images, particularly in the context of his controversial activity of 'emblem maker' for the Medici family. As Mario Biagioli has convincingly argued, 'because Medici patronage rewarded marvels that would fit the discourse of the court but not scientific theories or research programs, Galileo tended to present the satellites of Jupiter not as astronomical discoveries supporting a new cosmology but as dynastic emblems, and himself not as a discoverer, but only as the mediator of an encounter.'[39] Using these artifacts to flatter his patrons and present himself in the guise of a philosopher, rather than a scientist, Galileo transformed the ultimate goal of his scientific discovery, turning the satellites of Jupiter into Medici dynastic emblems. In so doing, by rhetorically lessening his role in the discovery of the stars, Galileo paved the way for *Medicea Sidera* to become mythological entities instead of truthful illustrations of celestial bodies. In contrast with epistemic images, they turned out to be 'nothing but Medici fetishes'.[40] More notably, Galileo revealed how these finely crafted emblems served as vehicles of ideologically oriented Medicean promotion.

By means of both strengthening and diminishing the epistemological and probative functions of the visual during the various phases of his career, Galileo unveiled the paradoxical nature of images, posing unprecedented challenges to the ultimate credibility of the representable. In parallel, the dialectical relationship between vision and technology in the age of Galileo further complicated and revolutionized man's understanding of the terrestrial and celestial worlds. On a visual level, pre-Galilean material, such as the charts based on Aristotelian models, was suddenly replaced by illustrations conceived in the light of astronomical observations enabled by the telescope. As far as the credibility of written sources was concerned, it was the Lincean Academy in particular that, drawing on Galileo's pioneering experimental method, 'contributed to the erosion of textual authority by suggesting that scientific method was found in nature'.[41] To sum up, as chiefly emerges from David Freedberg's, Paula Findlen's and Alina Payne's analyses, 'Galileo's was an inaugural moment of conflict and disjunction, a coming of age of the cognitive problems vision had posed for some time – a moment of recognition that sight was both a strength and a weakness – and it was tied to the sciences and to instruments.'[42]

Produced in the wake of the turmoil generated by the Galilean revolution, Tornioli's *Astronomers* reveals a meticulous pictorial strategy to reassert the credibility of the visual. Addressing the issue of the potential crisis of images, the painter cleverly takes part in the artistic debate through the representation of a scientific topic, succeeding in empowering the visual arts and, more importantly, the role of painting in the process of knowledge-making. A first example of the artist's rhetorical devices to convey his claims is offered by the representation of the phases of the moon in the upper right corner of the *Astronomers*. Highlighted by Copernicus's deictic gesture, the chronologically different phases of the satellite are presented simultaneously, by virtue of the pictorial prerogative of condensing temporal and spatial values.

In order to discuss this strategy, we should first consider the illustrations of the moon spots included in Teofilo Gallaccini's treatise *Monade Celeste* (*plate* 3). Gallaccini, a Sienese polymath and a close friend of Tornioli's, sketched these images to complement his account of the astronomical observations carried out by Galileo in August 1633. Galileo was residing in Siena in the palace of the archbishop Ascanio II Piccolomini and he repeatedly used the telescope from the Piccolomini Loggia, in the presence of Gallaccini and other members of the Intronati Academy. As Payne has argued, 'the scene on the Piccolomini loggia must have been very close to that

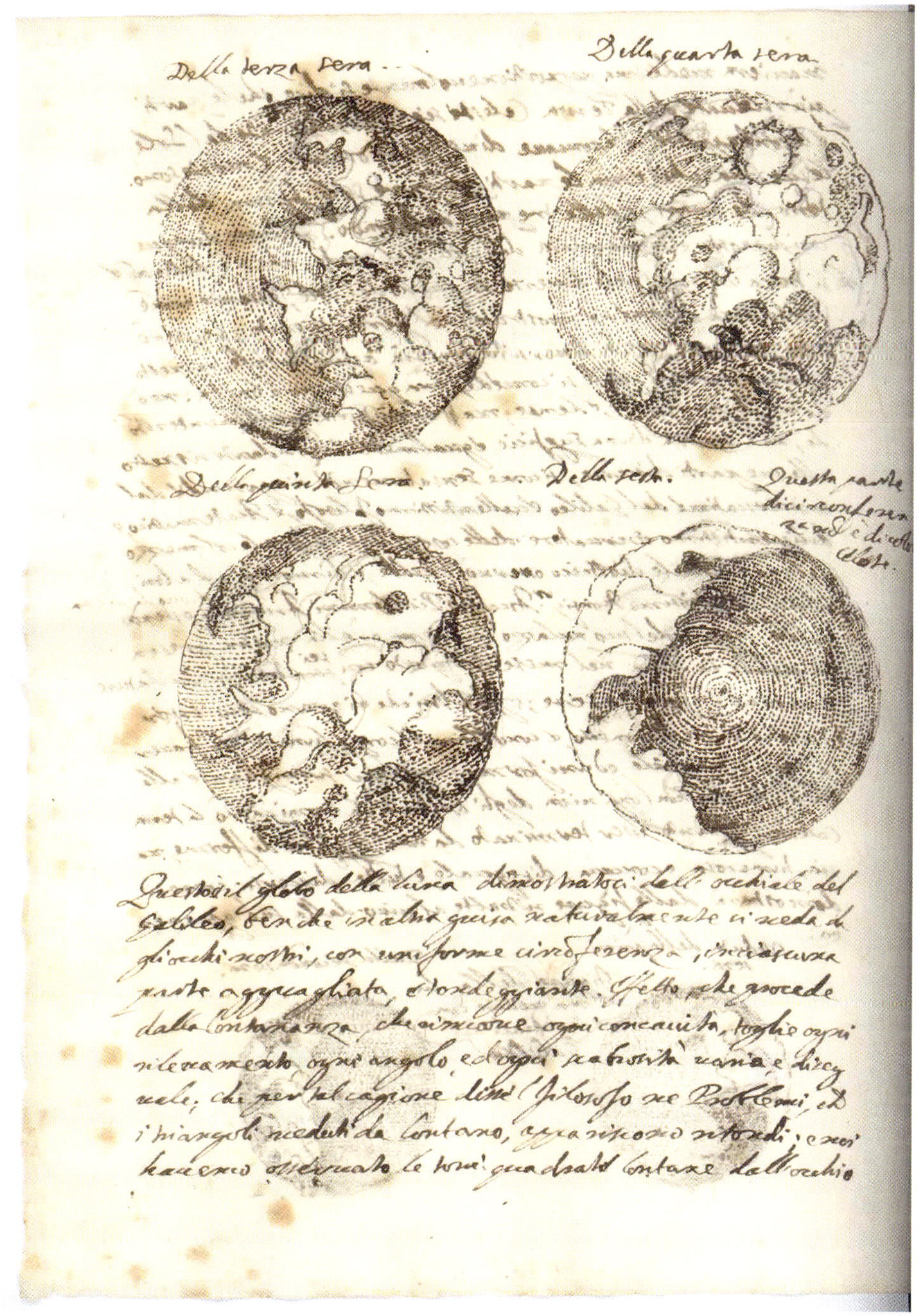

3 Teofilo Gallaccini, *Monade Celeste*, 1633–41. Siena: Biblioteca Comunale degli Intronati (L. VI. 31, f. 97v). Photo: Biblioteca Comunale Intronati di Siena

depicted by Tornioli in his *Astronomers*' and 'what was on display beyond the moon and beyond Galileo's finely ground lens was *sight itself*'.[43] Gallaccini explains that the six illustrations included in the manuscript immortalize six different observations of the moon by means of Galileo's telescope. Praised as an 'excellent philosopher, and mathematician, and most attentive observer of celestial things', Galileo argued that the lunar body does not have an equal or uniform surface ('*superficie uguale ed uniforme*'), but a 'deformed surface, varied, full of bruises, unequal and diverse' ('*una superficie difforme, varia, tumorosa, e concava, diseguale e tutta diversa*').[44]

To some extent, Gallaccini's use of 'didactic' images to illustrate his treatise could be related to Galileo's use of depictions of the moon spots in the *Sidereus Nuncius* to exemplify the productive alliance between written and visual sources. Confronted with the same scientific topic, Tornioli could rely on pictorial strategies only, but he nonetheless appropriates devices from textual rhetoric. The inclusion of chronologically separated phases of the moon allowed him to prove – and praise – the painter's capacity of controlling and visually 'possessing' nature. It might be argued that Tornioli used what Lorraine Daston has defined as 'synoptic effects'. Thanks to this pictorial expedient, early modern painters could achieve a synthesis of different scientific data, for the painted surface necessitated a compression of different phenomena (or observations of a single phenomenon) and so the perception of them 'at a glance'.[45]

In Tornioli's canvas the range of 'synoptic effects' is not limited to the record of celestial bodies. A powerful historical synthesis brings together the most eminent intellects of antiquity and the modern era. If the binary nature of the composition is aimed at stressing the duality between Aristotelian doctrine and the Galilean New Science, the ultimate unity granted by painting is dense in meaning and potential. Animated by a humanistic aspiration to universality, Tornioli reinforces the dialogue between the characters by means of squeezing the figures in a frieze-like composition, 'in the guise of a human chain' animated by intense gesture and lit by a Caravaggesque chiaroscuro.[46] This epistemological synthesis embraces a number of disciplines and methods. The simultaneous presence of Alexander the Great, Aristotle, Copernicus and Galileo testifies to the cross-historical capacities of painting. Moreover, the personifications of Geography and Astronomy summarize the two spheres explored by humankind. Whereas Alexander was the emperor of the known world, Galileo paved the way to a renewed scientific exploration and conquest of the celestial cosmos. In the super-historical and trans-disciplinary space of the painting, different systems of knowledge are not at odds, but rather merge in a truly universal reflection upon the possibilities of art. For this reason, the text-based and geocentric theories of the ancient and the nature-centred and heliocentric method of the Galilean age are brought together to celebrate the power of human *ingegni*, in a virtuous progression of achievements and discoveries.

Only painting could provide the physical and epistemological space for this otherwise impossible dispute, in which the transitory – and equally impossible – moment of the dialogue is eternalized in a great *summa* of human knowledge. At a time when, as seen in Galileo's career, scientific illustrations could be turned into mere court emblems, Tornioli empowered the visual by means of exploiting in every possible way its 'synoptic' potential. Moreover, the synthesis at work in the *Astronomers* shows that painting could not only visualize knowledge, but, thanks to its 'humanistic' historical and disciplinary collation, could create knowledge as well.

If Galileo's *Dialogue Concerning the Two Chief World Systems* of 1632 generated an irresolvable contrast between the Ptolemaic and the Copernican systems, in Tornioli's painting the central figure of Copernicus orchestrates a dialogue 'between' the two chief world systems. This synthesis enhances Galileo's 'sensed experiences', for the visual enables a comprehensive sight and understanding of natural, historical and philosophical phenomena. On this ground, the potential contrasts between art and science are reconciled in a strengthened bond, reviving the literary *topos* of *ut pictura natura* with a new fundamental alliance between disciplines in the edification of human knowledge. In Tornioli's *Astronomers*, the visual proximity of Galileo's and the painter's portrayals seems to allude to this reunification under the sign of a shared aspiration to universality.

Furthermore, the enigmatic trio formed by Galileo, Tornioli and the young astronomer who is pointing the telescope downwards should be reinterpreted in light of the painter's professional and ideological claims. The painter's choices should be linked to Galileo's controversial legacy with regard to visual knowledge and to the cultural transformations which occurred in the aftermath of the scientist's death. The revolutionary content of the New Science was not threatened by the censorship of the Church only. As Biagioli has shown, a breach in the credibility of Galilean findings was produced by the scientist himself, who performed the role of 'courtier' in his association with the Medici family. The case of the satellites of Jupiter testifies to the accurate and yet dangerous reinterpretation and reinvention of Galileo's own scientific discoveries, ineluctably deprived of their content of truth to serve as gracious self-fashioning tools for the Medici dynasty. In Seicento court society, patrons gave rise to a paradoxical connection between the celebration of the New Science and the search for self-celebration. As a result, Galileo's legacy was distorted and reshaped in accordance with a new set of rhetorical strategies and individual ambitions. The members of the Spada family were not immune to this process. As clearly emerges from the analysis of the *oggetti* and *concetti* included in Tornioli's *Astronomers*, they combined an overt support of Galileo's theories with a narcissistic praise of their enlightened artistic and scientific patronage, programmatically emerging as promoters, if not co-founders of a new 'golden age'.

In Tornioli's canvas, the iconographic features of the young astronomer may relate to this critical aspect of Galileo's heritage. The character's odd use of the telescope pointed downward makes the tool impractical, possibly betraying that this scientific instrument had, by this time, become a precious court 'toy' and a diplomatic gift to reinforce political alliances. Even if Copernicus is pointing at the phases of the moon, the astronomer fails to acknowledge the celestial phenomena and his downward-looking gaze prevents him from taking active part in the dispute. If this is so, the pose of the forward-leaning figure is connected to the impossibility of achieving or sharing knowledge.

By contrast, Tornioli's self-inclusion in the canvas is to be associated with the other side of Galileo's legacy. Even if it could be interpreted as an act of *hubris*, the insertion of the artist's effigy should also be understood within the circumstances of Tornioli's specific claims for the role of art. In Tornioli's point of view, the defence of the visual as a valid epistemological tool had immediate consequences for his profession, touching upon the artist's possibilities of contributing to collective knowledge. While the self-referential astronomer, a modern Narcissus, visualizes the court detour of Galileo's legacy, Tornioli's self-portrayal simultaneously testifies to art's possibilities of keeping the Galilean spirit alive, by means of pursuing and spreading the truth. Engaging the bystander in the scientific dispute, Tornioli's active gaze acclaims the artist's prerogative to convey a layered knowledge at a glance.

At a time when science and art were running the risk of being at odds, opposed in the polarities of real knowledge and crafted propaganda, Tornioli exploited his *Astronomers* to take part in this vital debate and to construct a defence and a praise of his profession. Taking advantage of a series of compositional, iconographic and rhetorical devices, the painter strengthened the bond between art and science under their mutual contribution to the cultural and moral advancement of mankind. Empowered by the artist's *ingegno*, painting rises like a phoenix from the ashes, regaining a chief role in the process of knowledge-making.

Notes

I would like to express my profoundest gratitude to Sheila McTighe, Scott Nethersole, Ben Thomas and Genevieve Warwick for their suggestions on this text.

1 For an updated bibliography on the canvas see Marco Ciampolini, *I pittori senesi del Seicento*, Siena, 2010, III, 885–6. The painting was purchased by Virgilio Spada in 1645; on the records of payment see Roberto Cannatà in Dante Bernini, ed., *Laboratorio di Restauro 2*, exh. cat., Palazzo Barberini, Rome, 1988, 178. For an insight into Virgilio's architectural patronage see Minna Heimbürger Ravalli, *Architettura, scultura e arti minori nel Barocco italiano. Ricerche nell'Archivio Spada*, Florence, 1977; Klaus Güthlein, 'Die Fassade der Barnabiterkirche San Paolo in Bologna. Ein Beitrag zur Tätigkeit der Architekten Girolamo und Carlo Rainaldi für die Familie Spada', *Römisches Jahrbuch für Kunstgeschichte*, 17, 1978, 125–55; Klaus Güthlein, 'Quellen aus dem Familienarchiv Spada zum römischen Barock. 1. Folge', *Römisches Jahrbuch für Kunstgeschichte*, 18, 1979, 173–246; Klaus Güthlein, 'Quellen aus dem Familienarchiv Spada zum römischen Barock. 2. Folge', *Römisches Jahrbuch für Kunstgeschichte*, 19, 1981, 173–243; Joseph Connors, *Borromini and the Roman Oratory: Style and Society*, Cambridge, MA and London, 1980; Joseph Connors, 'Virgilio Spada's defence of Borromini', *The Burlington Magazine*, 131: 1031, 1989, 76–90; on the patron's collecting activity see Roberto Cannatà and Maria Lucrezia Vicini, *La Galleria di Palazzo Spada. Genesi e storia di una collezione*, Rome, 1992, 72–84.

2 On this etching see Erwin Panofsky, *Galileo as a Critic of the Arts*, The Hague, 1954.

3 On this caricature drawing (anonymous artist, *Bernardino Mei, Niccolò Tornioli and Francesco Rustici*, black charcoal and white chalk on paper, 144 × 289 mm, Florence, Gabinetto Disegni e Stampe, 4837 S), see Ciampolini, *Pittori Senesi*, III, 877.

4 This iconography, quite common in the seventeenth century, is less diffused in Late Antiquity. However, possible ground for comparison is offered by the Farnese Cameo in the Museo Archeologico of Naples. See Carla Alfano, ed., *Alessandro Magno. Storia e mito*, exh. cat., Palazzo Ruspoli, Rome, 1995, 318, n. 115.

5 On this painting, attributed to Tornioli in the nineteenth century, see Ciampolini, *Pittori Senesi*, I, 139.

6 This interpretation would in a sense prove Anna Matteoli's neglected hypothesis that the female figure on the left is a personification of Earth, Gea, discussed in Anna Matteoli, 'Due inedite pitture barocche su temi biblici', *Bollettino dell'Accademia degli Euteleti*, 68, 2001, 153.

7 See Alina Payne, *The Telescope and the Compass: Teofilo Gallaccini and the Dialogue between Architecture and Science in the Age of Galileo*, Florence, 2012, 2 and 4.

8 On this painting see Luigi Salerno, *I dipinti del Guercino*, Rome, 1988, 304–5.

9 For the links between Tornioli's and Guercino's canvases see Claudia Cieri Via, ed., *Immagini degli dei. Mitologia e collezionismo tra '500 e '600*, exh. cat., Fondazione Memmo, Lecce, 1996, 162–4, and Marco Bona Castellotti, Enrico Gamba and Fernando Mazzocca, eds, *La ragione e il metodo. Immagini della scienza nell'arte italiana dal XVI al XIX secolo*, exh. cat., Sala Pietro da Cemmo, Crema, 1999, 118.

10 The tale of Endymion, narrated by Apollodorus (*The Library*, I, VII, 373–82) and Lucian (*Dialogues of the Gods*, 19), was reinterpreted by Pliny the Elder (*Naturalis Historia*, II, 43) and Nonnus (*Dionysiaca*, 41, 379). On this philological reconstruction see Cieri Via, *Immagini degli dei*, 162–4. According to Denis Mahon and Luigi Salerno, the work was commissioned by don Lorenzo Medici (1599–1649) in 1645 and subsequently donated to Camillo Pamphili, while Miriam Di Penta argued that it originally belonged to Alessandro Argoli and was given to a member of the Pamphili family as a sign of gratitude; see Salerno, *Dipinti del Guercino*, 320; and Miriam Di Penta, 'Guercino's "Endymion", "Hercules" and "Artemisia" for Alessandro Argoli', *Journal of the Warburg and Courtauld Institutes*, 67, 2004, 233–45.

11 Giovanni Battista Marino, *L'Adone poema del C. Marino Con gli Argomenti del Conte Sanvitale, e l'Allegorie di Don Lorenzo Scoto. Aggiuntovi la Vita dell'Autore con alcune sue lettere facete, e le Tavole delle cose notabili*, Paris, 1627, X, 43: 'Del Telescopio à questa etate ignoto/ Per te fia, Galileo, l'opra composta,/ L'opra ch'al senso altrui, ben che remoto,/ Fatto molto maggior l'oggetto accosta./ Tu solo osservator d'ogni suo moto,/ E di qualunque hà in lei parte nascosta,/ Potrai senza che vel nulla ne chiuda,/ Novello Endimion, mirarla ignuda.' On the link between Guercino's canvas and these verses see Cieri Via, *Immagini degli dei*, 163. Another possible source for Guercino's painting is offered by Francesco Parisani's *Diana Schernita* of 1629, which describes Endymion in the act of contemplating Diana with a telescope. For this suggestion see Bona Castellotti, Gamba and Mazzocca, *La ragione e il metodo*, 118.

12 Paula Findlen, *Possessing Nature: Museums, Collecting, and Scientific Culture in Early Modern Italy*, Los Angeles, CA and London, 1996, 407.

13 On this matter see, for instance, Filippo Camerota and Mara Miniati, eds, *I Medici e le scienze. Strumenti e macchine nelle collezioni granducali*, exh. cat., Museo degli Argenti, Florence, 2008.

14 On Galileo's Rome see Mario Biagioli, *Galileo, Courtier: The Practice of Science in the Culture of Absolutism*, Chicago, IL, 1993, 245–65.

15 Federico Zeri, *La Galleria Spada in Roma*, Rome, 1952, 21.

16 Payne, *The Telescope and the Compass*, 4.

17 On the bibliography of Caravaggio's *Narcissus* see Lorenza Mochi Onori and Rossella Vodret, eds, *Galleria Nazionale d'Arte Antica Palazzo Barberini. I Dipinti. Catalogo sistematico*, Rome, 2008, 283. Stressing the affinities between the *Narcissus* and the *Astronomers*, Gianni Papi attributed the former to Giovanni Antonio Galli, with whom Tornioli was associated in the works in St Peter's; see Gianni Papi, *Spadarino*, Soncino, 2003, 155–60; Gianni Papi, 'Un "concerto" di Spadarino', in Pierluigi Carofano, ed., *Atti delle Giornate di Studi sul Caravaggismo e il naturalismo nella Toscana del Seicento*, Pontedera, 2009, 73.

18 On this issue see Mario Biagioli, 'Galileo the emblem maker', *Isis*, 81: 2, 1990, 230–58.

19 See Giuseppe Finocchiaro, *Il Museo di curiosità di Virgilio Spada. Una raccolta romana del Seicento*, Rome, 1999, and Alessandro Zuccari, 'Borromini tra religiosità borromaica e cultura scientifica', in Alessandro Zuccari and Stefania Macioce, eds, *Innocenzo X Pamphilj. Arte e Potere a Roma nell'Età Barocca*, Rome, 1990, 35–73.

20 See Giovanni Pietro Bellori, *Nota delli Musei, Librerie, Gallerie & ornamenti di Statue, e pitture, nei Palazzi, nelle Case, e ne' Giardini di Roma*, Rome, 1664, 32.

21 On Borromini's museum, described in the architect's inventory of 1667, see Paolo Portoghesi, *Borromini nella cultura europea*, Rome, 1964, 374–90; Stefania Macioce, 'La "chiocciola" di S. Ivo alla Sapienza', in Zuccari and Macioce, *Innocenzo X Pamphilj*, 75–96; and Joseph Connors, 'Borromini's S. Ivo alla Sapienza: The spiral', *The Burlington Magazine*, 138, 1996, 668–82.

22 Matteoli, 'Due inedite pitture', 153.

23 The list of Virgilio's medallions is published in Finocchiaro, *Museo di curiosità*, 194–203.

24 On this painting see Ciampolini, *Pittori Senesi*, III, 884.

25 Krzysztof Pomian, *Collectors and Curiosities: Paris and Venice, 1500–1800*, trans. Elizabeth Wiles-Portier, Cambridge, 1990, 2.

26 On these commissions see Lionello Neppi, *Palazzo Spada*, Rome, 1975, 175–87 and 189–201.

27 Neppi, *Palazzo Spada*, 191, E. 'The hand which emulates nature has tried here to paint the skies, it could not paint the sun. Not to be excluded from this work the Sun, out of its own effort and by means of its own light, painted itself and its own course.'

28 Neppi, *Palazzo Spada*, 192, H. 'The iron shadow of the iron gnomon has faded. Not iron, but golden is the new age of the world.'

29 Maarten Delbeke, 'Gianlorenzo Bernini as "la fenice degl'ingegni", or the History of an Epithet', *Marburger Jahrbuch für Kunstwissenschaft*, 32, 2005, 247.

30 Neppi, *Palazzo Spada*, 192, G. The verses are extracted from a poetic dialogue between the sun and the catoptric mirror. The Sun complains about the vulgarity of the material chosen for the reflection of his precious rays, the catoptric mirror, and for the confining of his light under the vault of a narrow gallery. These lines belong to the answer of the mirror: 'Therefore, a wise Prince, bright light of the College of Cardinals, invites You in his house; and the hand and the learned skillfulness of a great astronomer force me to fulfill the wishes of such a patron. And we both have to obey these two chiefly wise men.'

31 On the Latin inscription see Neppi, *Palazzo Spada*, 191, F.1.

32 See Lorraine Daston, 'Epistemic images', in Alina Payne, ed., *Vision and its Instruments: Art, Science, and Technology in Early Modern Europe*, University Park, PA, 2015, 13–35.

33 See Panofsky, *Galileo*. For an insight on Galileo's links to the arts and the artists see Lucia Tongiorgi Tomasi, 'Galileo, le arti, gli artisti', in Lucia Tongiorgi Tomasi and Alessandro Tosi, eds, *Il cannocchiale e il pennello. Nuova scienza e nuova arte nell'età di Galileo*, exh. cat., Palazzo Blu, Pisa, 2009, 21–43.
34 On the fortune and symbolic meaning of Galileo's portraits from the seventeenth to the nineteenth centuries see Carlo Maccagni, 'L'iconografia di Galileo Galilei: dal ritratto all'apoteosi', in Bona Castellotti, Gamba and Mazzocca, *La ragione e il metodo*, 55–61.
35 Chrysa Damianaki, *Galileo e le arti figurative*, Manziana, 2000, 10.
36 On the epistolary exchange between Galileo and Cigoli see Anna Matteoli, ed., *Macchie di sole e di pittura: carteggio L. Cigoli – G. Galilei, 1609–1613: con introduzione e note*, San Miniato, 1959.
37 On the fresco decoration see Friedrich Noack, 'Kunstpflege und Kunstbesitz der Familie Borghese', *Repertorium für Kunstwissenschaft*, 50, 1929, 191–231, and Anna Matteoli, *Lodovico Cardi-Cigoli pittore e architetto. Fonti biografiche, catalogo delle opere, documenti, bibliografia, indici analitici*, Pisa, 1980, 245.
38 Pamela H. Smith and Paula Findlen, eds, *Merchants & Marvels: Commerce, Science and Art in Early Modern Europe*, New York, 2002, 9. See also David Freedberg, *The Eye of the Lynx: Galileo, his Friends, and the Beginnings of Modern Natural History*, Chicago, IL, 2002.
39 Biagioli, 'Galileo the emblem maker', 253.
40 Biagioli, 'Galileo the emblem maker', 254.
41 Paula Findlen, *Possessing Nature: Museums, Collecting, and Scientific Culture in Early Modern Italy*, Los Angeles, CA and London, 1996, 207.
42 Alina Payne, 'Introduction', in Payne, *Vision and its Instruments*, 3.
43 Payne, *The Telescope and the Compass*, 27; Payne, 'Introduction', 2.
44 The passage describing the astronomical observation is included in the manuscript of the Biblioteca Comunale di Siena BCS, L. VI. 31, ff. 97–98. It was partly published in Giovanni Targioni Tozzetti, *Notizie degli aggradimenti delle scienze fisiche accaduti in Toscana*, Florence, 1780, 318–19. The whole treatise *Monade Celeste* is published in Alina Payne, ed., *Teofilo Gallaccini: Selected Writing and Library*, Florence 2012, 247–405 (see in particular 385–6). On Gallaccini's links to Galileo see also Annalisa Pezzo, 'Una rete erudita', in Giovanni Morolli, ed., *Siena 1600 circa: dimenticare Firenze. Teofilo Gallaccini e l'eclisse presunta di una cultura architettonica*, exh. cat., Santa Maria della Scala and Magazzini della Corticella, Siena, 1999, 67–9.
45 On this issue see Lorraine Daston, 'The empire of observation, 1600–1800', in Lorraine Daston and Elizabeth Lunbeck, eds, *Histories of Scientific Observation*, Chicago, IL, 2011, 81–114.
46 Payne, *The Telescope and the Compass*, 3.

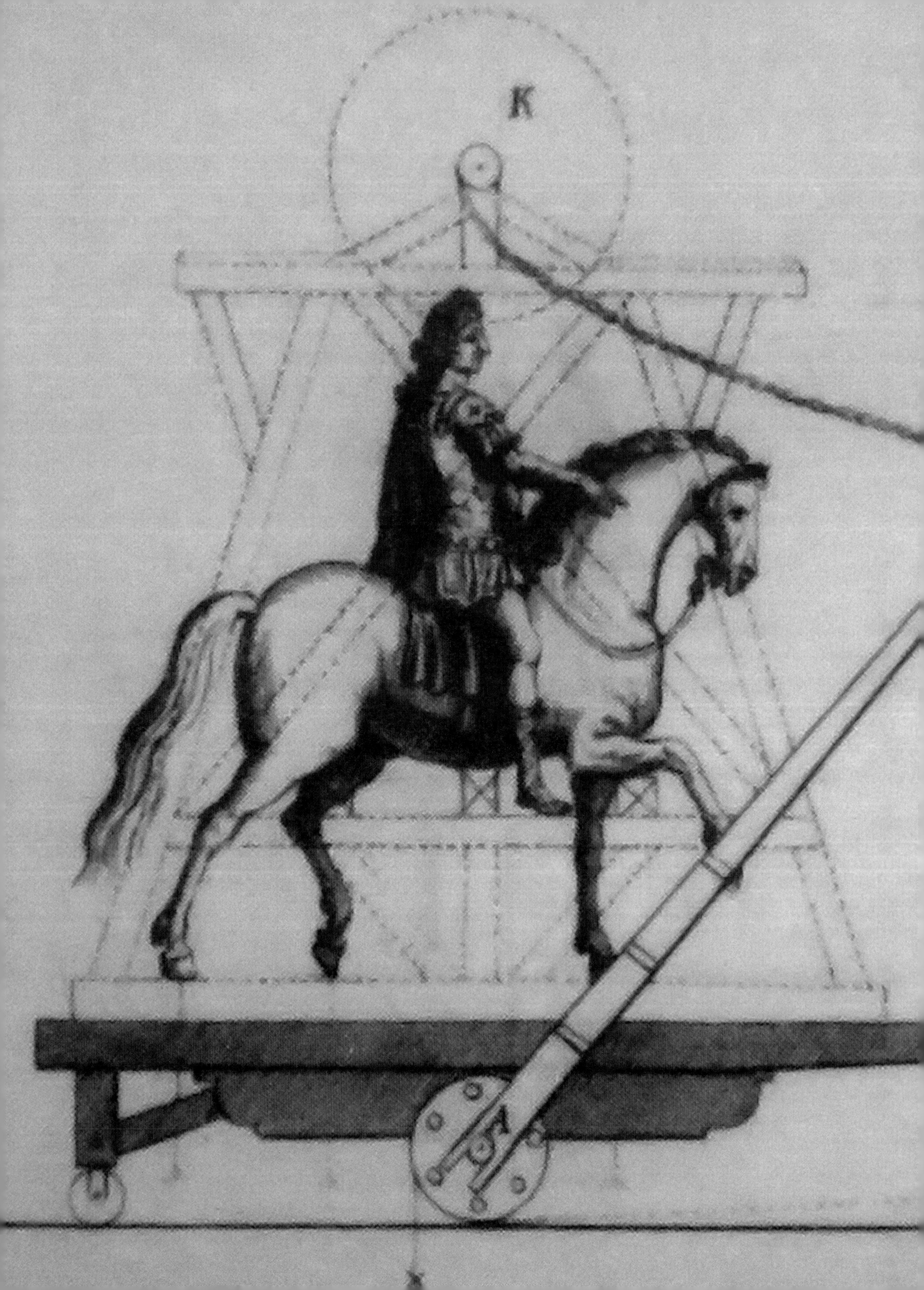
K

Chapter 7
The Monument to Louis XIV at the Place Vendôme (1699) as a Technical Achievement: A Question of Interest

Etienne Jollet

Detail from Franz Ertinger (engraver) and Thomas Gobert (draughtsman), *Moyen proposé pour élever la statue ...*, 1699 (*plate 10*).

Art and Technology in Early Modern Europe *Edited by Richard Taws and Genevieve Warwick*.

In the *Comptes des bâtiments du roi*, where all the sums spent for art and architecture ordered by French royal administration are kept, one notices for 1697 a payment to the painter René-Antoine Houasse for 'two pictures he has painted of the King's equestrian statue, to be printed' and another to Charles Simonneau for 'a plate he is currently engraving representing the King's equestrian statue'; and in 1699, to the same Charles Simonneau, a payment for a plate he engraved representing the same statue.[1] These commissions were hardly surprising at the end of a century during which so many prints of royal monuments had been made, from Henri IV's statue on the Pont-Neuf onwards. But Simonneau's print is peculiar, since it represents not the statue placed on its pedestal, but shows it instead in an intermediary state. Seen after casting but before being erected, and placed on rollers, it highlights the difficult task of moving the statue[2] (*plate 1*). Why this interest in the process of making and installation – an interest we can associate with technology, understood here as a kind of knowledge describing and prescribing a technical practice? Can we establish a relationship between the sitter's identity as the king of France, and technology, as a manifestation of his power, given that the whole process is carried out by the 'king's men', and since the gigantic statue is evidently a technical achievement? Or is this image, on the contrary, a way to create some distance from the idea of the king as omnipotent, by exhibiting his statue as an artefact? In this context, it is unsurprising, then, that Houasse's preparatory painting for the print was used for the cover of Peter Burke's influential book, *The Fabrication of Louis XIV*.[3] Here, instead of reducing these images to mere illustrations of the idea of royal propaganda, I would like to investigate the relationship between power and technology in terms of 'interest'. The great danger of any study of images of power is to implicitly assert that there is a power to images themselves, since such an approach has become a secularized form of magic, imparting to the art historian a legitimacy he or she desperately needs as a scholar specialized in the humanities, in a world which accords little value to his or her field. Hence studying the technologies of power, as we will do here with regard to Houasse's paintings and Simonneau's print, is a way to question the very notion of interest. My examination of this concept and its application centres on the choice of a very particular iconography – the display of the king's statue according to a specific technological register – revealing an interest which is linked to the logic of the sublime, evident here through the valorization of the statue's gigantism and the evocation of its casting. It corresponds to an art history that addresses the logic of attention – how and why a work attracts. Meanwhile, the technological process shown here, which concerns the branch of mechanics known as

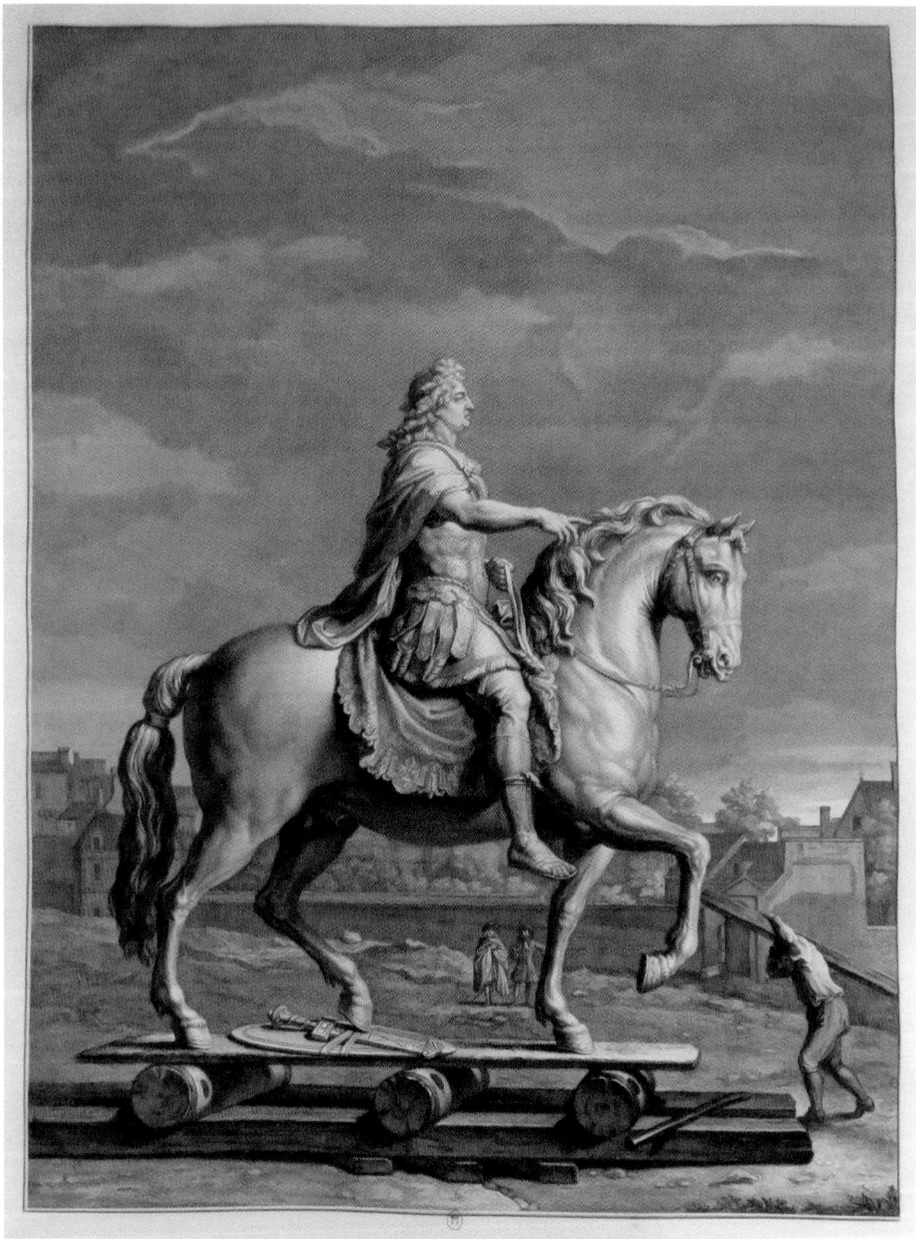

1 Charles Simonneau the Elder, *Equestrian Statue of Louis XIV on Rollers (Right Profile)*, 1699. Etching, 75.6 × 51.3 cm. Paris: Bibliothèque nationale de France, Département des Estampes et de la Photographie. Photo: BnF.

statics, offers a more discrete evocation of the recent progress of science and its interest for the country – interest having here a mainly economical meaning, as that which brings value. But a third meaning is present, when the question of interest is associated with the various actors involved in the process – in whose interest was it to order the pictures and the prints? Does it match any public interest? And in these conditions, is technology (or, here, technologies) a way to enhance the power of the royal image or a means to make it merely acceptable?

The Casting of a Colossus

The statue as represented in Simonneau's print is the result of a very long history, and the print cannot be understood without consciousness of this duration, which was widely understood in 1697 – to paraphrase Marvin Trachtenberg's 'building-in-time', a 'casting-in-time'.[4] Girardon's project seems to have been developed from the autumn of 1685 onwards, when a model of the pedestal was placed in the garden of the Hôtel de Vendôme and when the foundry was settled in the rue Neuve des Petits Champs.[5]

From January 1688 onwards, various operations took place: the construction of a cage for the model, using old metal, in which the casting was supposed to take place; and the placement of straw around it.[6] The foundry itself is first mentioned on 17 October 1688. Up to the final casting of December 1692, a long series of acts testifies to the process.[7] The casting itself took place on 31 December, under Robert de Cotte's supervision.

Simonneau's print represents a technological feat, one that was later reproduced in Germain Boffrand's book of 1743, which included prints showing the various steps of the casting.[8] Before then, the real issue is why such a step would be represented only for technical reasons, alluding to the work's creation. From the opposite perspective, the drawings kept in Robert de Cotte's collection at the Bibliothèque nationale de France belonged primarily to the sphere of reception thanks to their reproduction in print.[9] In fact, some meeting points between the two worlds exist, such as the statuette (more precisely the *réduction*) of the statue of Louis XIV made for Lyon by Martin Desjardins (1713) and kept in the Staatsmuseum in Copenhagen, where the statue appears complete with all the pipes by which the infusion of matter filled up the body.[10] The fact that this object was made and the fact it was kept may be considered as an early testimony of interest in the process itself.

Girardon's statue was certainly colossal, at least two or three times the actual size of the model.[11] In Simonneau's print, the figure at the right of the statue gives a sense of the scale, which is even clearer in the other picture. But in Girardon's case, there is a specific issue: the casting of a colossus at a time when the conception of the 'grand' – that which is large, but also important – was associated with the king, beginning with his common denomination as 'Louis le grand'. The large scale was also indicative of a conception of the sublime that prevailed in France after the publication in 1674 of Boileau's translation of the *Treaty of the sublime*, attributed in this period to Longin.

Reference to the characteristics of the colossus exists in the tradition of the royal statue in France. Here a Latin text, Statius' *Domitian's Horse*, was foundational. Its importance in the French tradition is underlined by the fact that it was translated and adapted to be a homage to the statue of Henri IV at the Pont-Neuf.[12] The first feature underlined by Statius was the *moles*, the mass of the colossus, which weighed so heavily on the ground that the ground resisted with difficulty.[13] This sheer quantity of metal was also stressed by the traveller Martin Lister when evoking Louis XIV's statue at the place Vendôme, 100,000 pounds of which were prepared and 80,000 actually used.[14] The second feature was the fact that the human figure and the horse were cast in one go, 'd'un seul jet'. This was considered a sign of technical prowess, since both figure and horse were cast together, which was never the case before, neither for Henri IV at the Pont-Neuf nor the statue of Louis XIII at the Place royale.[15] In fact, it was a 'tour de force', to borrow the French phrase used by Irving Lavin to describe the fascination during the Renaissance for the sculpture *ex uno lapide*, made from one block.[16]

Material Matters

The statue foregrounds 'solid material', to take François Lemée's definition of such a work's ideal properties, in his *Traité des statues* of 1688.[17] We may here stress the fact that the painter has chosen to create a contact, on the surface of the work, between the beam lifted by the worker and the leg of the statue, developing a contrast between non-elaborated material and one transformed according to an artistic shape. The value placed upon materials is inherent to other aspects of the print, such as the indication of brightness – the fact that on the statue, and above all on the horse, the gleams are translated thanks to highly lit zones. Bronze had been highly favoured at Versailles since the creation of the Parterre d'eau in 1685, and Louvois, Girardon, and Keller were active both there and in Paris, using bronze to facilitate the creation of a new Rome. As far as the statue is concerned, the material itself was considered noble, since it was given by the king himself. Both Houasse's pictures stress the materiality of the statue, especially perhaps the one which has not been engraved: here the glittering of the bronze is intense, thanks to the contrast created by the dark, almost nocturnal, light (*plate* 2 and *plate* 3). And it may work this way because this material was also a matter of some comment – 'the king's praise is an inexhaustible matter', wrote the *Mercure galant*.[18]

But simultaneously, in Houasse's pictures and Simonneau's engraving, the king is represented in a way close to the 'living statues' present notably in Rubens' paintings and used again later by Watteau. It seems to anticipate or evoke a 'Pygmalion effect', the statue becoming alive and descending from its support.[19] In fact, it may be associated with a larger vision, characteristic of sculpture, concerning the interplay between art

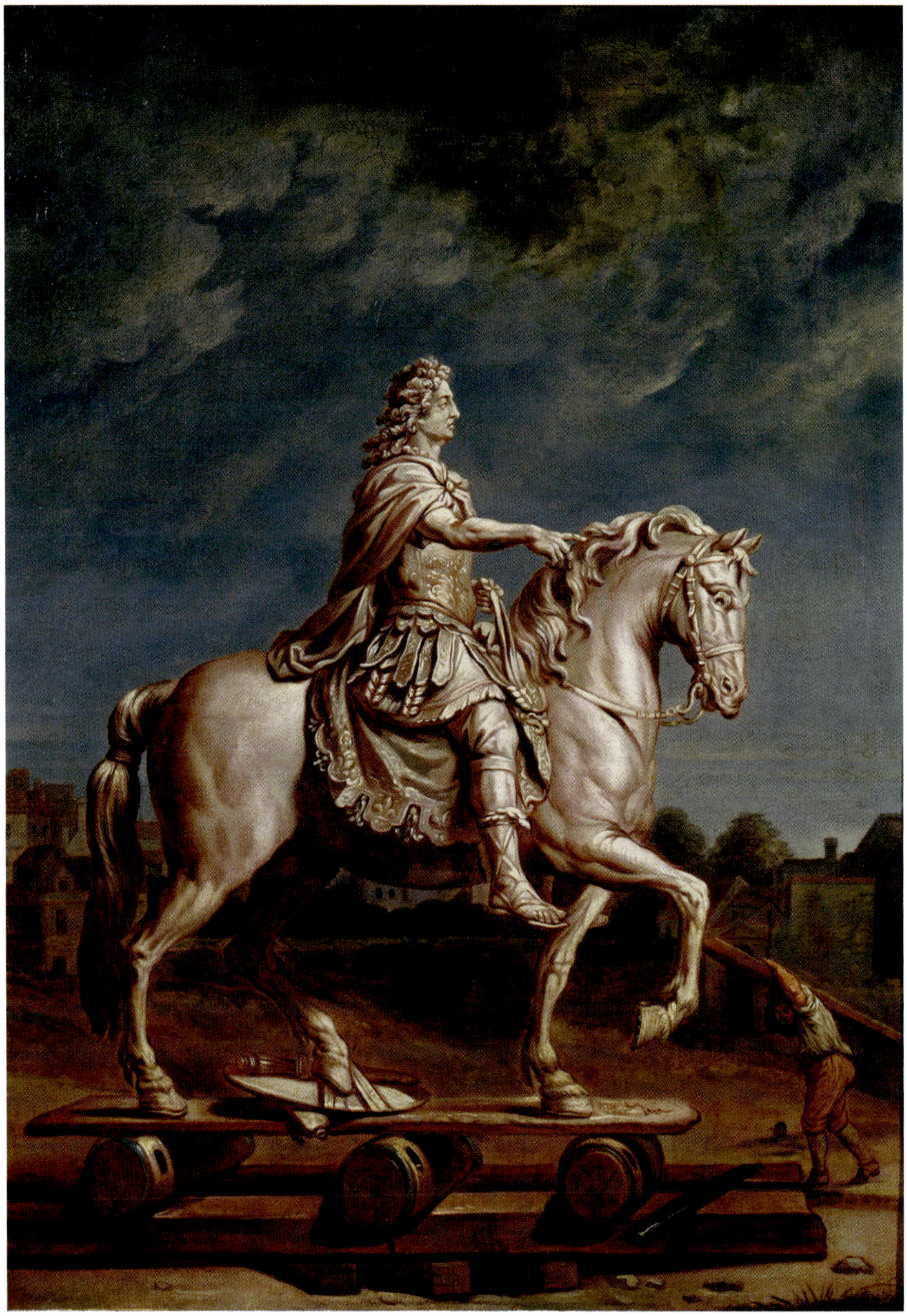

2 After René-Antoine Houasse, *Equestrian Statue of Louis XIV on Rollers (Right Profile)*, 1967. Oil on canvas, 72.5 × 54 cm. Paris: Musée Carnavalet. Photo: Bridgeman.

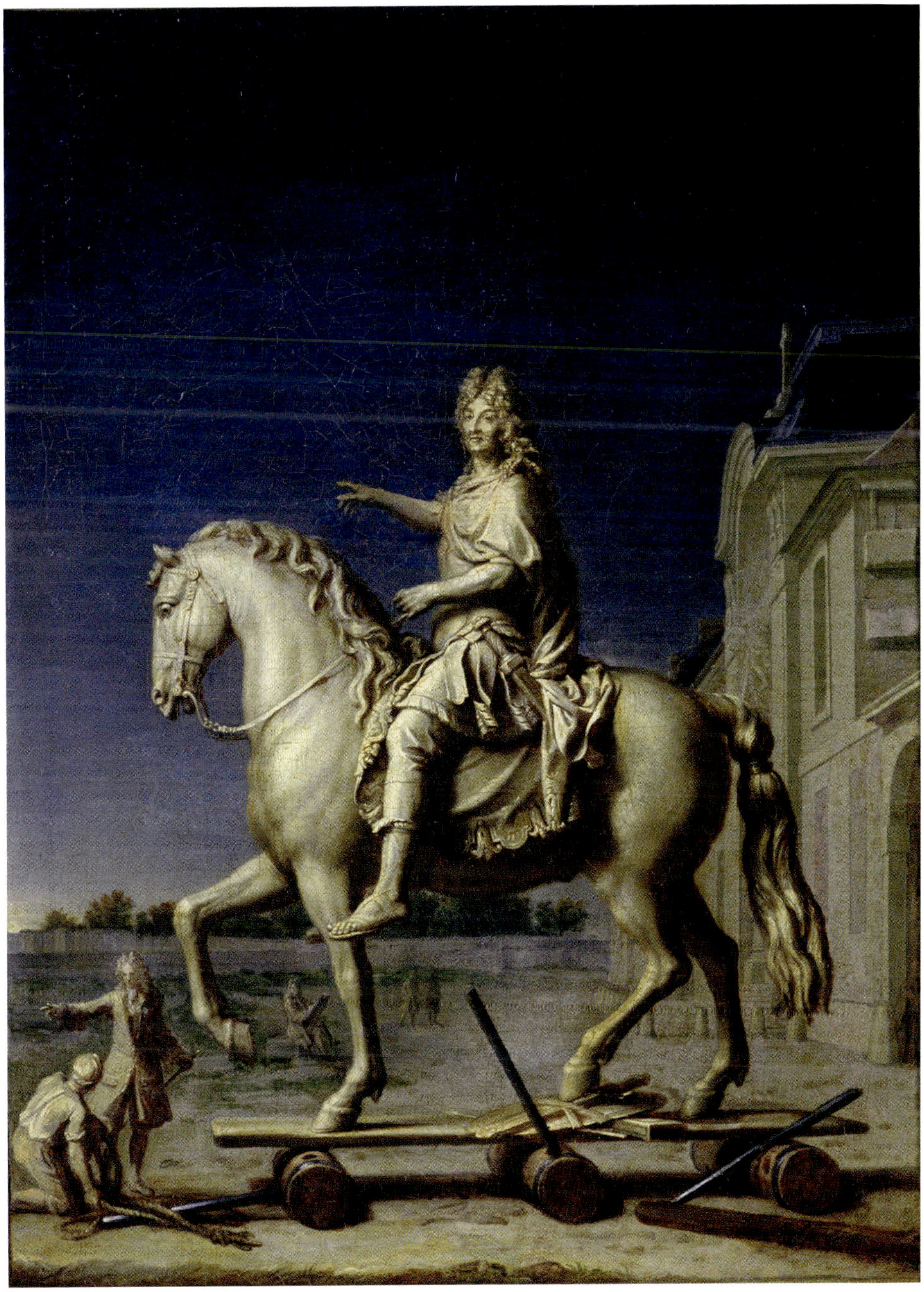

3 René-Antoine Houasse, *Equestrian Statue of Louis XIV on Rollers (Left Profile)*, 1697. Oil on canvas, 74 × 54 cm. Paris: Musée Carnavalet. Photo: Bridgeman.

and nature.[20] As Etienne Binet puts it in 1628, describing a bronze statue, 'the Metal has been entirely thrown from its own nature and has transported itself to an actual representation'.[21] This magical treatment was no longer the rule during the second half of the seventeenth century. In La Fontaine's fable, 'The sculptor and the statue of Jupiter', the eponymous sculptor asked himself whether he would produce a god, a bench, or a basin, thus thematizing and controlling the process of creation.[22] This new approach to materials was discernable in the representations associated with the casting of Louis XIV's statue, and especially in portraits of the caster, where it became a source of particular interest.

Keller's Works, Keller's Portraits

The caster was Balthazar Keller, the 'commissaire général de l'artillerie de France' (general commissioner of the French artillery).[23] A 'marché général', or general contract, was signed on 22 December 1683 for casting all the statues.[24] On 31 July

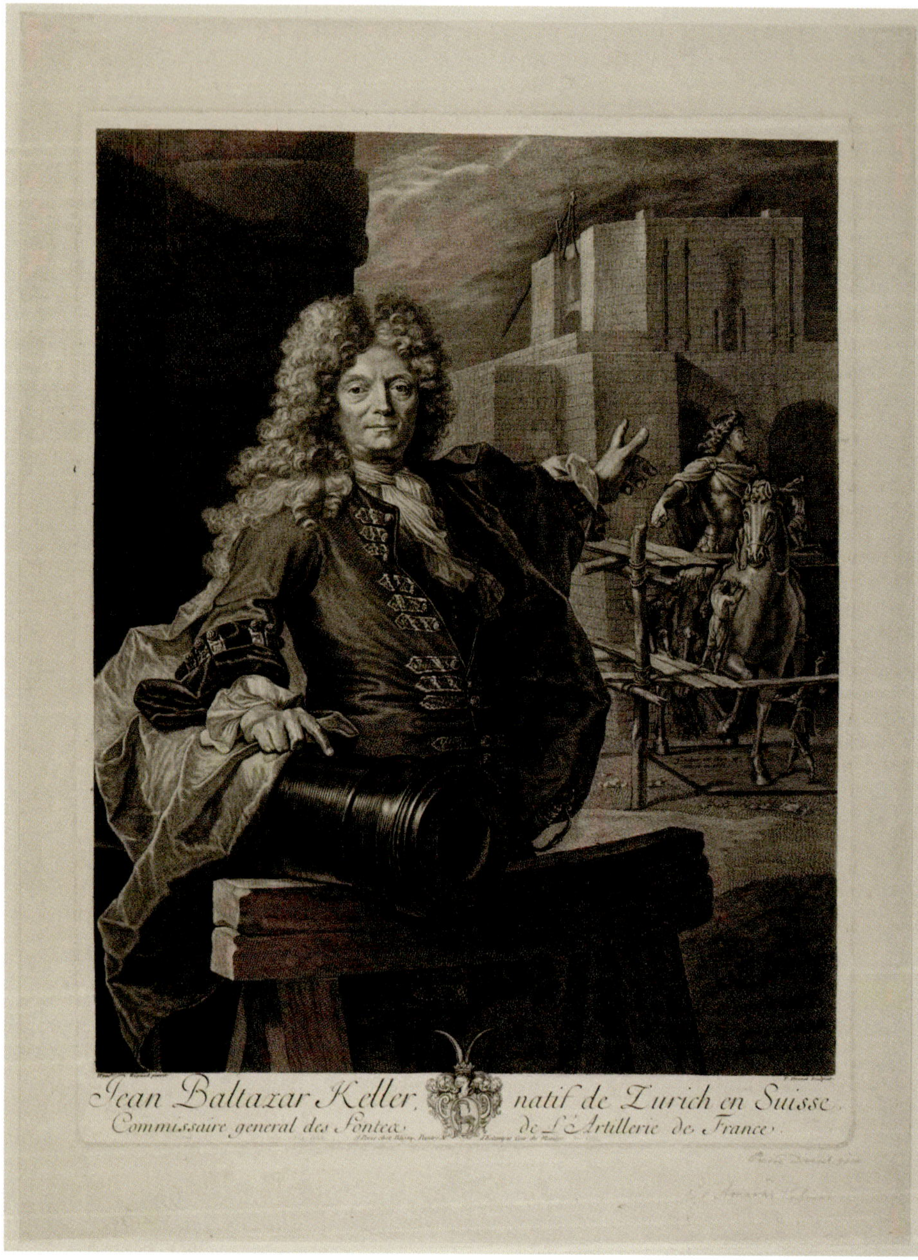

4 Hyacinthe Rigaud, *Balthazar Keller*, 1693. Etching, 44.8 × 35.3 cm. Cambridge, MA: Harvard Art Museums. Photo: Harvard Art Museums.

1692, Keller was asked to intervene.[25] The operation went well, and Keller was proud enough of the statue to ask, in 1693, the most reknowned portrait painter of the time, Hyacinthe Rigaud, to paint him with the statue of the king (plate 4). So far, so understandable. But what is very peculiar is the fact that the statue is shown being 'réparée', in the language of the time; in other words, being polished (to 'réparer' the statue was crucial after the casting, since small pieces of materials made the surface rough). This might have been a way to excuse the impression that the caster conceived the sculpture as his own creation. In fact, the importance and notoriety of the caster was enough to motivate the presence of his name in the legend of the most significant engraving produced for the inauguration of 13 August, by Langlois. But the difference is clear: in Rigaud's portrait of Keller, the statue is still a work in progress. The fact that the statue is in the process of being made is further underlined by the scaffold, while the building at the rear is the foundry, in the gardens of the convent of the Capucins, along the rue Neuve des Petits Champs.[26] This reference to specific works marks a difference from an earlier portrait.[27] In 1685, the forge is shown at the back, without any workers. It is a presentation which draws upon mythology: Vulcan is at work, in front of some sort of volcano. Drevet, when he engraved the portrait at the beginning of the eighteenth century, stressed the idea of process by writing on the base of the statue: 'cast in 1692 and erected in 1699'.[28] Later on, the valuing of process disappears, as in the portrait of Keller by Belle.[29] The situation is the same in Simonneau's print, where what is visually obvious is not the casting, but the potential movement of the statue, demonstrating a change in interest from material to mechanics.

The Mechanics of the King's Statue

Three features stress the fact that the statue, though static, is inscribed in a logic of movement. First, it is noticeable that the print faces in the same direction as the painting. This means that the orientation was important enough for the engraver to respect it, which meant the use of a mirror or other apparatus. Second, the rollers, displayed on the ground, and never represented before, are placed on beams that cut into the left-hand edge of the image, hinting to a possible continuity. Third, the deictic gesture of the figure on the right-hand side at the rear: his left hand, hidden behind the leg of the horse, indicates a direction to the right, so creating an oppositional tension and a sense of movement. Movement means here transportation, which was a crucial issue as far as royal statues were concerned, because of their weight. The year of the casting of the Parisian statue, a monument executed for Dijon became stuck in the mud in the village of La Brosse for twenty-seven years. In Paris, the transportation of the statue was a real challenge too.

Simonneau's print belongs to a tradition of representing such problematic transportation. The best known is transmitted to us thanks to the prints representing Domenico Fontana's stunning technological feat of lifting no less than four obelisks in Rome for pope Sixtus Quintus.[30] Under Louis XIV's reign, such achievements were represented and diffused by print. In Claude Perrault's translation of Vitruvius' *De Architectura*, published in 1673, book ten is illustrated with a view by Sébastien Leclerc of the construction of the east facade of the Louvre, the one featuring the colonnade.[31] Meanwhile, in the right-hand foreground of Adam François Van der Meulen's *La Construction de Versailles*, men lift a big block of stone, using the same lever as in Houasse's other painting.[32] The transportation of heavy stones affected too the relationship of the statue to the ground: a few weeks before the inauguration, the ground of the square had to be made as flat as possible, and resistant to the statue's weight.[33] Here intervenes a knowledge that was applied to a number of important projects – levelling – which had been recently developed in France, notably by Jean Deshayes, in *La théorie et la pratique du nivellement* of 1685. All these practices refer in fact to a great scientific transformation at the end of the seventeenth century: the development of mechanics both as a science and as a technique.

Fontenelle insists, in his eulogy of the physicist Amontons written in 1705, on the difference between two types of mechanics, one for machines and the other 'above', the 'sublime Mechanics, which have settled the universe'.[34] During the 1690s in France, the use of mechanics for machines – more precisely, as concerns our print, static mechanics – was still to be settled, although it was developed significantly during the century by scientists like Hérigone, Sauveur, and above all Roberval, whose work defines the discipline.[35] The Collège de France, unlike the University, valued a concrete approach to mechanics, as presented by Monantheuil in his lectures about mechanics as early as the end of the sixteenth century.

Numerous contemporary cases, such as the erection of the obelisk in Rome by Fontana, the building of the Pont-Neuf, or even the movement of Parisian porters when lifting heavy weights, bring to mind the posture of the worker in Simonneau's print. In fact, the association of this figure with mechanics is a reminder of the contemporary discovery of the notion of 'work', a word which designates any human activity, but also the special physical phenomenon associated with the use of lever and the displacement of heavy objects: the product of force and distance.[36] Here, the presence of the man lifting the beam may be associated with 'the body of the artisan' described by Pamela H. Smith in her examination of the relationships between art and experience in the Scientific Revolution, in which she stresses the role played by bodily labour in the experience of nature, causing a new orientation in the scientific attitude towards the world.[37] In this context, we may refer directly to a figure like

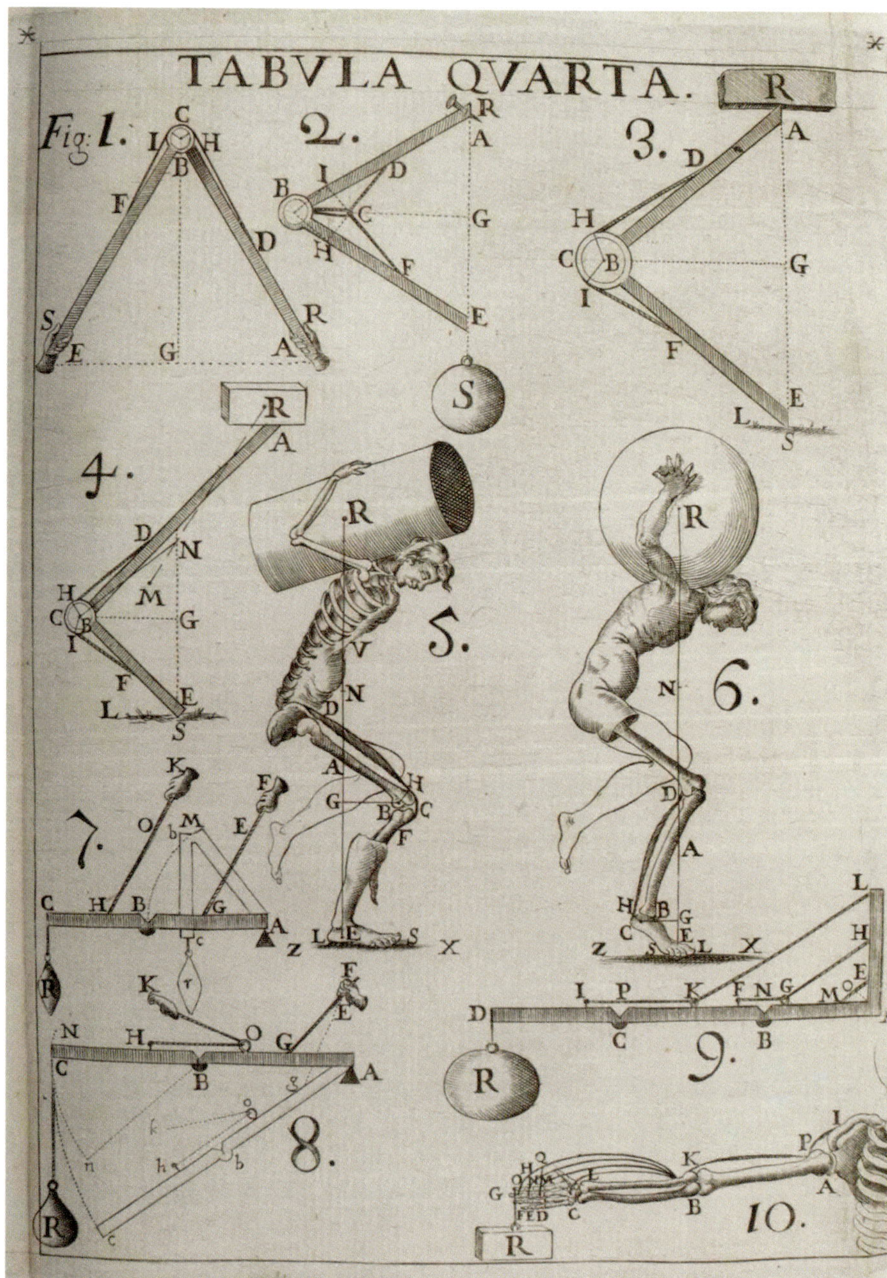

5 Giovanni Alfonso Borelli, 'Tabula Quarta' from *De Motu Animalium*, Rome, 1680. Etching, 21.6 × 16 cm. Paris: Bibliothèque de médecine, Université de Paris V. Photo: Bibliothèque de médecine, Université de Paris V.

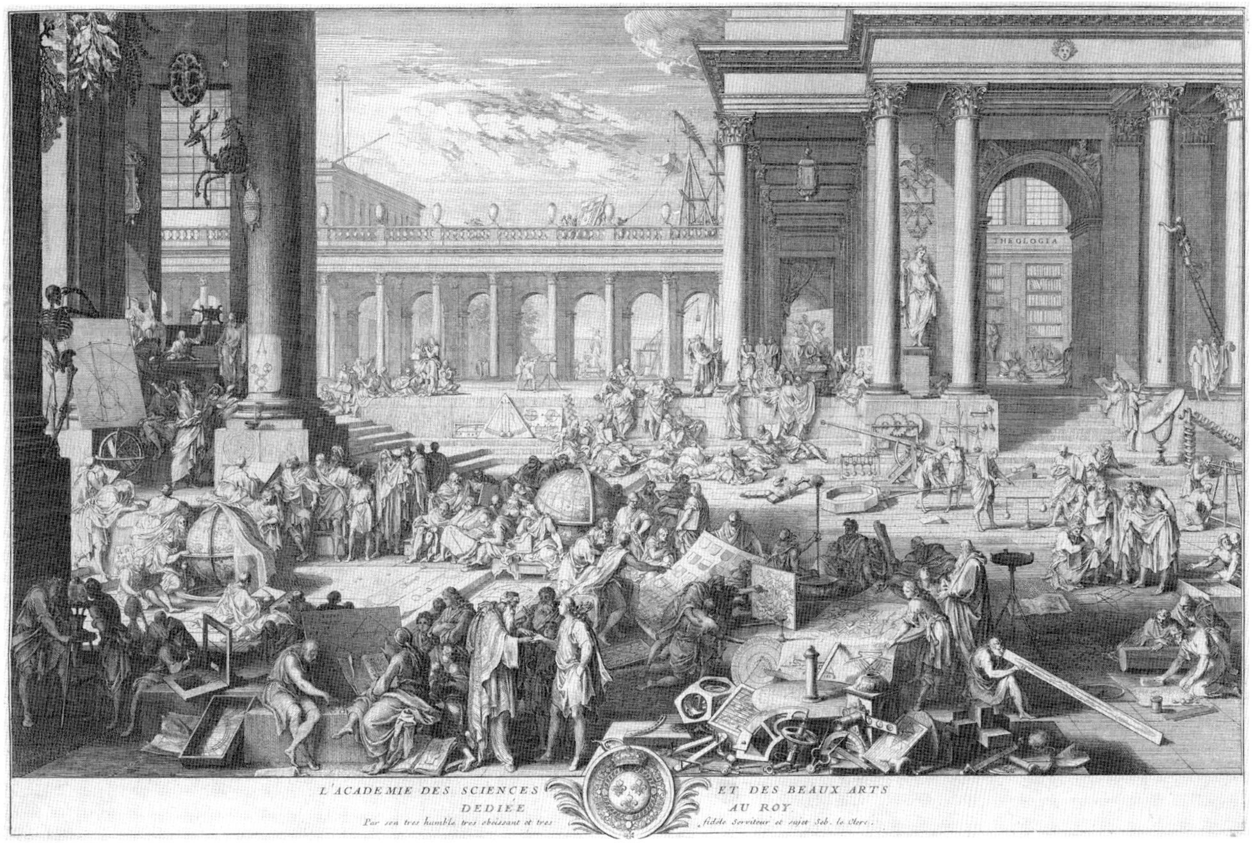

6 Sébastien Leclerc, *Académie des Arts et des Sciences*, 1698. Etching, 34 × 37.6 cm. Paris: Bibliothèque nationale de France, Département des Estampes et de la Photographie. Photo: BnF.

the one from Borelli's influential book, *De Motu Animalium* of 1680 (*plate* 5). But the valuation of useful science also belonged to a global movement in France which can be called neo-Colbertianism, after the death of Louvois in 1690, manifestations of which are for example the enquiry ordered for the duke of Burgundy, heir to the monarchy by the duke of Beauvilliers in 1697,[38] or the reorganization of the Académie des sciences in 1699, which highlighted the wish for usefulness of invention. In fact, no other deed in France was mentioned as frequently as Riquet's Canal du midi, the absolute technical masterpiece.[39] However, technical knowledge was favoured on a broader scale, and to be useful, this knowledge had to be transmitted: hence the role of illustration.

To understand Simonneau's print of the king on rollers in its relationship to contemporary knowlege, we have to consider two major issues. First, the problematics of the 'réduction en art', a stabilization of the description of technical processes thanks to written texts and illustration,[40] and second, what is now called 'visualization'.[41] New techniques were elaborated to make this visualization even more efficient.[42] In addition, important gatherings of knowledge were produced and standardized: the *Description des arts et métiers* began as early as 1693 with the preparation of a first volume about the art of the book (editing, printing and binding).[43] It included plates by Louis Simonneau, Charles's brother. The whole Simonneau family (Charles, Louis, his brother, Philippe) was involved in the illustration of the *Mémoires pour servir à l'Histoire naturelle des plantes*, another academic project under Denis Dodart's supervision. The link between the Académie des sciences and the Académie royale d'architecture was made notably through Philippe de la Hire, a member of the former institution from 1678, and later professor of perspective in the latter.[44] These relations are emphasized in Sébastien Leclerc's well-known print, the *Académie des sciences et des*

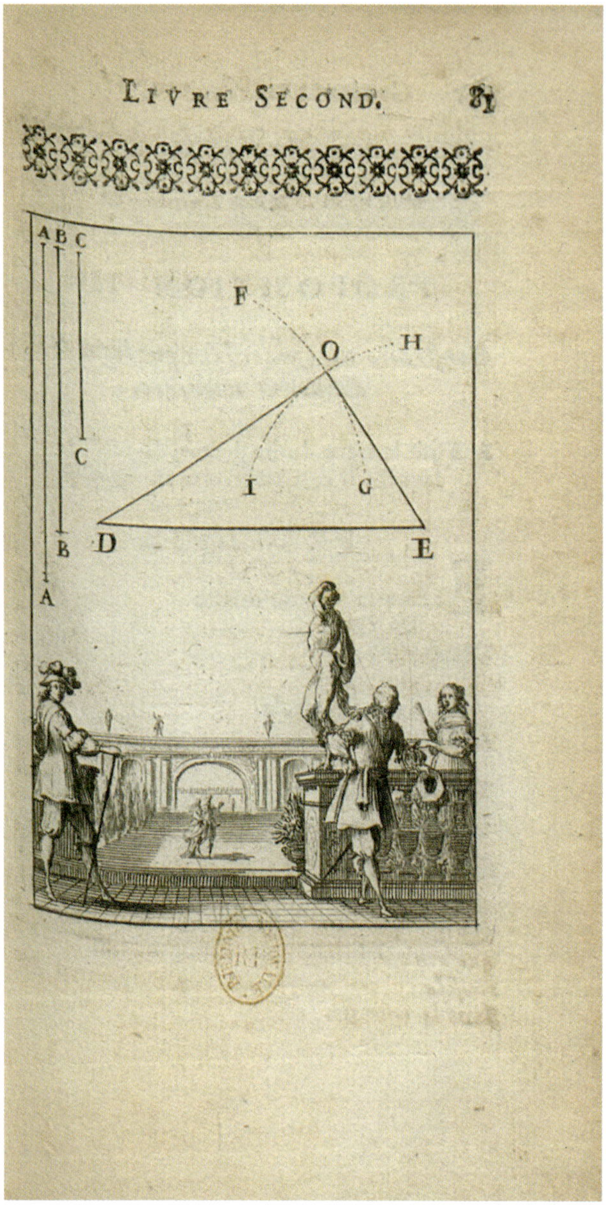

7 Sébastien Leclerc, *Pratique de la géométrie sur le papier et sur le terrain*, Paris, 1682, book 2, page 81. Etching, 12.7 × 18.7 cm. Paris: Bibliothèque nationale de France, Département des Estampes et de la Photographie. Photo: BnF.

beaux arts (1698). In the background of this image a royal equestrian portrait – a painting – is associated with a statue, the marble of which is being worked upon, while a crane lifts a block just above (*plate* 6). One can find the various features present in Simonneau's print: an iconography of the king; the process; and the problem of the material, presented in an analytical way. Here is already present the rivalry between architects and engineers described by Antoine Picon in relation to the eighteenth century.[45] Nonetheless, the problems of representing mechanics at the end of the seventeenth century are complex enough on their own to lend meaning to our print.

For example, one can find there the simultaneous presence of figurative and geometrical representations, a consequence of the mathematization of statics. Hence the strange association, in Sébastien Leclerc's *Pratique de la géométrie sur le papier et sur le terrain* (1682), of lines in the plane and of characters composing a scene at the bottom of the page (*plate* 7).[46] Here, the two dimensions at work in Simonneau's print, the logics of forces and processes on one side, the visible world on the other side, are separated, making their relationship an enigma. But this kind of visualization affects not only mathematics but also physics, as in the stunning prints by Bernard Picart for the English edition of Alberti, with its surprising bodies suspended in the air.[47] All this belongs to the long tradition of the *theatrae mechanicum*, or the representation of machines, which is as old as the print.[48] This tradition was in fact visible in Paris in the years preceding the execution of the print, at an exhibition held by Jean-Baptiste Picot in 1683.[49]

The major issue, at this point, is the confrontation between non-rational and rational approaches – in other words, the question of secularization in the images themselves. At the centre of this is, again, the value placed upon process – that is, in fact, the confrontation between the Renaissance logic of *natura naturans* and the new idea of technological process.[50] The best example of the survival of the older model might be what happened in 1687, in Poitiers, the day after the inauguration of the king's statue. The Jesuits organized a feast, using the technique of the 'enigmatic picture'.[51] On the face of things, this represented Hercules leaving Hell, dragging behind him a chained Cerberus. Yet it represented in fact the statue leaving the quarry, while the furies which appeared to be insulting him were the tools used by the sculptor; Jupiter was the king, Themis the magistrates, and Iris the admiration of the people.[52] A second enigma stressed Amphion, the one who built a city, and Thebes, with his lyre.[53]

What is at stake here is a valuing of process which is still in the register of the marvellous, defined as the accepted intertwining of natural and supernatural. It must be noticed that there were also technical feats that were appreciated just for pleasure, as in the case of the statue for the duke of Richelieu by Gobert, balanced on its two rear legs and moveable by a mere flick of the finger.[54] The Countess of Auneuil published in 1702 *La Tyrannie des Fees Detruite, ou l'Origine de la Machine de Marly*, where the furies, who behaved badly, are used 'as new Danaids', who 'fill without ending monstruous machines'.[55] The furies, 'working themselves to prepare their tortures', placed 'big beams on which they fastened wheels of a prodigious size'. At the end, the good fairy Serpente is addressed in these terms: 'I love the science you have, use it as usual to embellish those places honoured by the presence of the greatest king in the world ... find every day new inventions to please his eyes, join to nature all that the art of enchantment has the most beautiful.'[56] Here, technique ('science') is not only useful; it is valorized by the link created with the marvellous – as indeed at the royal court.

Here we might return, perhaps, to Jean-Marie Apostolidès' thesis concerning the 'roi-machine'.[57] A quotation from Félibien, as the epigraph to the second part of the book, stressed the seduction created by that which exceeds the strength of a normal person: the king is an astonishment, before becoming, in the second part of his reign, a mere mechanism.[58] The king is first active (the gesture of the hand), then passive (as an artefact); the king is first a figure, then becomes a structure. This dichotomy stresses the fact that what is at stake in such a representation of the monarch is the way in which he is at the same time the effect of the representation and the cause of it, since the people who ordered the work at the Direction générale des bâtiments did so on the king's behalf. But we are in a position of some distance from this mechanical model of power. Simonneau's print can certainly be considered as testimony to the pertinence of Apostolidès' model – the king's statue is presented as if a mechanical model. Yet on the other hand the fact that the process is shown leaves open the question of responsibility for its making, which may indeed be attributed to the king, the power of whom was beyond all recognition.

The question, concerning our print, is whether we are really entitled to speak of a 'machine', since for us a machine is used in a process, and here the king's statue in fact appears between two processes. At the beginning of the century, Dumoulin denominated Henri IV's statue as a 'machine royale',[59] a royal machine. Before the term was used to define a complex and big picture ('une grande machine', as people called it during the eighteenth century), Dumoulin stressed the functioning of things. Apostolidès' hypothesis is based, in fact, on a history of ideas, and behind all of this lies Descartes' 'animaux-machines'. A regime of necessity is present in these images and debates, corresponding to the legitimacy of power: the normal order, the natural order. But in fact Houasse's pictures and Simonneau's print are part of a story about various personal interests, from the people who ordered them to the people who were expected to admire them.

Beyond the Interest in Technology

Decades of work about Louis XIV have allowed us to take some issue with the idea that something like propaganda exists. There is a propagation of images, certainly, but nothing like a consistent discourse about the way that people should behave. The 'fabrication of Louis XIV' is in fact the accumulation of various interests, which are not, in all certainty, centred on the king's effigy. In Simonneau's print, two figures at the rear look at the statue. They are our equivalents, as spectators, on the other side, so their behaviour concerns ours, and even more, may contribute to a feeling of belonging to a community, particularly since there are two of them. The statue interests them, in the sense that their attention is focused on it, or him – that is, the king. This is the main issue. What is actually to be seen here, apart from the technical aspects stressed earlier? The beauty of the statue? In Houasse's second painting, an artist is represented, drawing the sculpture. Is this testimony of a desire to stress the pertinence of an aesthetic approach to the sculpture, or to the king himself? But even this question of the 'king himself' is not clear, since we must take into account the 'king's two bodies' Ernst Kantorowicz taught us to distinguish: the person, Louis-Dieudonné, and the current figure of the French monarchy. It might then be worth to speak not of an ideological operation as a whole, but to study precisely whose interests are simultaneously present in this image. We may have to see the monument instead as an opportunity, used for diverse personal interests, and to speak about 'interessment' in the sense in which it is used in banking, as a form of appropriation.

Charles Simonneau's print is the result of a series of decisions and choices. By whom? As very often during the Ancien Régime, it is impossible to give a name here. Nevertheless, it can, for the purposes of this discussion, be the 'surintendant des bâtiments' himself, Colbert de Villacerf. This representation of the king concerns him on both levels, as the person responsible for the various items made for the king, but more particularly, for the king's representation, for establishing the king's glory. For this task, a specialist known for his equestrian portrait of Louis XIV was chosen, for Houasse's work of 1679 was celebrated.[60] But this does not explain the interest in the creative process. Can we associate it with Houasse himself, or identify an interest in process manifested before in Houasse's career, in the same official context? We might refer here to his *Minerva teaching the art of sculpture to the people of Rhodes* (plate 8).[61] The work, the iconography of which is very rare, belongs to a series of works engaging with the figure of Minerva ordered for the Billiard Room at the Grand Trianon. The theme refers to Pindarus' seventh *Olympians*, maybe by way of Blaise de Vigenère's version of it in his *Images, ou tableaux de platte peinture de Philostrate Lemnien* of 1597: 'The green-eyed goddess gave them also from her part to surpass in skill all the inhabitants of the earth: in such a way that their alleys and crossroads produced already works that ressembled living animals. Which was something very pleasant to see.'[62] The insistence on process is particularly important in Houasse'work, since the goddess

8 René-Antoine Houasse, *Minerva Teaching the Art of Sculpture to the People of Rhodes*, 1689–93. Oil on canvas, 122 × 207 cm. Versailles: Musée national du château et du Trianon. Photo: RMN.

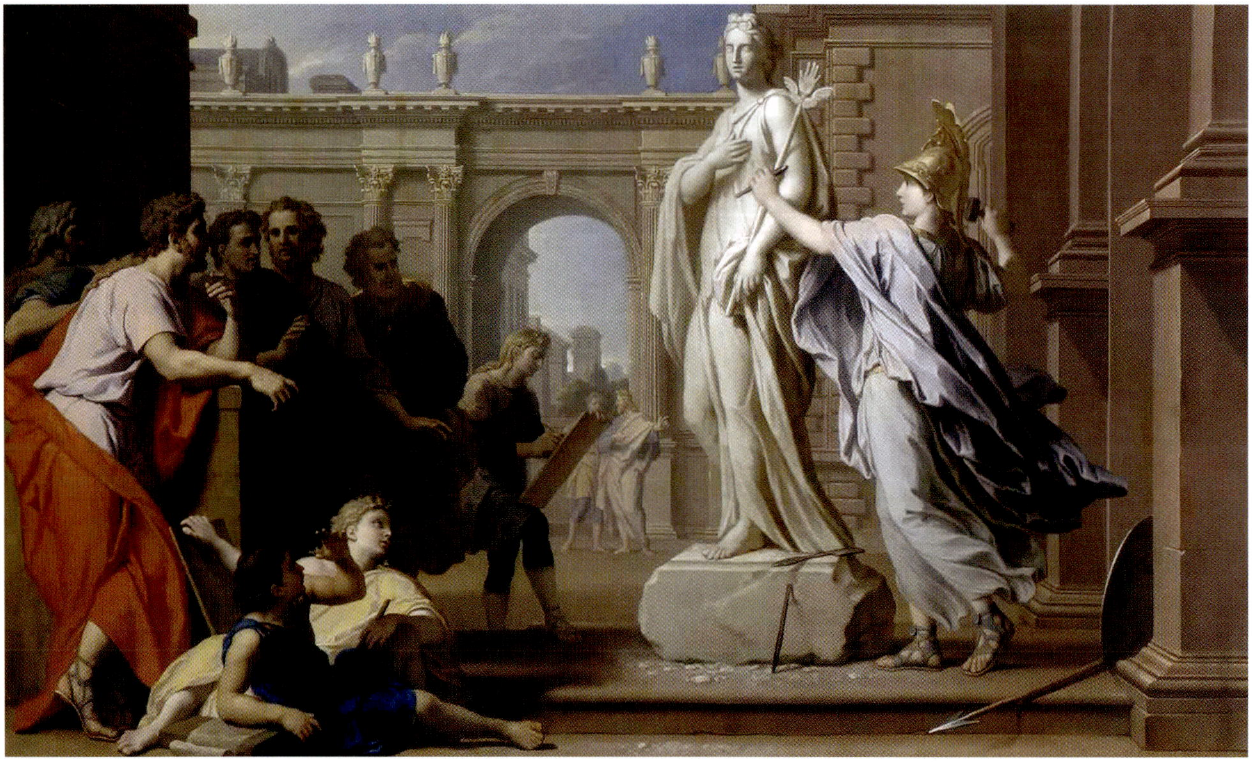

is represented working, which is not mentioned at all in Pindarus', Philostratus' or Vigenère's texts; represented are the consequences of the act of creation, the transformation of the material, via the visible bits of marble on the ground. What makes Houasse's insistence on valuing process even more obvious is that, as Matthieu Lett has proved, the sculpted figure represents Invention, according to the type defined by Cesare Ripa and translated by Baudouin. The artist stresses in fact the counterpoint of the technological dimension. The form he creates – the one drawn by the young man sitting at the rear in the other painting at the Musée Carnavalet – is very close to the one present in the Minerva painting.

Charles Simonneau the Elder, the author of the print, had already worked for the royal administration. Prints were part of a complex system of diffusion of the richness of art in France but also, directly, of the image of the king. Simonneau's print of the king on rollers appears often in a volume offered as a gift (mainly for diplomatic purposes), where it appeared alongside representations of famous ideological creations, such as the 'escalier des Ambassadeurs', but also of celebrated actions such as the second conquest of Franche-Comté, also by Simonneau.[63] In this context, Simonneau's king on rollers appears at the limit between the two logics, showing the result and the process; a renewed version of the old distinction between *natura naturata* and *natura naturans*, brought together for the king's glory. In fact, Simonneau's print and Houasse's pictures are not the only examples of the association of power and process. In the frontispice of the *Médailles sur les principaux événements du règne de Louis le Grand* (1702), published by the Académie des inscriptions, Simonneau has represented, after Antoine Coypel, a coining press at the rear (*plate* 9). This frontispiece is structured by the same logic of presenting a step of the technical execution, showing interest in the process of creation itself, this time in the workplace. The tradition of visits to the sculptor's studio as one of the pleasures of people of quality is well known.[64] Can we speak of a 'poetics of the workplace',[65] outside, in public space? Louis XIV himself paid visits to his statues in progress, visiting this one on 30 January and 10 May 1687.[66]

So much for the process, then. But what is seen in Simonneau's print is not the actual statue. The head of the king is shown in perfect profile and not, as it was in reality and in Houasse's other painting, with the head bent to the left. Consequently, fidelity to the model is not really the issue. Rather, the overall visual impact seems to be more important. The profile places the king apart from the actual situation, an effect that is accentuated further in the print, thanks to the clouds which have been added by Simonneau. These effectively resituate the king in the sky, in a dramatic atmosphere, among various layers and types of clouds, on the top of which the king's figure appears gesturing as a ruler and thus ordering even nature. The clouds make a sort of a 'z' from the head of the horse to the upper-left angle, indicating a movement towards the top of the work, while the intermediate part functions as a canopy above the king, the face of whom is lit in a very intense manner. In the paintings, Houasse had further stressed the strange nocturnal atmosphere of the scene, with the use of a deep blue for the sky. Is there any relationship between this image and a nocturnal viewing of the royal monument scheduled by le maréchal de la Feuillade for the monument of the place des Victoires? In 1697–99, the statue was lit thanks to lanterns placed at the corners of the square, a presentation that took place in a context when light at night in major cities remained a major challenge in France, even following La Reynie's creation of public lighting in Paris from 1667.[67] Even in the print, the contrast between the two worlds is extreme, since there is not, as in the other painting, a high building – and especially a religious one – to constitute a link between them.

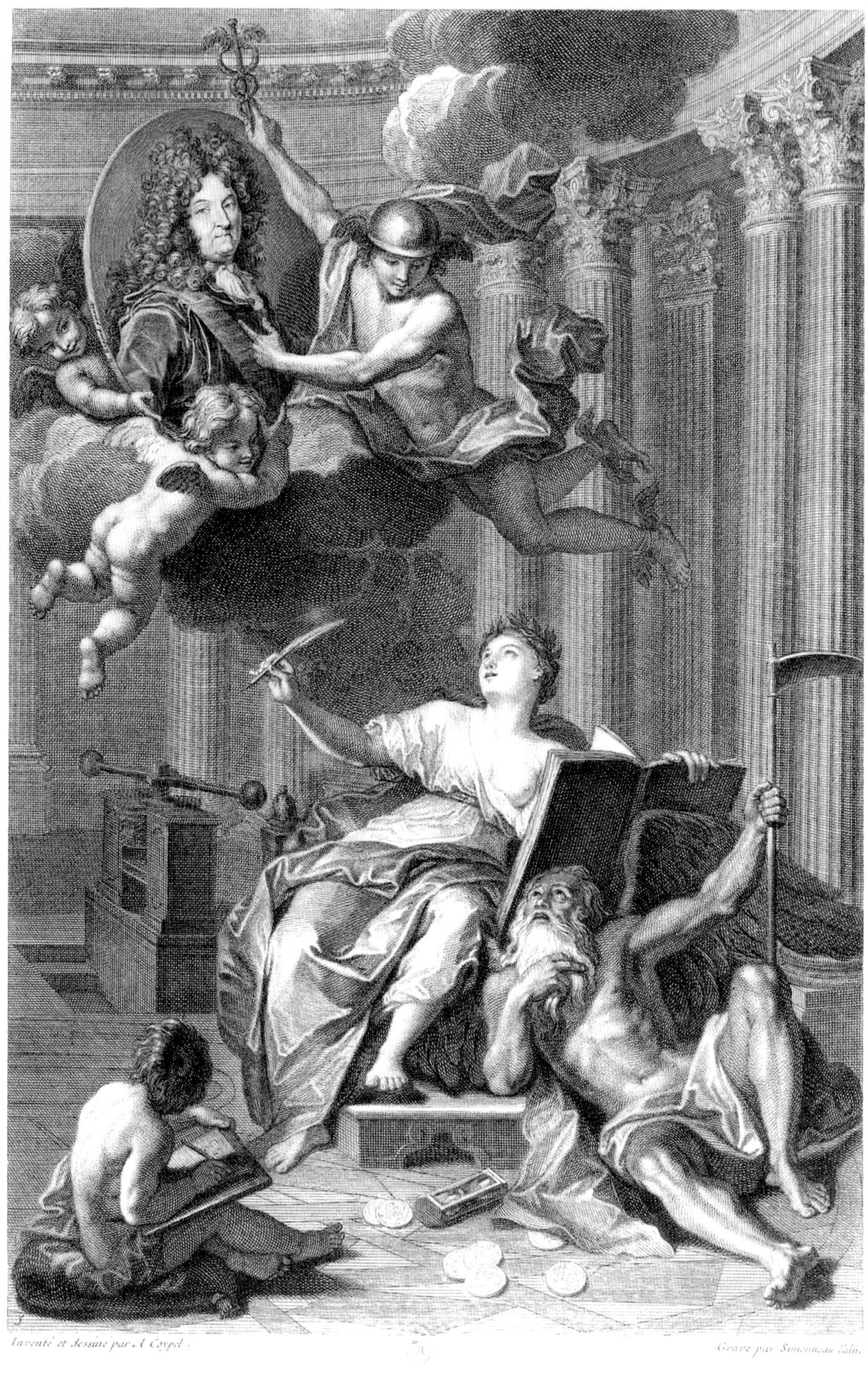

9 Charles Simonneau after Antoine Coypel, *Médailles sur les principaux événements du règne de Louis le Grand*, Paris, 1702, frontispiece. Etching, 44.5 × 28.5 cm. Paris: Bibliothèque nationale de France, Département des Estampes et de la Photographie. Photo: BnF.

So, where then is Louis XIV, in terms of his relationship to reality – a reality which is here underlined thanks to the technological apparatus displayed? Is he in the sky or on earth? Is he real or not? Such questions were in fact asked about this statue in September 1699 by a protestant opponent, Gueudeville. In his *Esprit des cours*, Gueudeville recounted Philip II of Spain answering to a 'skilled Milanese worker' who proposed to place his statue on the doors of all the Duchy of Milan that he didn't want a statue on earth, but that he wanted instead somebody who could place one in the sky. Gueudeville added that he was persuaded that the king thought there was something 'solid' in this answer; that if a statue in the sky was indeed impossible, what Philip wanted was to place himself there.[68] So what we see here is a spectacular figure, evidently of sufficient interest to be diffused thanks to the print; but spectacularized to the point of modifying its appearance, fixing a transient image in eternity, above the present place and the present situation.

The print, an initiative of the Direction générale des bâtiments, participated too in the valorization of process. The casting was announced by the semi-official *Gazette*.[69] Later on, close to the inauguration, other printed periodicals also mention the extent of the ceremony's preparation.[70] The readers of the *Histoire journalière* of 28 July 1699, on the date of 20 July, learn that 'last Thursday' (i.e. 16 July), 'they began to move the statue', which had already been 'lifted from the ground in some sort of cage or frame built around it, to be dragged on rollers up to the pedestal',[71] the publication stressing the fact that 'several persons of quality were there to see the beginning of this enterprise'.[72] On 30 July, the same gazette writes that 'the equestrian statue of the King attracts a lot of spectators, to see the way it is guided for the last four days'.[73] But, we may ask again, where does the interest reside for either 'persons of quality' or normal people: in the king, the statue, or the moving process itself?

In fact, the crowd's interest failed to prevent the monument from being mocked.[74] The question of the reception of the royal monuments in France is still very difficult to apprehend, since it is hard to establish a link between the pamphlets that exploited any type of situation to criticize the monarch, and the opinions of the majority of people. We have to rely on testimonies like those produced by diplomats, such as a very important passage of a letter by Daniel Cronstrom, who wrote in 1698 from Paris to Nicodemus Tessin in Sweden that 'between us, since the great degeneration in which His Majesty has found himself for several years, the spirit of praise and one can say of adulation that reigned in this kingdom has not only diminished but has changed entirely from white to black … People would like to be able to raze all the monuments erected and efface the tributes written during the life of the king.'[75]

Even worse, the king himself seemed to want to get rid of them too. A letter from Madame de Maintenon to Hardouin-Mansart of 28 July 1698 indicates that the king said that Louvois 'has it made almost in spite of himself; that all these ministers

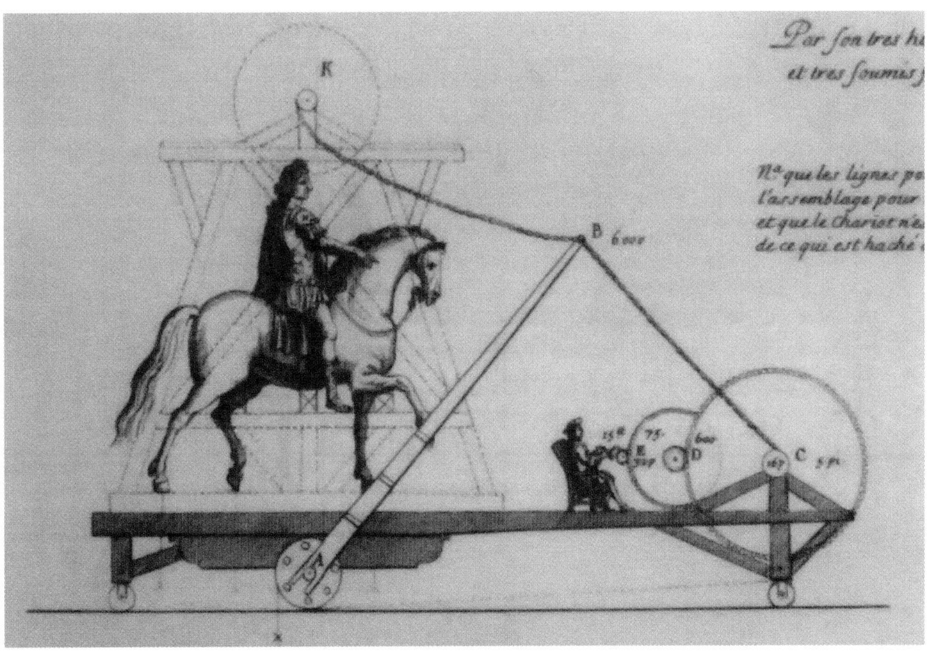

10 Franz Ertinger (engraver) and Thomas Gobert (draughtsman), *Moyen proposé pour élever la statue …*, 1699. Etching, dimensions unknown. Paris: Musée Carnavalet. Photo: Musée Carnavalet.

wanted to do something which would be an honour for them in the future; that they succeeded in presenting him to the public as somebody who likes these kinds of vanities; that I was a witness of the grief M. de Louvois and M. de la Feuillade had caused him on this point.'[76] The historical context in 1697 was very different from that of 1686, when the monument erected by La Feuillade created a major diplomatic crisis in Europe.[77] The Peace of Ryswick had just been signed, and Pontchartrain, in a letter from May 1699 to the Académie des Inscriptions, insisted on the fact that the pedestal must be simple: 'nothing, in a word, that resembles the reliefs, slaves, and inscriptions of the statue of the Place des Victoires'.[78] The simplicity of the figure of the king, first in its form as an artefact, second as an object deprived of any ornament, is what is offered in Simonneau's print.

What we can call an instrumentalization of the monument in favour of displaying technology is illustrated in a spectacular way by Gobert's project for the erection of the statue in 1699 (*plate 10*). According to this image, one man, sitting in a comfortable armchair, can achieve the task by turning a wheel in front of him. The print itself is a declaration of love for the king, but is infected throughout by the application of physics, based on 'the proportion of the numbers of time and weights'. An analogy is made with the king's governance: '... it is by this just balance that he regulates with much wisdom all the movements in the distribution of the

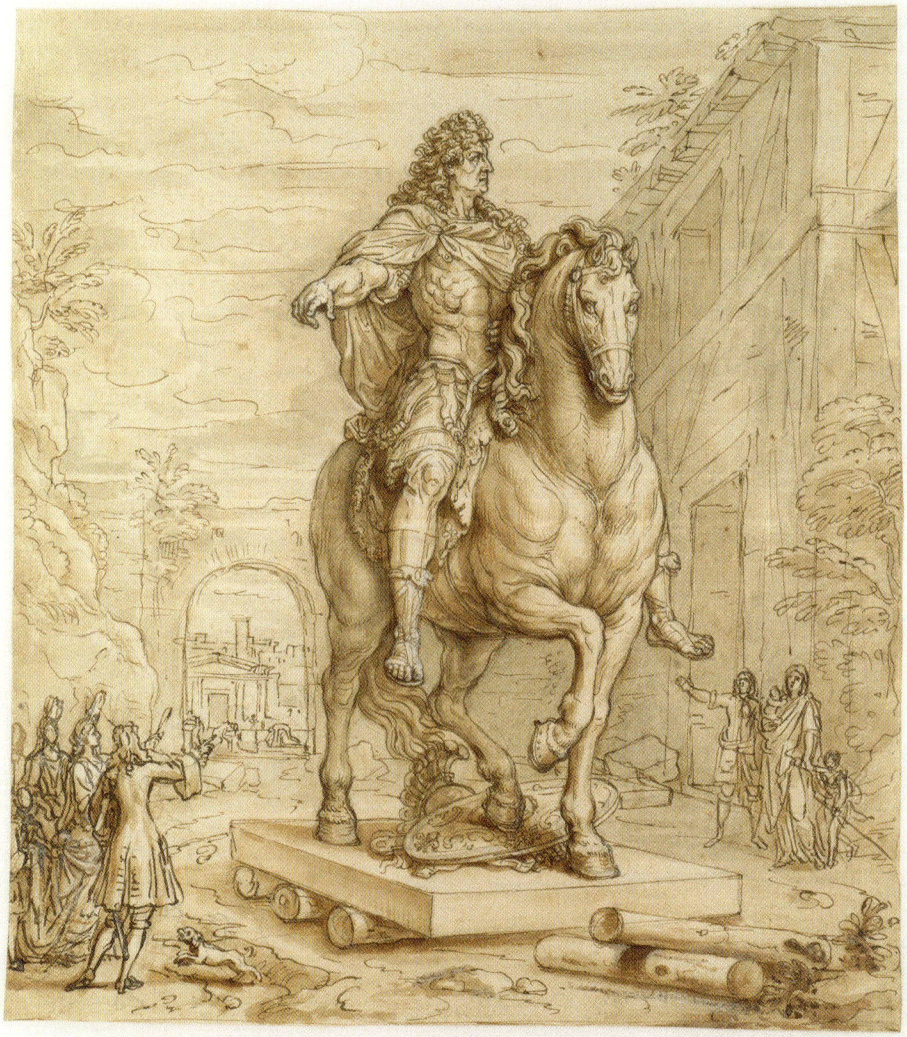

11 Unknown artist, *Equestrian Statue of Louis XIV by Girardon*, c. 1699. Pen, brown and black ink and wash, 31 × 26.6 cm. Weimar: Goethe-Nationalmuseum. Photo: Goethe-Nationalmuseum.

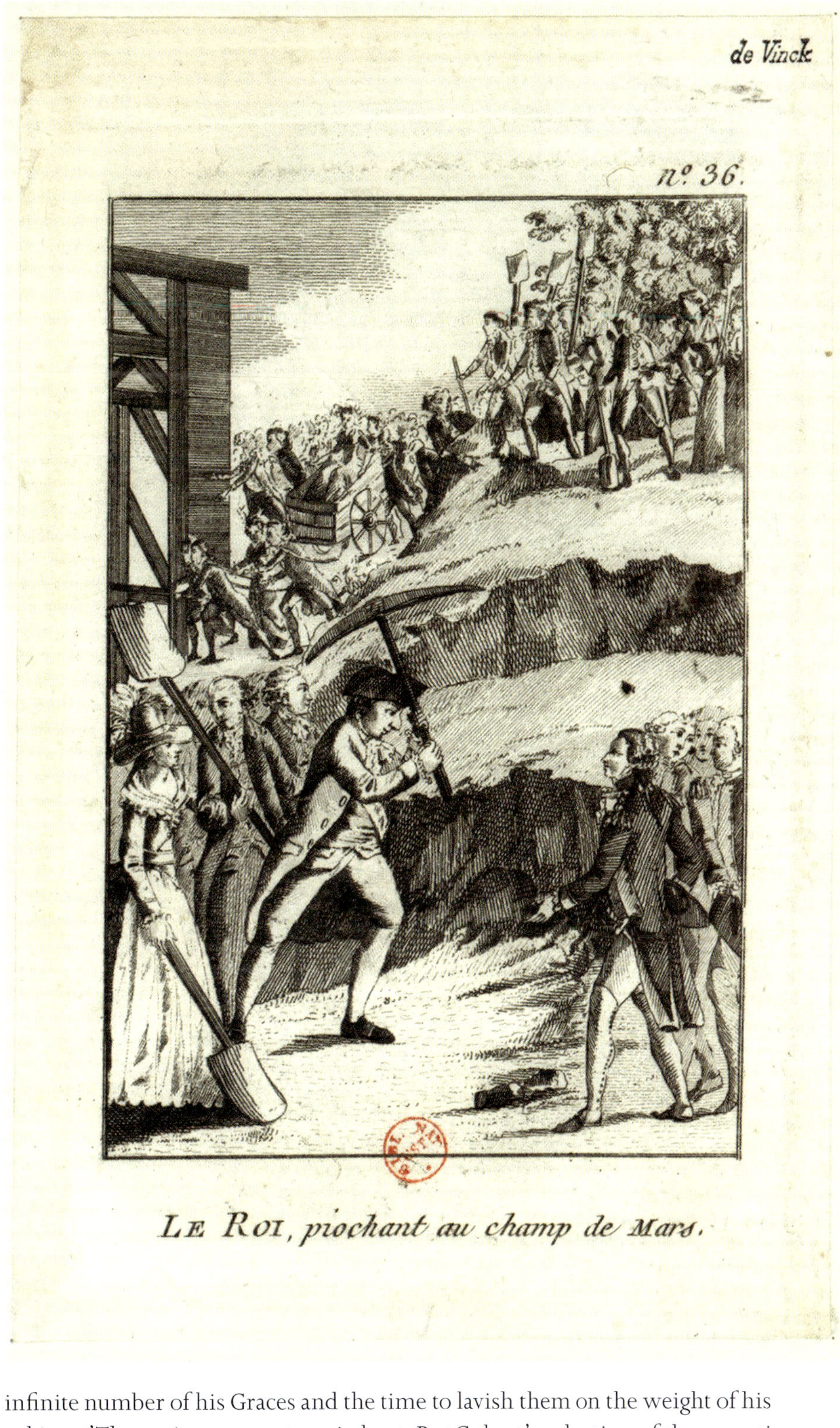

12 Unknown artist, *Le roi piochant au Champ de Mars*, 1790. Etching, 12.5 × 8.5 cm. Paris: Bibliothèque nationale de France, Département des Estampes et de la Photographie. Photo: BnF.

infinite number of his Graces and the time to lavish them on the weight of his subjects.' The project was not carried out. But Gobert's valuation of the operation itself was problematic for Gueudeville, who referred with heavy irony to the fact that the monument 'must have been placed on its pedestal thanks to the most beautiful secrets of Mechanics', accentuating in this way the material dimension of the tribute to the king.[79]

Even with the presence of a technological apparatus, or maybe because of it, Girardon's statue of Louis XIV might be considered among the 'caprices', or deliberately imaginary settings, of the period. It appears again in a series of five drawings kept in the Goethe-Nationalmuseum in Weimar, the author of which is unknown.[80] These drawings offer a diversity of angles of view on the statue and its presentation: two of them, numbers 760–4 (*plate 11*) and 760–5, show the statue on rollers. The statue is set in different urban landscapes characterized by an atmosphere of absence of decorum. In drawing number 760–5 the houses seem to refer to Italian urban landscapes, with a half-timbered house and, on the left, some open ground, as in a semi-deserted city such as was Rome in this period. At the rear, an arch may allude to the one that was to appear at the rear of the 'place de nos conquêtes', but appears here as something irregular and old, offering a perspective on a Roman-type facade of a temple. Nonetheless, the costumes of the people on both sides of the image are French, and it is easy to identify the 'Fontanges' headdresses, characteristic of the end of the century. In drawing number 760–4 the contrast concerning the localization is exacerbated by the opposition between a man dressed in an oriental style and, on the right at the rear, a classical temple, a common trick that evoked a play between the image of the king and the scale of the whole world (in the form of the *étranger*, the foreigner as the witness of the diffusion of his celebrity). In this way, the statue appears in different places, with or without rollers: in the three other drawings, the base is directly on the ground, as if it makes no difference, just as it appeared in Tardieu's engraving of Le Sueur's drawing, which adorned Boffrand's publication of 1743. This presentation might be seen as a way to value the statue as a statue, and not as part of the monument. It may in turn be associated with the presentation of statues in François Perrier's *Segmenta* of 1638. This very important collection, the first collection of prints representing the masterpieces of antique sculptures published in France, shows the statues extracted from their actual context and placed in a landscape – for example, the *Wrestlers* appear on the grass, with a triumphal arch and the Colosseum behind them. In this way, Louis XIV's statue could be included fully in the long tradition of sculpture; a marvel, but also doomed to decay. This sense of the fragility of everything, including the technological knowledge necessary to achieve such a statue, was used by Boffrand to legitimate his own enterprise. Focusing on the technical process, the prints were used to 'give a great light to the discourse', but also because 'they make present in every place the works of the most remote Nations, they perpetuate them in the future: without them they would fall again into the nothingness from which the great Artists have pulled them out.'[81] Here the communication is from one artist to another, as Boffrand says he gave his memoirs and drawings to Lemoyne, in the name of the tradition of art, for posterity.[82] This monument of technical knowledge was, according to Boffrand, intended for the public, as befitted the actions of a 'good Citizen'.[83] Was this as disinterested as it appears? Maybe: the implication is that although Louis XIV is dead, the monarchy and the casting of giant statues must go on. And the print, which circulates in both space and time, is a monument of a monument – or of the process of making and installing the monument.[84]

As he stands, on rollers, in Houasse's pictures and Simonneau's engraving, the king may create new interest among his people but also among prominent foreign rulers by showing that such a magnificent statue was made for him, according to a long and complicated process. Demonstrating an unheard-of casting method and a feat of statics, it exposed the techniques central to the greatness of his reign. But this presentation was also a way to show the statue as an artefact, thus preventing the future monument from being criticized for creating an idol. In this context, the reasons for the valorization of the idea of process, which includes sculpture and even painting, is clear enough. Whatever the value, the king stands above everything: the workplace, the houses, the workers, the viewers below. He appears in the sky, where his profile appears simultaneously eternal and vivid. This was in accord with expectations for his behaviour regarding the politics of monuments: ignoring them, accepting them, or giving them to the cities, just as the place Vendôme was given to the city of Paris just before the inauguration (in actuality it was not much of a gift, given the expense of the erection and accompanying ceremonies). But the 'prevost et échevins' and the people were supposed to be as interested – in fact more than interested, grateful – as the officials who received, in the years to come, Simonneau's print as a diplomatic present. Might we consider this propaganda stressing technical achievement, or technical achievement itself being praised on its own terms? The answer to this question was never actually given through words, for we will never know the real level of interest. All we can do is note the continuous representation, throughout the eighteenth century, of the technical execution of the royal monument. Regarding, first, the casting: on 6 May 1758, first-rank figures of the Versailles court were present when the oven was opened after the casting of Louis XV's statue. The artist, Bouchardon, and the caster Gor were greeted, and the scene was engraved in Mariette's book about the statue.[85] For the erection on the pedestal: in 1713, the installation of the statue in Lyon was engraved, as was later that of the statue of Louis XV by Bouchardon in Patte's *Monuments*.[86] We might consider too the feat that took place in Russia when the engineer Carburi de Ceffalonie succeeded in moving a huge rock to be used as a pedestal for Falconet's statue of Peter the Great. We may even discern the vestiges of these concerns lasting into the French Revolution, most notably the construction by the people of the huge basin on the Champ de Mars for the Fête de la Fédération on 14 July 1790, during what was called the 'Journée des brouettes', the day of the wheelbarrows. The king of France, Louis XVI, came himself and did some digging, albeit reluctantly (*plate 12*). Prints of this scene act in the people's interest, in both meanings of the term: as something raising attention, and as something valuable for them, the king sharing the creation of some sort of common ground. Equally, such prints were, theoretically, in the king's interest too, demonstrating a technically simple action, shared now with his people. It just proved not to be very politically effective.

Notes

1 Respectively: '17 mars–29 septembre: au sieur Houasse, peintre, pour deux tableaux qu'il a peints de la statue équestre du Roy, pour les faire graver (4 p.) … 840 livres', *Comptes des bâtiments du roi*, vol. 4, col. 187 (année 1697, Diverses maisons royales); '14 avril–27 octobre: au sieur Simonneau l'aîné, sur une des planches qu'il grave représentant la statue équestre du Roy (3 p.) … 500 livres', *Comptes des bâtiments du roi*, vol. 4, col. 193–4 (année 1697, Gravure); '12 avril: à Simonneau l'aîné, graveur, à compte d'une planche qu'il a gravée, représentant la statue équestre du Roy … 200 livres', *Comptes des bâtiments du roi*, vol. 4, col. 485 (année 1699, Gravures). Simonneau's print is to be found in the Bibliothèque nationale de France, in the N5 series. One painting by Houasse, a view of the right-hand side of the statue which corresponds to the print, is in a private collection in Paris, a copy of which exists at the Musée Carnavalet, Paris (inv. P 630; E 2773; oil on canvas, 72 × 54 cm). A second work, a view of the left-hand side, is also at the Musée Carnavalet (inv. P 156; E 2391; oil on canvas, 74 × 56 cm). I thank Matthieu Lett, who has just submitted his PhD thesis on René-Antoine Houasse, for this kind indication and others concerning the rest of the text.

2 See Anne Nadeau, 'Charles Simonneau: un graveur de l'entre-deux siècles. Un aperçu de la gravure d'interprétation de 1667 à 1727', in Pierre Wachenheim, ed., *La gravure: quelles problématiques pour les temps modernes?*, Paris, 2009, 35–44.

3 Peter Burke, *The Fabrication of Louis XIV*, New Haven, CT and London, 1994.

4 Marvin Trachtenberg, *Building-in-Time: From Giotto to Alberti and Modern Oblivion*, New Haven, CT, 2013.

5 Jules Guiffrey, ed., *Comptes des bâtiments du roi*, vol. II, col. 775–6, 978, 981ff., for the payments for the model of the equestrian statue.

6 *Comptes des bâtiments du roi*, III:79, 25 janvier 1688, for the cage of the model of the equestrian statue.

7 *Comptes des bâtiments du roi*, III:83, 17 octobre 1688 up to III:707, 14 décembre 1692.

8 Germain Boffrand, *Description de ce qui a été pratiqué pour fondre en bronze d'un seul jet la figure équestre de Louis XIV, élevée par la ville de Paris dans la place Louis le Grand, en 1699…*, Paris, 1743, 4.

9 See François Fossier, *Les Dessins du fonds Robert de Cotte de la Bibliothèque nationale de France: architecture et décor*, Rome, 1997.

10 The *réduction* was made by Roger Schabol.

11 The statue stood at 19 feet 6 inches.

12 Nicolas Bergier, *Le Cheval de Domitian, extraict du premier livre des 'Bocages' de P. Statius Papinius, et mis en français pour servir de parallèle au cheval de Henry le Grand, posé sur le Pont Neuf à Paris, au mois d'aoust l'an 1614*, Paris, 1614.

13 See Statius, *Silvae*, book I:9; English trans. J. H. Mozley, Loeb Classical Library, 1928, 5: 'Quae superimposito moles geminata colosso/ Stat Latium complexa forum?' ('What mighty mass redoubled by the huge form surmounting it stands gathering to itself the Latian forum?') and 10: 'Vix sola sufficiunt, insessaque pondere tanto/ Subter anhelat humus: nec ferro aut aere laborat,/Sed Genio' ('Scarce doth the soil hold, and the ground pants beneath the pressure of so vast a weight').

14 Martin Lister, *A Journey to Paris in the year 1699*, London, 1699, 26–7.

15 See Lister, *Journey to Paris*, 27, and the *Mercure historique et politique*, September 1699, 289. According to the latter source, the statue was 'was cast in one go, that is to say, the horse and the figure together'.

16 Irving Lavin '*Ex Uno Lapide*: The Renaissance Sculptor's *Tour de force*', in Matthias Winner *et al.*, *Il cortile delle statue*, Mayence, 1998, 191–210.

17 François Lemée, *Traité des statues*, Paris, 1688, 8: 'Et quand les ouvrages qui en sortent sont d'une matière solide, on les appelles des Statues, des Simulacres et des Signes.'

18 *Mercure galant*, June 1685, 1: 'Je vous l'ay dit bien des fois, Madame, l'Eloge du Roy est une matière inépuisable.'

19 See Kenneth Gross, *The Dream of the Moving Statue*, Ithaca, NY, 1992; and Victor Stoichita, *L'Effet Pygmalion. Pour une anthropologie historique des simulacres*, Genève, 2008.

20 See Michael Cole, *Cellini and the Principles of Sculpture*, Cambridge, 2002.

21 Etienne Binet, *Essai des merveilles de nature*, ed. 1628, 333, 'La façon de louer les Statues': 'Le Métal s'est entièrement jetté hors de sa propre nature et s'est transporté à une véritable représentation.'

22 'Un bloc de marbre était si beau /Qu'un Statuaire en fit l'emplette./ Qu'en fera, dit-il, mon ciseau?/ Sera-t-il Dieu, table ou cuvette?' Jean de La Fontaine, *Fables*, book IX, fable 6.

23 Aline Magnien, 'Les bronzes "Keller"', *Bulletin de la Société de l'histoire de l'art français*, 1996, 37–63; Lorenz Seelig, 'Keller, Balthasar (Quelair, Quailer)', in *Neue Deutsche Biographie*, vol. 11, 1977, Berlin, 429ff.

24 See Arthur de Boislisle, 'Notices historiques sur la place des Victoires et sur la Place de Vendôme', *Mémoires de la société de l'histoire de Paris et de l'Île de France*, 15, 1888, 1–272, 122; Colbert de Villarchef to Keller, 31 July 1692, BN, ms. Fr. 7801, fol. 408.

25 Keller had just been asked to cast a pedestrian statue, also made by Girardon, for a triumphal arch in Tours. See letter from M. de Miroménil, 'intendant' at Tours, to the 'contrôleur général des bâtiments, 22 août 1691', in Arthur de Boislisle, *Correspondance des contrôleurs généraux des finances avec les intendants des provinces* (1683–1715), Paris, 1874–97, 3 vols, I: 247, doc. 978.

26 See Fossier, *Les Dessins du fonds Robert de Cotte*, plan 39, 2, 185–6.

27 See Stéphan Perreau, *Hyacinthe Rigaud, Le peintre du Roi*, Montpellier, 2004, 161, fig. 126.

28 'Fondue en 1692 et érigée en 1699.'

29 Alexis Simon Belle, *Portrait of Balthazar Keller*, 1703, Versailles, Château de Versailles et de Trianon.

30 Domenico Fontana, *Della Trasportatione dell'Obelisco Vaticano* ..., Rome, 1590.

31 Claude Perrault, *Les dix livres d'architecture de Vitruve*, Paris, 1673, livre X, ch. XVIII, 390, note 4, comprises a very long text concerning the tension of the cables 'necessary to make move the machine used to erect the big stones with which the pediment over the main entrance of the Louvre is covered'. Perrault designed a model permitting a perfectly horizontal hoisting of the stones. Plate LVIII, engraved by Edelinck, shows a machine in action in a deserted aera, whereas plate LIX by Etienne Gantrel shows three elevating machines. The process was spectacular enough to be noticed by a traveller such as Martin Lister, who describes it in these terms: 'It is a masterwork of art, equal to all that the Ancients may have done in such a way, to have succeeded in raising to this height two such big and fragile stones.' Lister, *Journey to Paris*, 52.

32 c. 1680, Londres, Royal Collections.

33 It was necessary to put criss-crossed beams beneath the pedestal, because of springs in the earth. See *Histoire journalière*, 'Du jeudi 28 mai 1699; de Paris le 22 mai'.

34 Bernard Le Bovier de Fontenelle, 'Eloge de M. Amontons', in *Histoire de l'Académie Royale des Sciences, Année 1705, Avec les Mémoires de Mathematique & de Physique, pour la même Année*, Paris, 1730, 150–4.

35 See Domenico Bertoloni Meli, *Thinking with Objects: The Transformation of Mechanics in the Seventeenth Century*, Baltimore, MD, 2006.

36 See Yannick Fonteneau, 'Penser le travail à l'époque moderne (XVIIe–XIXe s.): introduction et perspectives', *Cahiers d'Histoire*, 110, 2009, 11–38.

37 Pamela H. Smith, *The Body of the Artisan: Art and Experience in the Scientific Revolution*, Chicago, IL, 2004.

38 See Louis Trénard, *Les Mémoires des intendants pour l'instruction du duc de Bourgogne, 1698, introduction générale*, Paris, 1975.

39 See Chandra Mukerji, *Impossible Engineering: Technology and Territoriality on the Canal du Midi*, Princeton, NJ, 2009.

40 Andrea Catellani, 'Images scientifiques de la première modernité: entre emblème et diagramme', in Maria Giulia Dondero and Audrey Moutat, eds, 'Techniques de transformation, transformation des techniques', *Visible*, 6, 2010, 71–90. For the 'réduction en art', see Hélène Vérin, *La Gloire des ingénieurs. L'intelligence technique du XVIe au XVIIIe siècle*, Paris, 1993.

41 See among others Caroline Jones and Peter Galison, eds, *Picturing Science, Producing Art*, New York, 1998.

42 See for example Henri Gautier, *L'art de laver ou la nouvelle manière de peindre sur le papier*, Paris, 1687.

43 Madeleine Pinault-Sorensen, 'Les dessinateurs de l'Académie royale des sciences', in Christiane Demeulenaere-Douyère and Eric Brian, eds, *Réglement, usages et science dans la France de l'absolutisme (Actes du colloque, Paris, 8–10 juin 1999)*, Paris, 2002, 147–67. Around the abbé Bignon, the coordinator, were three scientists: Des Billettes; Jaugeon, responsible for the arts of printing; and father Sébastien Truchet. They recruited three draughtsmen: G. Quineau, Louis Simonneau and Pierre de Massart dit de Rochefort. They engraved around 165 plates from 1695 to 1718.

44 See Denis Lanier and Jean-Pierre Le Goff, 'L'héritage arguésien. Le fils spirituel: Philippe de la Hire', *Cahiers de la perspective, points de vue*, 5, 1991, 61–78.

45 Antoine Picon, *Ingénieurs et architectes au siècle des Lumières*, Paris, 1988.

46 See William B. McGregor, 'The authority of prints: An early modern perspective', *Art History*, 22: 3, 1999, 389–430.

47 Bibliothèque nationale de France, Ed 56 c petit fol., p. 2: 'Trois Planches de la manière de suspendre des Statues par des Poulies, d'après des Esquisses de ***, ornées de Paysage par B. Picart, pour le même livre 1726'. Two years later, in 1702, Bernard Picart illustrated Jean-Baptiste Delespine's text 'Pour la pratique des Forces mouvantes' ('For the practice of the moving forces'), which included the lifting of an equestrian statue.

48 See Luisa Dolza and Hélène Vérin, 'Figurer la mécanique: l'énigme des théâtres de machines de la Renaissance', *Revue d'Histoire Moderne et Contemporaine*, 51: 2, 2004, 7–37; Wolfgang Lefèvre, ed., *Picturing Machines: 1400–1700*, Cambridge, MA, 2004; Janet-Louise Clarke, 'The Guenegaud Theatre, 1673–1680 and the machine plays of Thomas Corneille', PhD thesis, Warwick University, 1988.

49 *Explication des modèles des machines et forces mouvantes, que l'on expose à Paris dans la rue de la Harpe*, Paris, 1683. See too Arthur Birambaud, 'L'exposition de modèles de machines à Paris en 1683', *Revue d'histoire des sciences et de leur application*, 20, 1967, 141–58.

50 See Cole, *Cellini*.

51 See Jennifer Montagu, 'The painted enigma and French seventeenth-century art', *Journal of the Warburg and Courtauld Institutes*, 31, 1968, 307–35.

52 *Relation de ce qui s'est passé à l'érection de la statue du roi, dans la ville de Poitiers, le vingt-cinquième jour d'août 1687*, Poitiers, 1687, 20–1.

53 *Relation de ce qui s'est passé à l'érection de la statue du roi*, 20–1.

54 La Moureyre, 'Thomas Gobert, architecte des bâtiments du Roi et de Monsieur, ingénieur, inventeur ... et sculpteur', *Gazette des Beaux-arts*, 116, 1990, 67–93.

55 Countess of Auneuil, *La Tyrannie des Fees Detruite, ou l'Origine de la Machine de Marly*, Paris, 1702, 85–6.

56 Auneuil, *La Tyrannie des Fees Detruite*, 87.

57 Jean-Marie Apostolidès, *Le Roi-machine. Spectacle et politique au temps de Louis XIV*, Paris, 1981.

58 See André Félibien's quotation, p. 154, from the *Description du château de Versailles*: 'Une machine par ses mouvements surprend et charme les spectateurs, et surpasse les effets ordinaires de la nature. Ainsi Sa Majesté par ses vertus et ses actions héroïques étonne et ravit tous ceux qui en sont les témoins, et surpasse les forces anturelles et la porée ordinaire des hommes.'

59 René Du Moulin, *Triomphe royal de Henry le Grand sur la statue de bronze dressée sur le pont des Hernis. Dédié à Monsieur Bartolini Gentil-homme florentin, agent de monsieur le grand duc de Toscane*, 1614, 4.
60 This work is at the Musée des Beaux-arts, Arras.
61 Matthieu Lett dates the execution of the work between 1689 and 1 September 1693.
62 See Paris, chez Abel Langelier, 938.
63 See Georges Duplessis, 'Le cabinet du Roi. Collection d'estampes commandées par Louis XIV', *Le Bibliophile français*, 1869, 87–105.
64 William McAllister Johnson: 'Visit to the salon and sculptors' ateliers during the Ancien régime', *Gazette des Beaux-arts*, 120, 1992, 17–35.
65 See 'Poétique du chantier', special issue of *Ligieia*, 101/104, 2010; especially Basile Baudez, 'La représentation du chantier urbain à l'époque moderne. Quelques pistes de réflexion', 88–95.
66 For the visit of 30 January 1687, see the *Mercure galant*, February 1687, 70–1: 'He declared himself satisfied but said, because it was not easy to see the figures so high, he had to climb to see them from close; and that it was too late, that he will come back' ('S.M. parut très contente de la disposition et de ce qu'elle put voir des figures, dont il était malaisé de juger tout à fait, parce qu'étant fort élevées, il falloit monter pour les voir de près: de sorte qu'elle dit qu'il étoit trop tard, et qu'elle les verroit une autre fois').
67 Craig Koslofsky, *Evening's Empire: A History of the Night in Early Modern Europe*, Cambridge, 2011. The same public lighting was decreed for the rest of France during the same year 1697.
68 *L'esprit des cours*, September 1699, 370–1.
69 *Gazette*, 10 January 1693, 24.
70 See *Histoire journalière de ce qui se passe de plus considérable en Europe. Avec privilège de Nosseigneurs les Etats généraux de Hollande et de Westfrise (et les armes de Hollande)*, à La Haye, chez Jean-François Du Four, 10 July, 13 July.
71 *Histoire journalière*, 'Du mardi 28 juillet 1699. De Paris le 20 juillet'.
72 *Histoire journalière*, 'France, de Paris le 20 juillet'.
73 *Histoire journalière*, 'Du jeudi 30 juillet 1699. France. De Paris le 24 juillet'.
74 See Bibliothèque de l'Arsenal, MS 3128, fol 196 V; Lister, *Journey to Paris*, 28–9.
75 Daniel Cronström to Nicodème Tessin the Younger, 14/24 October 1698, in *Les Relations artistiques entre la France et la Suède (1693–1718). Extraits d'une correspondance entre l'architecte N. Tessin le Jeune et D. Cronström, publiés par R.-A. Weigert et C. Hernmarck*, Stockholm, 1964, 27, 206.
76 Thierry Sarmant, *Les Demeures du Soleil. Louis XIV, Louvois et la surintendance des bâtiments*, Seyssel, 2003, 498: when Louis de Pontchartrain, 'contrôleur général des Finances' and 'secrétaire d'Etat de la Maison du roi', suggested pulling down the primitive facades to build the square according to a new scheme, 'le roi répondit que Louvois l'avoit fait presque malgré lui; que tous ces Messieurs les ministres vouloient faire quelque chose qui leur fît honneur à l'avenir; qu'ils avoient trouvé moyen de le donner au public comme aimant ces vanité-là; que j'étois témoin des chagrins que M. de Louvois et M. de La Feuillade lui avoient donné là-dessus' (letter to the Archbishop of Paris, 28 July 1698, *Correspondance de Mme de Maintenon*, Paris, 1865–1866, IV, 242–3).
77 See Henrik Ziegler, *Der Sonnekönig und seine Feinde. Die Bildpropaganda Ludwigs XIV. in der Kritik*, Petersberg, 2010.
78 Quoted by Boislisle, 'Notices historiques', 198.
79 *L'Esprit des cours de l'Europe*, September 1699, 363ff.
80 See David Mandrella, Hermann Mildenberger, Benjamin Peronnet and Pierre Rosenberg, *De Callot à Greuze. Dessins français des XVIIe et XVIIIe siècles des musées de Weimar*, Berlin, 2005, no. 35ff.
81 Boffrand, *Description*, 3.
82 Boffrant, *Description*, 3–4.
83 Boffrand, *Description*, 4.
84 See Félibien, introduction to the first volume of the *Tableaux du Cabinet du Roi*, ed. 1677, 8: he praises the prints 'since it is through them that posterity will see one day, thanks to pleasant figures, the history of the great actions of this august monarch, and that from now on the most remote people take advantage just like us of the new discoveries made in the academies established by His Majesty for Sciences and Arts'.
85 Jean-Denis Lempereur and Pierre-Jean Mariette, *Description des travaux qui ont précédé, accompagné et suivi la fonte en bronze, d'un seul jet, de la statue équestre de Louis XV le Bien -Aimé, dressée sur les mémoires de M. Lempereur, ancien échevin, par M. Mariette …*, Paris, 1768.
86 Pierre Patte, *Monuments érigés à la gloire de Louis XIII*, Paris, 1765, plate 13.

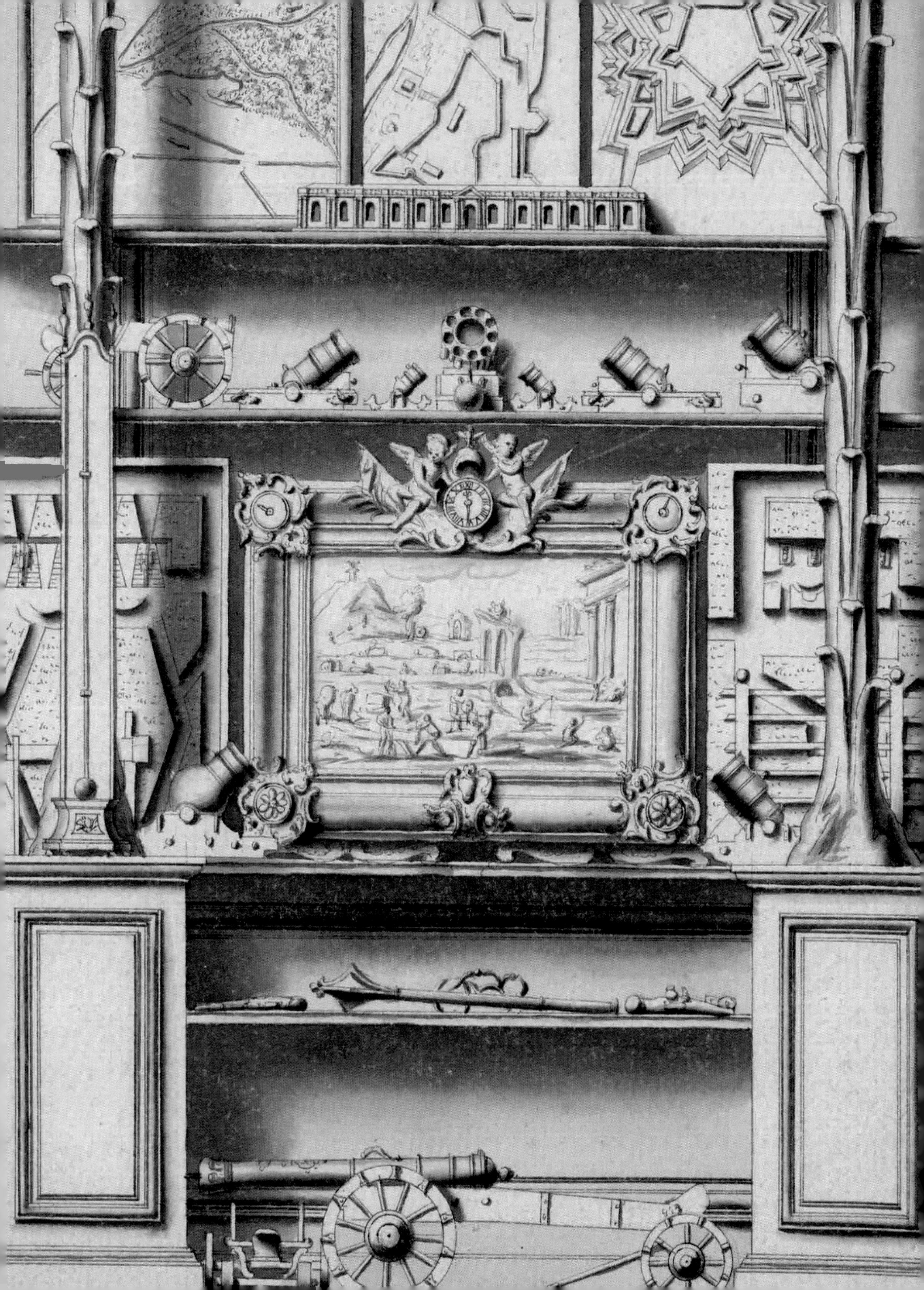

Chapter 8
A Clock Picture as a Philosophical Experiment: The Tableau Mécanique in the Physics Cabinet of Bonnier de la Mosson

Hanneke Grootenboer

A Moving Image

In 1739, an obscure painter of nature scenes and still lifes of fish by the name of Nicolas Spayement created a classical landscape populated with figures clad in early eighteenth-century workers' attire (plate 1). The humbleness of their clothing is countered by the stateliness of the neo-classical architecture arising on the right, which vaguely recalls the pastoral scenes of Poussin and Claude Lorrain. However, the representation of so many human figures engaged in manual labour certainly sets Spayement's painting apart from such Arcadian compositions. Backed by a hilly landscape, men prepare large slabs of stone that have been quarried from a stone pit. Various figures transport materials towards a distant city along a meandering pathway, while women do the laundry in a small stream running diagonally through the composition, and a fisherman sitting nearby hopes for a good catch. Even the animals are appropriately diligent: on the left, cattle are duly grazing and chicken frantically picking at the ground, while at the foot of the hillock children are doing their share by bouncing on a seesaw. The hilltop windmill seems to have been included to demonstrate that not only creatures but machines, too, contribute to the daily round of work.

Approaching this picture at the Musée des Arts décoratifs in Paris today, one notes the questionable quality of its execution and its rather limited colour scheme. The scene's many lovely details notwithstanding, the figures are rather roughly outlined and as a result do not blend very well into the composition as a whole. Incongruously, this poorly rendered landscape has been mounted in a pompously sculpted gilded frame, heavily decorated with garlands and crowned with two putti flanking a built-in clock. In truly rococo fashion, the frame has overtaken a fair bit of the artwork proper, its width being almost one-quarter that of the picture plane. The frame's ornateness contrasts sharply with the mediocre painting, until we realize that its heaviness is not purely decorative but rather serves a functional purpose: when we turn this picture around (plate 2), we see that it carries part of a complex wind-up mechanism that is attached to the back of the painting.[1] As it turns out, this is not a painting but a clock. The enamelled white faces inserted in the frame indicating the hours, minutes, and, on the left, the days of the week are not ornamental elements but essential parts of the object as a whole. More than just a painting that features a clock in its frame, this unusual artefact hides an ingenious surprise. Rather than chime on the hour, when wound up using a key that fits the holes in the middle of the bottom portion of the frame, this tableau of working people jumps instead into action. The rough outlines of various pictorial elements are not indicative of the artist's lack of skill, but are in fact the edges of the cut-out wooden figures and animals,

Detail of Jean Courtonne and Jean-Baptiste Courtonne, *Cabinet de Bonnier de la Mosson, The Physics and Mechanics Cabinet*, 1739–40 (*plate 7*).

Art and Technology in Early Modern Europe *Edited by Richard Taws and Genevieve Warwick.*

sixty in total, that move within the picture in accordance with the particular activity in which they are involved. A hybrid object, neither a picture nor a real clock, and not quite an automaton in the common sense of the term, this is no doubt one of the most animated paintings of its period. What exactly is it? Is it a genuine tableau mouvant, a rudimentary moving image antedating the invention of technologies like the zoetrope that, in turn, anticipate cinema? Is this what painting looks like when time – literally – is added to it?

Unique as this object may be, it is embedded in the intertwined history of technologies of clockmaking and automata that had been developed extensively in the preceding centuries. This interlaced history manifested itself in major scientific highpoints such as the astronomical clock with a mechanical cock installed in 1350 in the cathedral of Strasbourg, eventually followed by numerous other examples, such as the famous parade of jacquemarts and automata in the clockworks adorning St Mark's in Venice.[2] Similar crossovers can be found in the elaborate table clocks with moving figures popular in Germany in the sixteenth and seventeenth centuries, or in the French and Swiss gold watches that grew ever thinner over the course of the seventeenth and eighteenth centuries, and were often embellished with moving parts or erotic imagery hidden beneath camouflaged covers. The rare fusion of clock and painting that we see in a tableau mécanique (as it was called) was often created specifically for cabinets of curiosity; however, few survive. They serve as precursors to the *spectacles mécaniques* of Pierre and Henri-Louis Jaquet-Droz, and later the *Eidophusikon*

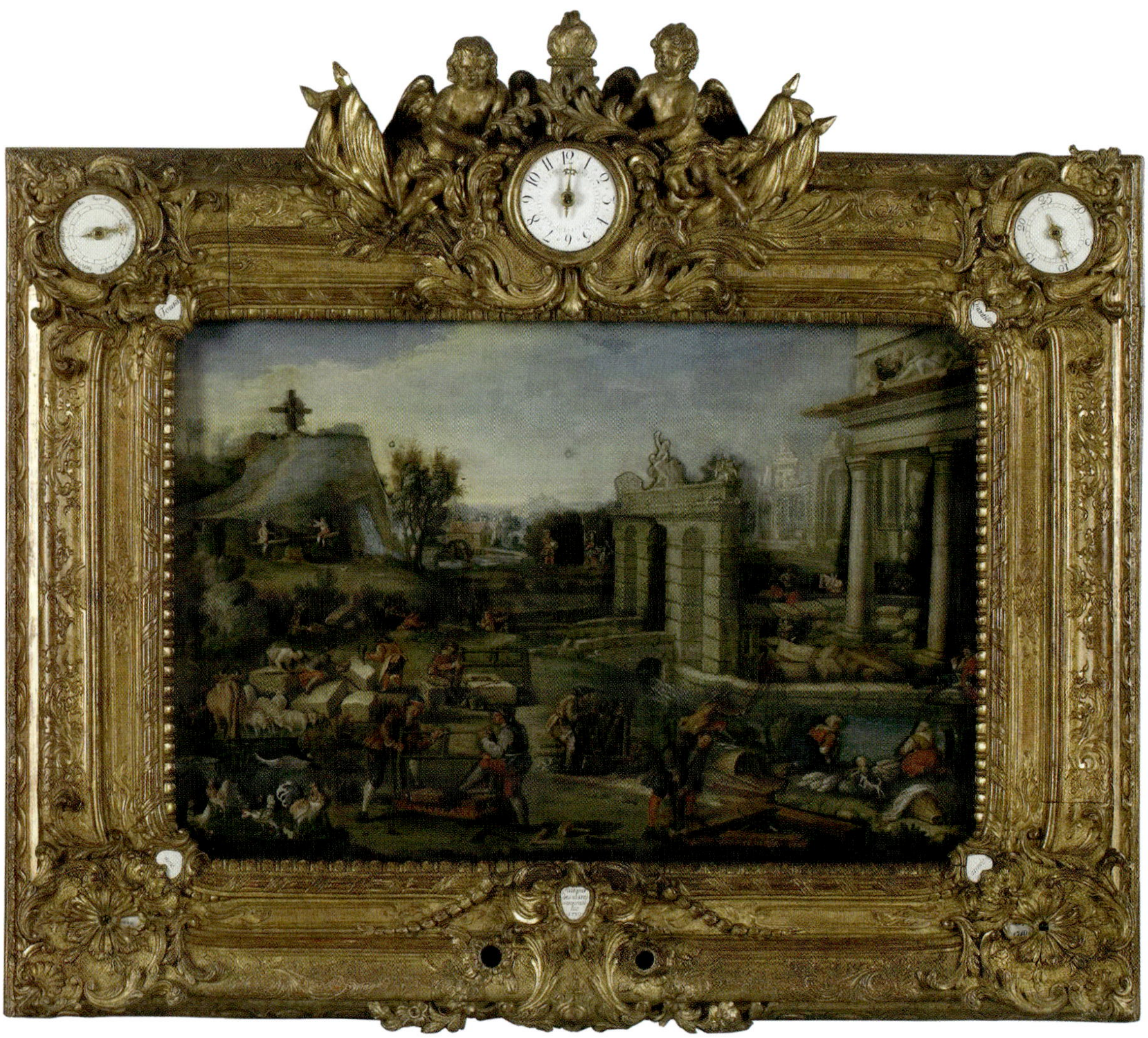

1 Nicolas Spayement and Desmares, *Tableau Mécanique*, 1739. Oil on wood in gilded frame, 79 × 91.5 cm. Paris: Les Arts Décoratifs, Paris. Photo: Les Arts Décoratifs, Paris/Jean Tholance.

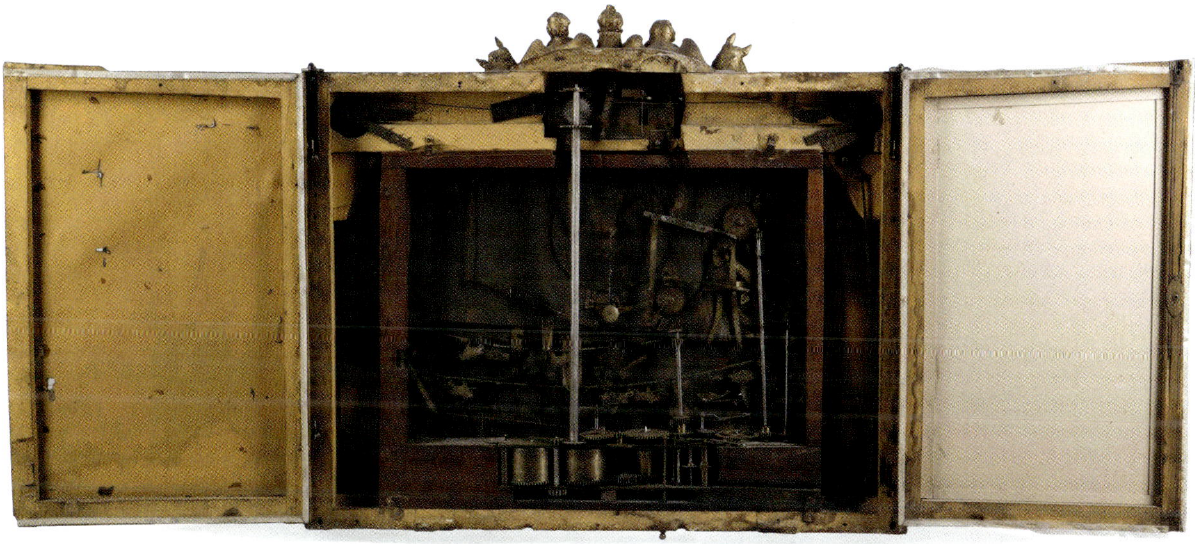

2 Reverse Side of Nicolas Spayement and Desmares, *Tableau Mécanique*, 1739, showing the clockwork mechanism. Photo: Les Arts Décoratifs, Paris/Cyrille Bernard.

of Philippe Jacques de Loutherbourg discussed by Ann Bermingham elsewhere in this volume. As text on two slabs of wood within the composition makes clear, Spayement worked closely together with a certain Desmares, a 'mécaniste' of Versailles, who produced the wind-up clock mechanism for this unique artefact.

This fusion of clock and painting was very timely. It was produced in a period when the French were becoming increasingly temporally literate, particularly in Paris.[3] In addition to the Horloge du Palais and the numerous clocks visible in local neighbourhoods, the postal system had become highly regularized, and transport by passenger carriages took place according to strict timetables. The introduction of the pendulum clock in domestic spaces further contributed to an emergent temporal mentality, ruled by punctuality and schedule, that was significantly different from the understanding of time based on the sun's position or the sound of church bells. Exhibited in the Louvre in the same year Spayement and Desmares's tableau mécanique was made, the series *Four Times of Day* by Nicolas Lancret (1690–1743) shows us in four separate images this new life regulated by clocks and watches that dictated activities typical for morning, afternoon, or evening (*plate 3*). Later in the century, Louis-Sébastien Mercier's chapter on 'Les Heures du Jour' in his *Tableau de Paris* (1782-88) recounted the sounds, sights, and smells of the city at various moments during the day.[4] However, Mercier complained that the pendulum clock that now decorated every mantelpiece had a regrettable effect on one's experience of time, for 'there is nothing so sad as to contemplate a pendulum: you see your life slipping away, so to speak, and the movement warns you of all the moments that are being taken away and which will never return'.[5] Many others commented about this change in mentality caused by the visibility of timepieces, among them Ferdinand Berthoud (1727–1807), clockmaker-*mécaniste* to the king, who wrote various entries on timekeeping for the *Encyclopédie*. He called the truly innovative clockmaker an *architecte-mécanicien*, a theorist and mechanician at once. In this context, Spayement and Desmares had embarked on an ambitious project that was clearly meant to delight, but also to instruct. Whereas many paintings of the period reflect new conceptions of time, often in allegorical compositions, this tableau mécanique stands out because of its presentation of time as defined by work, not leisure as we see in Lancret's series. The movability of the figures has ingeniously been made to represent – we might even say theorize – time as *lived*, as event rather than concept. It was partly the tableau's pseudo-status as an automaton that had enabled its makers to make such a theoretical statement.

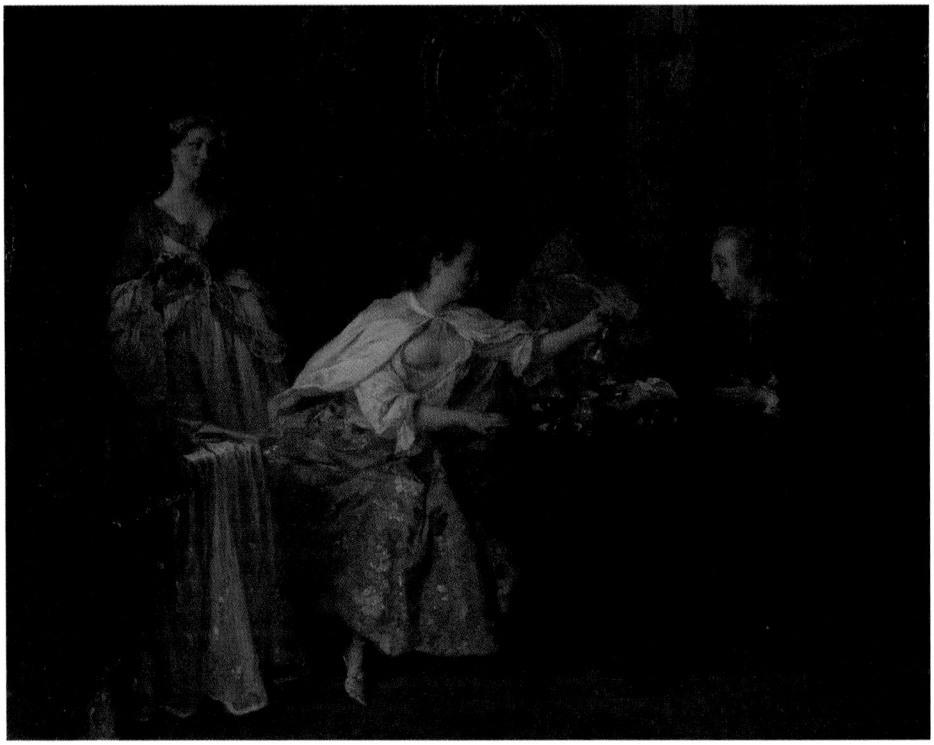

3 Nicolas Lancret, *The Four Times of Day: Morning*, 1739. Oil on canvas, 28.3 × 36.4 cm. London: National Gallery (Bequeathed by Sir Bernard Eckstein, 1948). Photo: © The National Gallery, London.

Spayement and Desmares's tableau mécanique was manufactured at the start of what Minsoo Kang has termed the automaton craze.[6] Its production coincides with an exhibition at the Hôtel de Longueville in Paris of Jacques de Vaucanson's automatons in 1738–39. Despite the hefty entrance fee of three *livres*, crowds flocked to see Vaucanson's creations, in particular a moving duck flanked by two androids playing a flute and pipe-and-tambour. Not only did the duck move around as if it were alive, but it simulated the working of a real bird's body by taking a grain from the hand, swallowing it, and – after a short pause to heighten the tension for its viewers – defecating. Parisians were flabbergasted. Claiming to the Académie Royale des Sciences that he wanted to demonstrate 'les solides principes de mécanique' (evidently in addition to making a profit), Vaucanson was subsequently praised by Voltaire as a 'rival of Prometheus' and by Diderot as the archetype of mechanical genius.[7] Writing in 1854, Hermann von Helmholtz was still celebrating Vaucanson's duck, which he considered *the* scientific marvel of the previous century.[8]

Though initially made for public entertainment, Vaucanson's inventions were not mere trifles. They resonated widely in scientific and philosophical discourse. Vaucanson insisted that the mechanism animating the duck's wings had not been modelled after previous automata but was based on close observation of the anatomy of actual birds.[9] His marvel was as much a scientific as a philosophical experiment – we might even call it a materialized hypothesis. Even though the mechanical digestive system of Vaucanson's duck was manufactured and its excrement fake, it implied that a bird's digestive system operates like a machine. As such, the duck had a wide-ranging impact and was considered a serious contribution to the then-current debate among physiologists as to whether digestion was a chemical or a mechanical process (making a clear case for the latter).[10] In his famous 'Man a Machine' essay of 1748, Julien Offray de La Mettrie went even further when he declared that the bodies not just of animals but of humans were machines. Referring to Vaucanson's duck, he presented the human digestive system as having a wind-up mechanism and cast the human body as a whole as a clock, 'the structure of it fram'd with such exquisite workmanship and dexterity, that if the wheels which serve to mark the seconds should stop short ... [it] would still continue to move on in its natural course'.[11] Only when the wheels are rusty or out of order, La Mettrie advised, should the mechanism stop, just as obstructions in blood vessels stop the beating of the heart. The analogy of the body with a clock was not La Mettrie's invention. Initiated by Descartes, the clock metaphor was prominent in philosophical debates about materialism as well as what was called 'mechanical philosophy'. Leibniz, Hobbes, and Pascal all commented on the idea that the dynamic complexity of the world was powered by a system analogous to a clock mechanism, and that the human body's various organs are comparable to the wheels and levers found inside a watch.

In this essay I am interested in the extent to which Spayement and Desmares's clock picture is a variation on the clock metaphor in artistic-technological terms, a kind of 'philosophical instrument' like Vaucanson's duck that was capable of making a specific contribution to scholarly debates precisely through its unusual form. I am particularly keen to explore this object's initial meaning and function as it was originally commissioned by Joseph Bonnier de la Mosson for his immense collection of curiosities. As we will see, within that collection it was conceived very much as an instrument – and not a painting – having found its place in Bonnier's Mechanics and Physics Cabinet among optical devices and architectural models. I seek to understand what this unusual artefact – an exception even within the exceptional history of automata, precisely because of its hybrid nature and marginal position – can reveal to us about the invisible interstices between the mechanical and the philosophical, between craftsmanship and thought.

The Clock Picture as a Functional Object and 'Problem' in Bonnier's Cabinet

In July 1744, Joseph Bonnier de la Mosson died unexpectedly at the age of forty-two, leaving behind a massive *kunstkammer* collection, unique in the wide range of subjects it embraced.[12] Praised by all who saw it as one of the most remarkable assemblages of curiosities of its time, exceeded only by the collection of the French king, it included devices and instruments relating to astronomy, anatomy, and hydrology, and counted among its many items a working miniature theatre, a library, and an extensive assortment of shells. Son of the treasurer of Languedoc who had been one of the richest men of his generation, Bonnier de la Mosson inherited his father's vast fortune in 1726, and used most if not virtually all of it to amass his extraordinary collection and to house it in nine richly decorated rooms (designed by architects Jean Courtonne, Jean-Baptiste Leroux, and François Debias-Aubry), arranged *en enfilade* on the first floor of the old Hôtel du Lude (demolished in 1861), on the rue Saint-Dominique, in the heart of Paris. Only a few of the exquisite bespoke cabinets have survived, now adorning the walls of the library in the Muséum national d'histoire naturelle. Jean-Marc Nattier portrayed Bonnier de la Mosson at ease in this cabinet, resting an open book on his knee while looking out – deeply contentedly, it seems – over the impressive number of objects displayed on shelves along the walls, a sample of which we see right behind him (*plate* 4). The painting was completed in 1745, after the death of its sitter, and was exhibited at the Salon in the Louvre

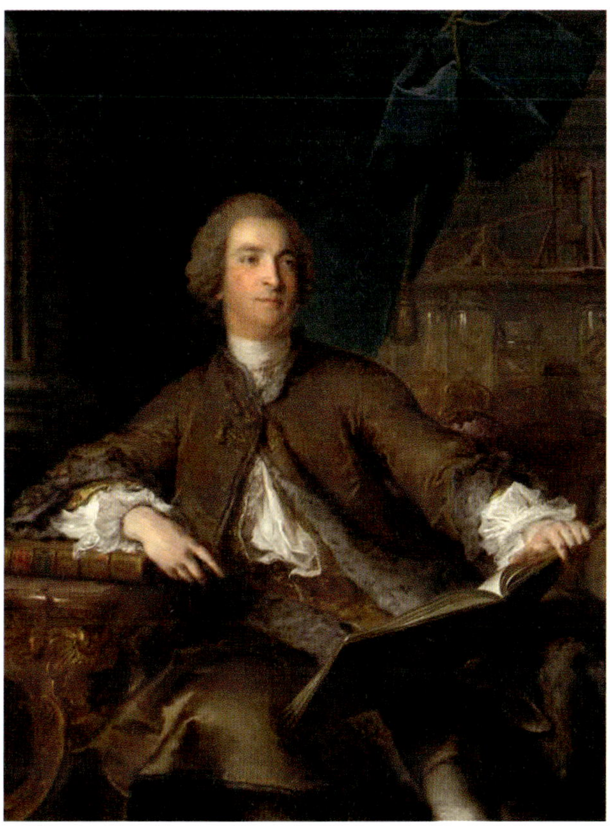

4 Jean-Marc Nattier, *Joseph Bonnier de la Mosson*, 1745. Oil on canvas, 138 × 105 cm. Washington, DC: National Gallery of Art (Samuel H. Kress Collection). Photo: National Gallery of Art, Washington, DC.

in the following year. An annotated catalogue for the sale of the entire collection was drawn up by none less than Edme-François Gersaint, dealer in art and luxury articles and a familiar name in well-to-do Parisian circles, best known to us today through Jean-Antoine Watteau's painting of his shop sign in 1720–21.[13] The scope of Gersaint's catalogue indicates the scale of the collection as a whole, running to more than 234 pages and containing 966 entries. Categorized by room, the list of items follows a route through the apartments at the Hôtel du Lude that Gersaint claims he had been given by Bonnier himself.[14] Browsing the catalogue today is like embarking on a virtual tour through what must have been an absolutely spectacular gallery.

Let's walk with Gersaint for a moment to get a clear sense of this superb collection. Opening the pages of the catalogue, we start in the anatomy room, where we glance over human and animal skeletons, perhaps frown at a mummy's head, and bend over some wax models. We proceed through the chemistry cabinet, where we admire the intricate technology demonstrated by scale models of fountains and furnaces, and go straight to the pharmacy room, where lining the shelves are hundreds of empty faïence jars, indicating that this room had not yet been finished when its owner died. We continue to the adjacent 'cabinet de drogue', housing nearly eight hundred jars filled with mineral substances, dried herbs, and other extracts meant for medicinal compounds (we can imagine the smell of herbal fragrances and, presumably, alcohol). On we go through the lathe cabinet, chock-full of tools essential for clockmaking and woodworking attached to panels inserted in the wall, a common elite interest at the time, to arrive at three consecutive natural history rooms. Their cabinets represent the *kunstkammer* collection as we have come to know it: stuffed birds and large glass jars containing mammal parts in preserving fluid – or snakes, as we see in Nattier's picture – fill the shelves behind the cabinets' glass doors. We may imagine how visitors might have looked around and quickened their pace through this well-trodden field of knowledge to arrive at the virgin territory of the gallery's *pièce de la resistance*: the Cabinet de Mécanique et de Physique. It is here that we find Spayement and Desmares's tableau mécanique.

The section on the Physics Cabinet is the longest of the catalogue, and Gersaint, well aware of how exceptional it was, portrays the objects and instruments in lively detail. This room must have filled visitors with awe, as it housed some of the most recent inventions, state-of-the-art instruments, and exceptional artefacts, along with the usual mathematical measuring apparatuses and optical devices. Gersaint also calls this gallery the 'machine room', and indeed Bonnier had conceived an unrealized plan to have a life-size automaton of a guard open the door for visitors and close it behind them.[15] Instead, upon entering, we first encounter two impressive life-size mannequins, although without movable parts: 'a wild Indian' (a Native American) carrying various belts to which tools are attached, and a Chinese man in full traditional attire (items 597 and 598). They are placed near a cupboard filled with small replicas of building devices, including a scale model of the famous machine constructed to lift the two gigantic slabs of stone making up the façade of the Louvre's

5 Detail of Jean Courtonne and Jean-Baptiste Courtonne, *Cabinet de Bonnier de la Mosson, The Physics and Mechanics Cabinet*, 1739–40, showing plate 8, part of the Physics cabinet. Pen and ink on paper, 37 × 195 cm. Paris: Bibliothèque de l'Institut National d'Histoire de l'Art (collections Jacques Doucet). Photo: INHA/RMN-Grand Palais.

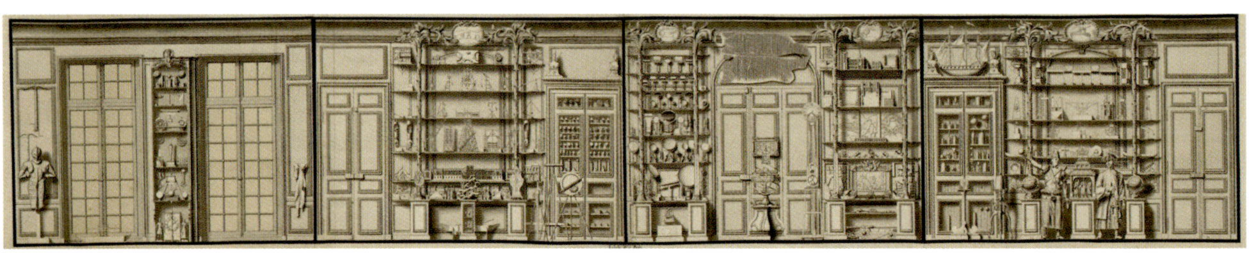

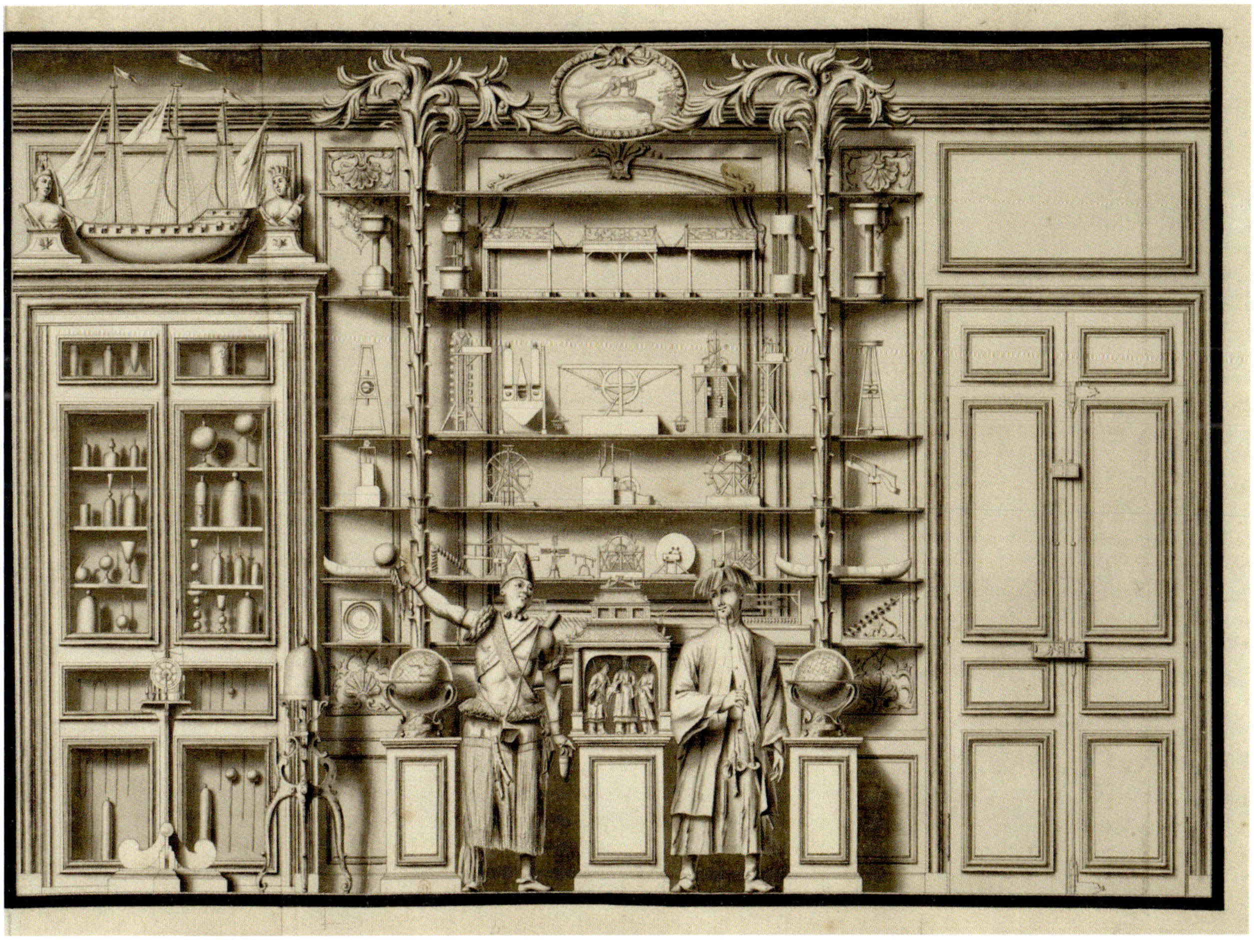

6 Detail of Jean Courtonne and Jean-Baptiste Courtonne, *Cabinet de Bonnier de la Mosson, The Physics and Mechanics Cabinet*, 1739–1740, showing plate 8, part of the Physics cabinet. Pen and ink on paper, 37 × 195 cm. Paris: Bibliothèque de l'Institut National d'Histoire de l'Art (collections Jacques Doucet Photo: INHA/RMN-Grand Palais.

main entrance (item 409). Here Gersaint is quick to identify a selling point: this device appears in full working order in a well-known 1677 print by Sébastien Le Clerc of the Louvre under construction. Following the catalogue as our tour guide, we are asked to come to a full stop in front of item 618, our tableau mouvant, whose description runs to more than two pages. The length and quality of the text reveals Gersaint's unbridled enthusiasm for what must have been – even in this remarkable cabinet – an extraordinary object. Gersaint paid great attention to the various figures, describing in meticulous detail the different tasks they perform. He was particularly passionate about the accuracy of the mechanism's movements with regard to the operations being carried out. Just as Vaucanson had studied the wing motion of ducks when designing his automaton, it appears that in refining the machinery of the tableau Desmares had examined not just the construction of previous automata but also the actual movements of human and animal bodies, in an attempt to mimic them. Admittedly, Gersaint was first and foremost a dealer of pictures, and though this unusual clock-painting must have filled him with delight, he may have written his loving entry primarily to get the highest price for it.[16] Indeed, the tableau mouvant sold for a whopping 1,151 *livres*, one of only three objects in this cabinet to fetch a price over 1,000 *livres*.

The tableau mouvant was evidently conceived to belong in the Physics Cabinet. It was not considered part of Bonnier's extensive picture collection, also recorded in the catalogue, which included work by Quentin Metsys, Pieter Neefs, Alexandre-François Desportes, and the contemporary artist Jean Raoux, who also carried out work at the family's Château de la Mosson near Montpellier. Yet the fact that it was a decorated

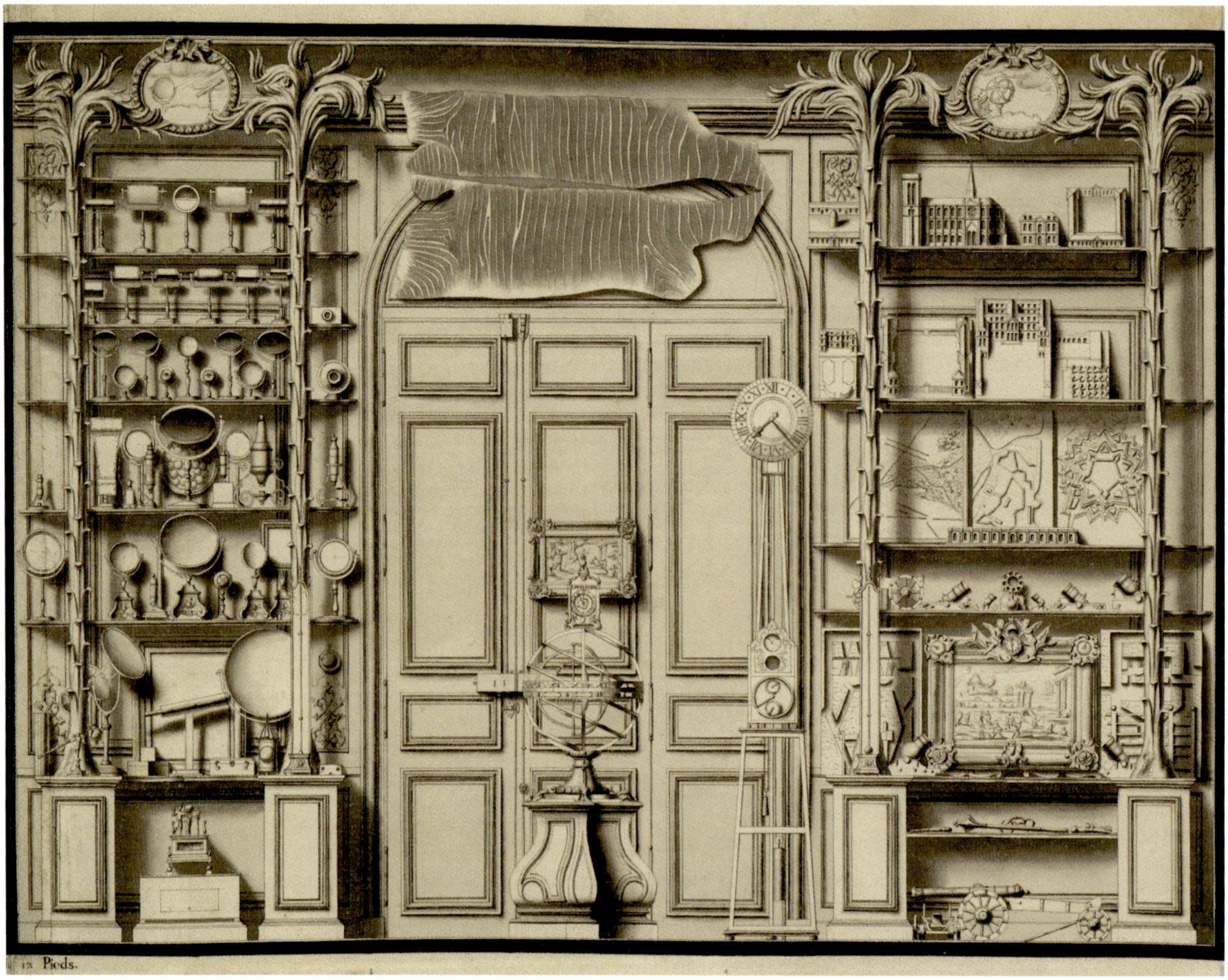

7 Detail of Jean Courtonne and Jean-Baptiste Courtonne, *Cabinet de Bonnier de la Mosson, The Physics and Mechanics Cabinet*, 1739–40, showing plate 8, part of the Physics cabinet. Pen and ink on paper, 37 × 195 cm. Paris: Bibliothèque de l'Institut National d'Histoire de l'Art (collections Jacques Doucet). Photo: INHA/RMN-Grand Palais.

object, or at least rendered with great sophistication, accorded with the high quality of the other objects, most of them produced at considerable expense. The question arises as to how this picture-clock was actually placed with other objects. Did it hang on a wall as if it were a painting (as it does now), partly fulfilling a decorative role, or did its sole significance lie in its status as a functional object?[17]

This answer to this question can be found by looking at a series of eight pen-and-ink drawings of the cabinet's original layout made by Jean Courtonne, who had also been involved in the design of the *kunstkammer* as a whole (*plate* 5). The drawings are beautifully rendered in meticulous detail.[18] Studying the drawings with Gersaint's catalogue in hand, a largely complete image of the entire space emerges before our eyes, as we are able to identify virtually all the objects we see. If we look at the sheet with the depiction of the Physics Cabinet, we meet the figures of the Native American and the Chinese man at the entrance, followed by shelves full of scale models of construction devices (*plate* 6). We immediately recognize the Louvre's lifting machine, various Claude glasses in different sizes, optical instruments, and even the wax models of eyes on the ledge second from the top. To the right of the door overhung by a zebra hide, on a shelf filled with architectural maquettes and surrounded by models of cannons, we find our clock-painting. It is standing proudly on two typically rococo legs (now removed), accentuating its status as a functional object rather than as a picture. There is a clear distinction between the picture we see in Courtonne's drawing, hanging at eye level between the doors, and the tableau mécanique on the shelf, which visitors had to bend over to see (*plate* 7). The tableau mouvant, completed in 1739, was probably one of the last objects to be added to these shelves, for Courtonne died in January 1739, and it was left to his son Jean-Baptiste to complete the drawings the following year.

There was yet another – rather fictionalized – depiction of the Physics Cabinet, commissioned in 1734 as part of a series of four paintings by Jacques de Lajoue (two of which survive in the collection of the Alfred Beit Foundation).[19] Facing a wide-open, church-like space, the now familiar cabinets with lenses, mirrors, and Claude glasses are surrounded by built-in shelves filled with models of construction implements and other devices that in many cases correspond with those found in Gersaint's catalogue (*plate* 8). The large pendulum clock, in Courtonne's drawing placed directly left of the tableau mécanique, has been moved to centre stage and is given an elaborate niche of its own, its face now framed by a flowery crown. Various bronzes, absent from Courtonne's record, give the orderly assemblage a dynamic twist. Lajoue took the liberty of transforming the learned man's cabinet into a rococo fantasy of a scholar's dream, embodied in an ancient figure who, surrounded by various hefty volumes, leans against a column in the left corner and looks up from his geometry book towards the vault of this cathedral of learning. This is likely the classical architect and author Vitruvius, who from the vantage of the past envisions the riches of engineering genius that the eighteenth century – and Bonnier – have acquired. He even seems to look beyond the present and into the future: behind the pendulum clock's fringed face, the cabinet's wall seems to have dissolved to reveal a pale blue sky. In the diffuse pastel sunlight emerges a hazy construction site with figures busy climbing ladders or hoisting materials. Two columns and part of a triglyph lie on the ground, and are probably about to be lifted by a crane whose model we see on the shelf. The theory and practice of the mechanics of building are represented in this otherwise rather frivolous painting, suggesting that Bonnier's collection, anchored in classical thought, stands at the foundation of a rising city that symbolizes the nation or society writ large. Clever support for this concept is provided by the temporal diagonal running from the architect of antiquity in the bottom left-hand corner through the present-day space of the cabinet in the middle and extending towards the future-in-the-making that we see on the right. Lajoue has deliberately placed the pendule in the centre, as it serves as a theatrical repoussoir as well as a strong visual metaphor. Running past the central pendule, the diagonal line connects various means of imagining time, from the depths of the past to the measurable present and on into the immensurable, infinite future.

8 Jacques de Lajoue, *Physics and Mechanics Cabinet of Bonnier de la Mosson*, 1734. Oil on canvas, 80 × 68 cm. Previously in the Alfred Beit Collection, Ireland, sold at Christie's, June 2015, present location unknown.

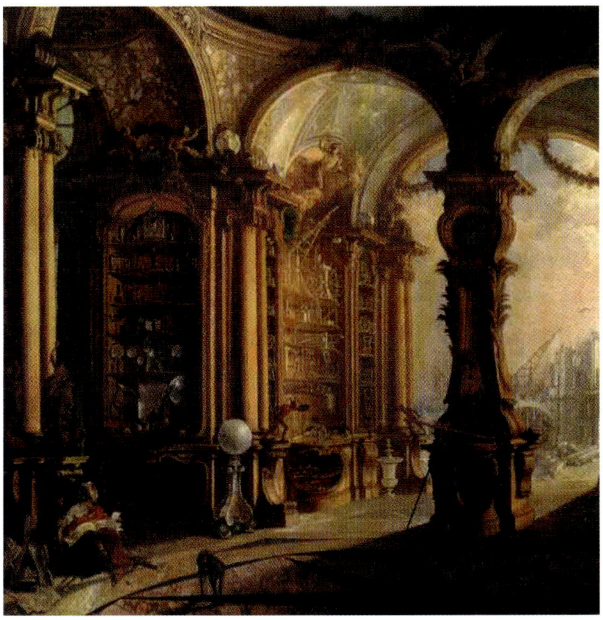

The intricate allegorical play of Lajoue's painting is only deepened when we note its original function as an overdoor in one of the natural history rooms. In Courtonne's drawing we find it there accompanied by a gazelle's skull (*plate* 9). In an amusing and remarkably self-aware *mise-en-abîme*, Bonnier offers his visitors a signpost, a preview or teaser of what is to come once they pass through the doorway. One wonders what other playful yet meaningful caprices could have been found in Bonnier's cabinets. Was the tableau mécanique meant to play a similar serious joke on time, its measurement, and ideas of progress via the patent connection of a clock mechanism with a representation of construction and building? If we assume for a moment that Lajoue's painting can be read as a statement on progress and the advancement of technology while remaining a decorative element, we can raise the question of the tableau mécanique's significance within the thematic strands woven by Bonnier through his cabinets. In his description of the Physics Cabinet, Gersaint states that it addresses various problems in mechanics, statics, arithmetic, and geometry. Consequently, we might ask whether the tableau mécanique – neither a model nor an instrument or a device in the strict senses of these words – can be categorized as a (philosophical) problem as much as an example of mechanical ingenuity. Part of the answer lies in the tableau mécanique's resonance with Lajoue's temporal theme and involves the connection between the representation of labour in the painted surface and the clock mechanism behind it, which were deeply linked in what was then called mechanical philosophy. It might be hard to precisely determine Bonnier de la Mosson's vision of this approach, but we can safely assume that he knew all about it.

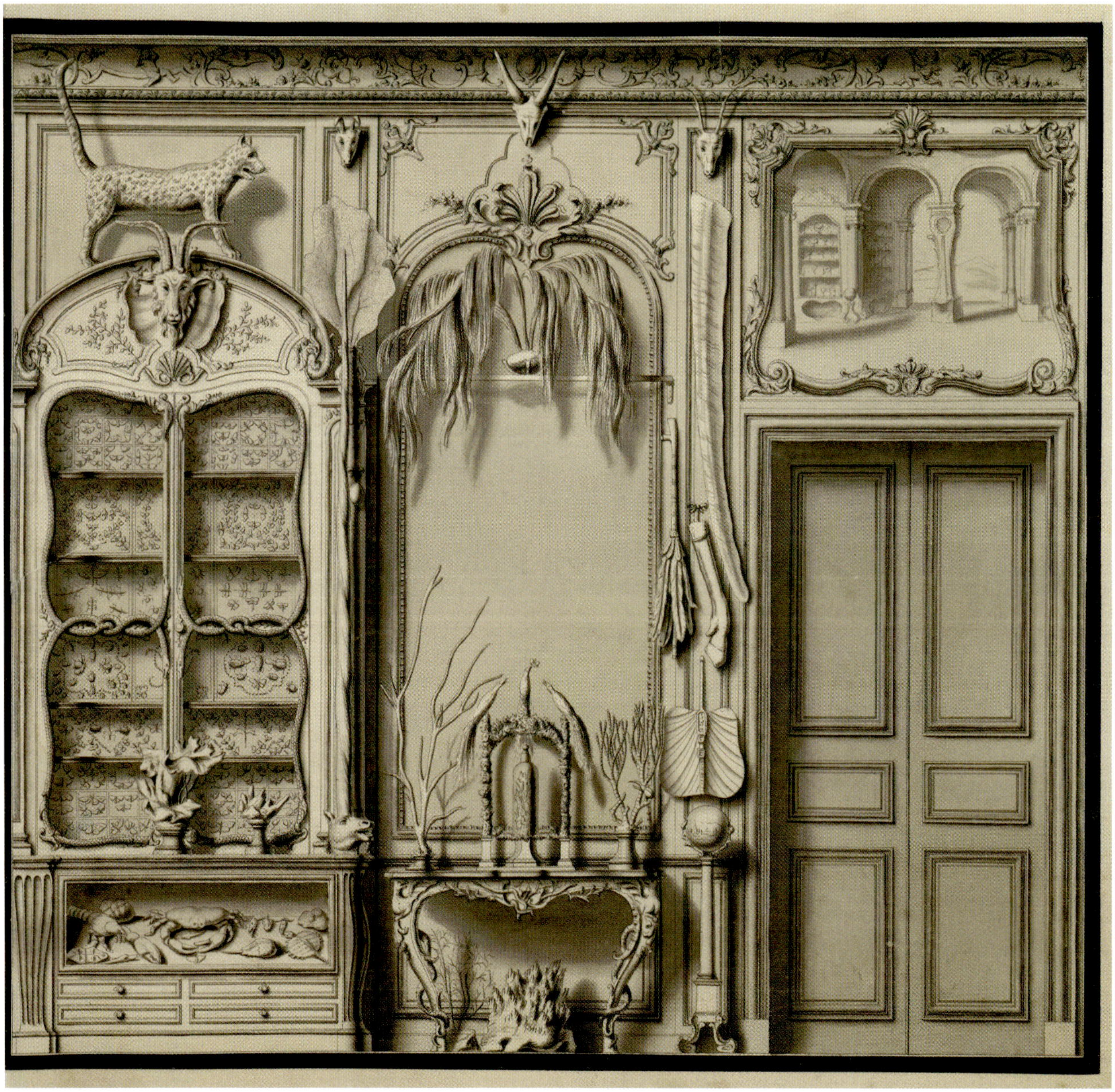

9 Detail of Jean Courtonne and Jean-Baptiste Courtonne, *Cabinet de Bonnier de la Mosson, The Physics and Mechanics Cabinet*, 1739–40, showing plate 7, part of second cabinet including the overdoor painting by Lajoue. Pen and ink on paper, 37 × 124 cm. Paris: Bibliothèque de l'Institut National d'Histoire de l'Art (collections Jacques Doucet). Photo: INHA/RMN-Grand Palais.

An Embodied Clock Metaphor

Mechanical philosophy is an umbrella term for a series of theories developed during the Scientific Revolution that are linked in their assumption that the world is not built of organisms, as Aristotle had thought, but of discrete units of moving matter that operate together like a machine.[20] Nature was considered to be mechanical in that it was made up of unthinking, inanimate matter set in motion as if a piece of machinery, such as, indeed, an automaton or a clock. This metaphor was derived from Descartes, who, fascinated by automata from an early age (he had even invented one of his own, a tightrope walker powered by magnets), spices up his oeuvre by repeatedly making references to them.[21] For instance, in the second of his *Meditations on First Philosophy* (1641), discussing the unreliability of the senses, Descartes wonders, 'If I look out of the window and see men crossing the square, as I just happen to have done, I normally say that I see the men themselves .... Yet do I see any more than hats and coats which could conceal automatons? I *judge* that they are men.'[22] Already in the fifth part of his *Discourse on the Method* (1637), Descartes had made his now infamous claim that animals were soulless vehicles of locomotion, echoing Aristotle's comparison of animals to 'automatic puppets' in his *Movement of Animals*.[23] A description in another of his works, *Passions of the Soul*, is particularly apt: 'Every movement we make without any contribution of our will – as often happens when we breathe, walk, eat, and indeed, when we perform any action which is common to us and the beasts – depends solely on the arrangement of our limbs and on the route which the spirit, produced by the heat of the heart, follow naturally in the brain, nerves, and muscles. This occurs in the same way as the movement of a watch is produced merely by the strength of its spring and the configuration of its wheels.'[24]

Descartes's *Treatise on Man*, written in the early 1630s but published posthumously in 1662, fully embraces the machine metaphor when it presents the human body as an automaton that houses all the different parts required for walking, eating, and breathing. Descartes describes the various organs – stomach, liver, heart, brain – one by one, laying them out in front of us (the text is illustrated) as if he were dismantling an automaton to show its operation by means of its parts. 'I should like you to consider', he writes at the end, 'that these functions follow from the mere arrangement of the machine's organs every bit as naturally as the movements of a clock or other automaton.'[25] In the ensuing international debate about Descartes' ideas, subsequent thinkers picked up on the clock metaphor. Leibniz and Wolff in Germany, Pascal and Malebranche in France, and Boyle and Hobbes in England considered the clock to be an archetypal machine, and as a consequence an excellent image to explain the workings not only of the body, but of the world as such. In the popular yet serious series of dialogues published in 1686 by Bernard de Fontenelle (1657–1757), the philosopher-narrator explains to his interlocutor, the Countess, that 'philosophy is now become very mechanical .... They will have the World to be in great, what a Watch is in little', upon which the countess wittily replies that she values the world all the more now that she knows it resembles a watch.[26] Robert Boyle in particular used the clock metaphor to explain the theological dimension of mechanical philosophy. In his *Excellence of Theology Compared with Natural Philosophy*, he explains that the world is 'a great piece of clock-work, the naturalist, as such, is but a mechanician'.[27] The difference really is only in the size of the parts that make up the engine. He praises God as the supreme artist for having framed 'so great and admirable an automaton as the world, and the subordinate engines comprised in it, for several ends and purposes'.[28]

Looking closely again at the tableau mécanique, we can see how literally the mechanization of the world-picture has been taken up by Spayement and Desmares, whose image of civilization – made up of nature and culture – hides an enormous clock mechanism. Instead of stating that the world consists of different parts, it actually shows, at the top of every hour, how its separate units move on their own as parts of an invisible engine. In his typically dry and systematic style, Kant would later describe mechanical philosophy as 'the method of explaining the specific differences of materials through the constitution and composition of their smallest particles in terms of machines', and this is precisely what is presented to us here.[29] The clock does not stand for a world picture, but it *is* a world picture, as much as the picture is a clock. Both aspects of the work – clock and picture – are further linked in the way each is put together, for they both depend on a careful attention to framing.

However, the tableau mécanique reveals not only the potential but also the limitations of the clock analogy, posing several problems with regard to the nature of the self-moving mechanism which imply that the course of the world is predetermined. Standing in front of this moving image, we may say, with Fontanelle's Countess, that we value the world all the more now that we are considering it as a clock. However, we also start wondering about the repetitiveness of the figures' movements, and if they might ever escape the limited scope of their programmed gestures. The picture suggests that just as the cow can only graze, and the chickens will forever continue to pick in the sand, so the women will wash their clothes, the fisherman fish, and the workers perform their labour in an endless, repetitive round.

The clock metaphor's determinist implications had already been recognized by Boyle, whose solution to the problem was to distinguish between those who see the world as a deceptive puppet theatre, and the more refined minds, who see it as:

> a rare Clock, such as may be that at *Strasbourg*, where all things are so skillfully contriv'd, that the Engine being once set a Moving, all things proceed according to the Artificers first design, and the Motions of the little Statues, that at such hours perform these or those things, do not require, like those of Puppets, the peculiar interposing of the Artificer, or any Intelligent Agent imployed by him, but perform their functions upon particular occasions, by vertue of the General and Primitive Contrivance of the whole Engine.[30]

For Boyle the essential difference was that, like the parade of figures on the Strasbourg clock, human beings carry out certain tasks not because they are moved by a greater causal agent holding the strings, but because they are compelled by the force of the engine as a whole.

The predetermination implied in the clock analogy was a hotly debated topic in the two decades leading up to the creation of Spayement and Desmares's tableau mécanique. The most extreme determinist was Christian Wolff (1679–1754), whose controversial ideas caused him to lose his professorship at the University of Halle in 1723, a result of the personal intervention of King Frederick of Prussia. According to many contemporaries, Wolff, who considered the clock analogy to be the foundation for his philosophy, had simply gone too far. Taking the analogy too literally, he needed to find a solution for the fact that a clock needs to be wound up, and therefore he claimed that the world was ruled by machinery steered from 'outside', implying a confinement for everyone within it. Wolff's downfall signalled a new direction for the automaton metaphor. Kant, by that time weary of the clock analogy, criticized it precisely for the lack of freedom it gave to the individual human being – who would end up as 'a marionette or an automaton like Vaucanson's, fabricated and wound up by the Supreme Artist'.[31]

The question arises as to the extent Spayement and Desmares' tableau mécanique, created in the interval between the end of the clock metaphor's heyday and the start of the automaton craze, reflects the intense contemporary debate around the clock metaphor. By its unusual composite of image, instrument, and clock, its conjunction of a medium of stillness and a mechanism of perpetual motion, it in fact seems to embody a 'problem', of the sort Gersaint claimed with regard to other items of geometry or hydraulics. Does this object, in many ways a kind of materialized metaphor, make a statement about the clock analogy? If we assume for a moment that it does, can we say then that it sides, for instance, with Boyle's enthusiastic embrace of this metaphor? Or, in contrast, does it anticipate Kant's critique of a clockwork world of puppetry? The answer lies in the various images of temporality that are hidden (or not so hidden) within the otherwise mundane scene. The various kinds of labour we have seen are coupled with different modalities of time. The to-and-fro movements of the two sowing men represent repetition, the cattle stand for the cycle of nature, the women doing household tasks are emblems of the tedious sequence of domestic work carried out each day. Idleness is embodied in the smoking figure on the extreme right, who personifies a *vita contemplativa* that counterbalances the *vita activa* of the other figures. Finally, the fisherman stands for patience. As in Lajoue's painting, the sequence of components when read from left to right articulates the idea of progress: from the stonemasons busy carving on the left, to the figures behind dragging slabs of stone on wheelbarrows over the bridge and beneath the arch to the construction site to the right. A sculptor sitting in the tympanum in the top right corner putting the final touches on a reclining female figure that embodies rest is the closing feature of this chain. Similarly, the river streaming diagonally through the scene connecting nature with culture symbolizes eternity as well as the ever-changing course of history. Heraclitus' famous dictum (widely known at the time) that everything changes and nothing stands still, and that no man steps in the same river twice, gains new meaning in the context of this picture, which thematizes time as much as it embodies it literally. And yet there is more. At the beginning of the *Treatise on Man*, Descartes mentions two other manmade machines in addition to a clock: the artificial fountain (models of which were on display in Bonnier's cabinet) and the mill. In the tableau mécanique, we see represented both a water-mill and a windmill, devices that offer a subtle comparison between animal movement and human labour and true machinery. They have the power to move on their own account, and each has its own temporality. All these separate parts are equal when they are set into motion the moment the hour-mark arrives. We could say that the clock-picture poses here the philosophical problem as to whether this conception of order is something to strive for, or whether it is on the contrary something to fear, as a form of indoctrination and oppression brought about by mechanics.

A Mechanical Monster

The most telling elements in the composition of the tableau mécanique can be read, perhaps, as making a gesture towards the future of the clock metaphor. As we have seen, at the bottom of the hill stands a seesaw, occupied by two children at play. A very simple mechanism, the seesaw signifies balance and its image would replace the clock metaphor in the next half of the century, when the debate on mechanical philosophy turned from comparisons of body and machine towards government and nation. The seesaw or equalizer would serve as a counterargument against the determinism of Boyle and to a certain extent Wolff. It is particularly significant that it is here occupied by children: in the far-reaching allegorical mode of reading so popular with Bonnier and his circle, this would of course mean that it represents the future.

In *Authority, Liberty & Automatic Machinery in Early Modern Europe* (1986), Otto Mayr demonstrates how the authoritarian system of ruling body and world evoked by the clock, dominant before the mid-eighteenth century, was slowly to be replaced by a notion of liberalism embodied by the figure of balance.[32] Terming this new view on government 'equilibrium thinking', Mayr argues that under the influence of the ideas of Isaac Newton and Jean-Jacques Rousseau (who was, in fact, the son of a clockmaker), complex dynamic systems were capable of being self-balancing or self-regulating. Equilibrium thinking culminated in the work of Adam Smith, in whose *Wealth of Nations* (1776) self-regulation is a central concept. It may be too much to say that the equalizer in the tableau mécanique somehow anticipates the equilibrium thinking that would soon replace the clock metaphor, but it certainly hints towards such thinking. The utopian image of a world order under an authoritarian system that this picture-instrument celebrates, where each worker carries out his task with clockwork precision, was under threat. It would soon be demolished by the extreme results of Smith's liberalism and replaced by the grim consequences of the Industrial Revolution that changed work into labour, concrete labour into abstract labour, and a day's worth of activity into exploitation, where, ultimately, balance would never be found without revolution.[33] Further support for the apparently inevitable consequences of the clock mechanism on society is found in the interesting leaps in Vaucanson's career. Having lost interest in his defecating, writing, and breathing automata, he wished to serve society and intended to modernize the French textile industry. His designs for an automatic loom never materialized during his lifetime, but they did eventually inspire Joseph Marie Jacquard to invent the mechanical loom around 1800. The punch-card system regulating the weaving operation originally devised by Vaucanson ultimately became a major step in the development of computer programming. In this light, we might wish to revisit the tableau mécanique as the embodiment of a worldview conditioned by clock technology, considering the transformation of this pretty picture of perfect working conditions in the hands of Karl Marx. Whereas Kant and others observed the shortcomings of mechanical philosophy in turning the world into a puppet theatre where humans were without free will, and whereas Smith and his followers would see liberalism as its solution, Marx fully realized the truly horrific consequences of the fusion of the authoritarian clock-regulated system with the free-market system of liberalism that was meant to overcome it. By evoking the same metaphors as his seventeenth-century predecessors, Marx brought the clock metaphor to its logical conclusion when he writes that instead of finding a balance, technological progress had produced a monster at the exact moment when its mechanism was at its most refined:

> An organized system of machines, to which motion is communicated by the transmitting mechanism from a central automaton, is the most developed form of production by machinery. Here we have, in the place of the isolated machine, a mechanical monster whose body fills whole factories, and whose demon power, at first veiled under the slow and measured motions of his giant limbs, at length breaks out into the fast and furious whirl of his countless working organs.[34]

A far cry yet from such monstrosity, the tableau mécanique contributed to the debate prior to Marx about the progress of technological advancement as well as its dangers. Its significance as a philosophical painting has only gained more relevance in the centuries after its making, and will likely continue to do so in the years to come.

Notes

1 For a more detailed description of the workings of the clock mechanism, see Marianne Roland Michel, 'Le Cabinet de Bonnier de la Mosson et la participation de Lajoue', *Bulletin de la société de l'histoire de l'art français*, 1975, 211–21.

2 A vulgarized version of this marvel appears later in the nineteenth century in the form of the more common 'clock pictures', usually inferior naive paintings of cityscapes featuring a church tower with a working clock inserted in it.

3 On temporal literacy in eighteenth-century Paris, see Charissa Bremer-David, 'About time: The hours of the day in eighteenth-century Paris', in Charissa Bremer-David, ed., *Paris: Life and Luxury in the Eighteenth Century*, Los Angeles, CA, 2011, 11–31.

4 Louis-Sébastien Mercier, *Tableau de Paris*, vol. 4, 1782, chapter 330.

5 Quoted in Bremer-David, *Paris*, 25.

6 Minsoo Kang, *Sublime Dreams of Living Machines*, Cambridge, MA, 2011, in particular chapter 3. See also Adelheid Voskuhl's splendid *Androids in the Enlightenment: Mechanics, Artisans, and Cultures of the Self*, Chicago, IL, 2013.

7 Quoted in Reed Benhamou, 'The automaton in eighteenth-century France', *Studies in Eighteenth-Century Culture*, 17, 1987, 92. For more details on Vaucanson, see Jessica Riskin's excellent essay 'The defecating duck, or the ambiguous origins of artificial life', *Critical Inquiry*, 29, Summer 2003, 599–633.

8 Hermann von Helmholtz, 'On the interaction of natural forces', in *Popular Lectures on Scientific Subjects*, trans. E. Atkinson, New York, 1885, 154. Also quoted in Thomas L. Hankins and Robert J. Silverman, eds, *Instruments and the Imagination*, Princeton, NJ, 1995, 223. A fascinating discussion of the significance of automata and their afterlife in nineteenth-century literature is given by John Tresch in 'The machine awakens: The science and politics of the fantastic automaton', *French Historical Studies*, 34: 1, Winter 2011, 87–123.

9 Benhamou, 'Automaton', 98.

10 Riskin, 'Defecating duck', 609. Various *méchanistes* continued to design prostheses for people with amputated limbs; see Benhamou, 'Automaton', 100–1.

11 Julien Offray de La Mettrie, *Man a Machine*, trans. by Marquis D'Argens, Dublin, 1749, 62.

12 On Parisian cabinets of curiosities, see Bettina Dietz and Thomas Nutz, 'Collections curieuses: The aesthetics of curiosity and elite lifestyle in eighteenth-century Paris', *Eighteenth-Century Life*, 29: 3, Fall 2005, 44–75.

13 See Andrew McClellan's article 'Watteau's dealer: Gersaint and the marketing of art in eighteenth-century Paris' in *The Art Bulletin*, 78: 3, September 1996.

14 Edme-François Gersaint, *Catalogue Raisonné d'une Collection considérable de diverses Curiosités en tous genres, contenues dans les Cabinets de feu Monsieur Bonnier de la Mosson*, Paris, 1744. For a detailed discussion of Bonnier's cabinet, see also C. R. Hill, 'The Cabinet of Bonnier de la Mosson (1702–1744)', *Annals of Science*, 43: 2, 1986, 147–74. The reference to Bonnier's preferred route can be found on page 154.

15 Roland Michel, 'Le Cabinet', 231.

16 '…ce qui augmente encore le mérite de cette piece, qui, indépendamment de ce dernier mérite, en a par elle-meme assez pour pouvoir piquer la curiosité d'amateur' quoted by Edouard Gélis in 'A propos d'un tableau méchanique du Musée des Arts décoratifs', *Bulletin de la Société de l'Histoire de l'Art Français*, 1923–24, 254–72. The quote can be found on page 265 in Gélis' article and in no. 171 of Gersaint's *Catalogue*.

17 See for more information on tricksy objects Mimi Hellman, 'Furniture, sociability, and the work of leisure in eighteenth-century France', *Eighteenth-Century Studies*, 32: 4, 1999, 415–45.

18 Courtonne's drawings have also been discussed in detail by Katie Scott, in *The Rococo Interior: Decoration and Social Spaces in Early-Eighteenth-Century Paris*, New Haven, CT, 1995, 169ff., and by Hill in 'Cabinet'.

19 For a comparison between Courtonne and Lajoue see Scott, *Rococo Interior*, 169–76.

20 For the role of automata in mechanistic philosophy, see Derek J. de Solla Price, 'Automata and the origins of mechanism and mechanistic philosophy', *Technology and Culture*, 5: 1, Winter 1964, 9–21; and Silvio A. Bedini, 'The role of automata in the history of technology', *Technology and Culture*, 5: 1, Winter 1964, 24–42. The classic study on automata is Alfred Chapuis, *Le monde des automates: étude historique et technique*, 2 vols, Paris, 1928.

21 Otto Mayr, *Authority, Liberty & Automatic Machinery in Early Modern Europe*, Baltimore, MD, 1986, 63.

22 René Descartes, *Meditations on First Philosophy with Selections from the Objections and Replies*, ed. John Cottingham, Cambridge, 2013, 43.

23 See Kang, *Sublime Dreams*, 117.

24 René Descartes, *The Philosophical Writings of Descartes* vol. 1, trans. John Cottingham, Robert Stoothoff, and Dugald Murdoch, Cambridge, 1985, 335.

25 Descartes, *Philosophical Writings*, 108.

26 Bernard le Bovier de Fontenelle, *A Plurality of Worlds*, trans. John J. Glanvell, London, 1688, in Leonard M. Marsak, ed. and trans., *The Achievement of Bernard le Vovier de Fontenelle*, New York, 1970.

27 Quoted in Mayr, *Authority*, 88.

28 Robert Boyle, *A Free Enquiry into the Vulgarly Received Notion of Nature*, n.p., 1686. Quoted in Kang, 113.

29 Quoted in Mayr, *Authority*, 57.

30 Robert Boyle, *The Works of Robert Boyle*, ed. Michael Hunter and Edward B. Davis, 14 vols, Charlotteville, 2003, vol. 10, 448.

31 Quoted in Kang, *Sublime Dreams*, 108.

32 Mayr, *Authority*, part two 'Liberal Systems', 139–200.

33 See for an in-depth discussion of time keeping during the Industrial Revolution E. P. Thompson's classic essay 'Time, Work-Discipline and Industrial Capitalism', *Past and Present*, 38, December 1967, 56–97.

34 Karl Marx, *Capital, Volume One: A Critique of Political Economy*, London, 1867, 417.

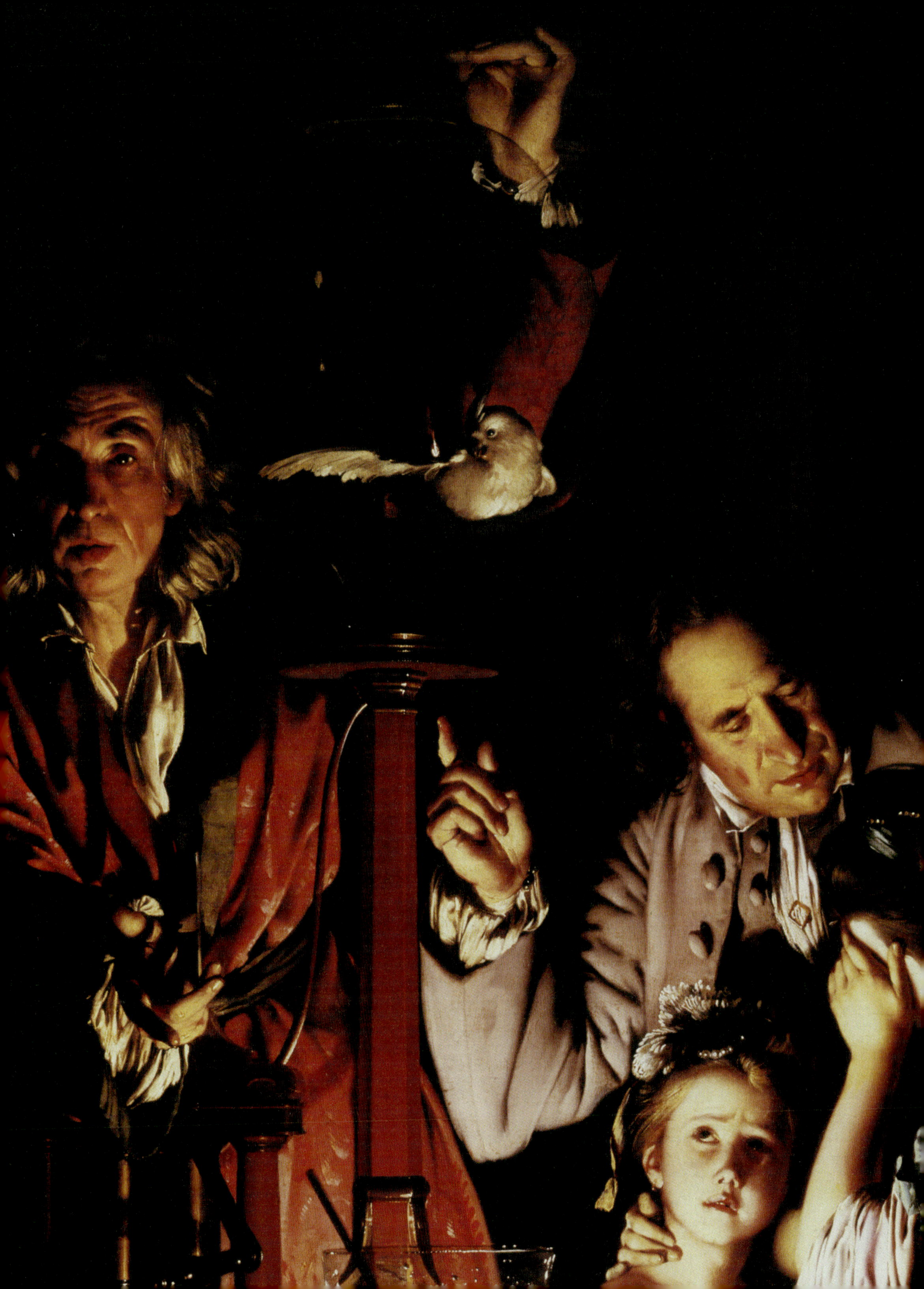

Chapter 9
Of Air Pumps and Teapots: Joseph Wright of Derby, John Singleton Copley and the Technology of Seeing

Bryan J. Wolf

On a dark night in 1779, a crowd of striking labourers, disgruntled with working conditions in the mills of Derbyshire, marched through the countryside intent on destroying the 'engines' of the Industrial Revolution. Their itinerary included manufacturing centres in Bolton, Manchester and Stockport. Their final goal was Cromford, site of Richard Arkwright's massive cotton factories. Arkwright declared Cromford to be in a state of siege and supplied his factories with hastily recruited defenders and a battery of cannon. The 'rioters' never reached their destination. They were dispersed by government troops sent from Liverpool. A period of severe repression followed, during which several labourers were sentenced to hang.[1]

Not long after, in the years around 1782–83, Joseph Wright of Derby also marched to Arkwright's mills. He too was interested in certain forms of redress, but not ones that the workers would have recognized or accepted. His goal was to paint Arkwright's cotton mills by moonlight, transforming what others perceived as a blot upon the landscape into a sublime and nocturnal image (*Arkwright's Cotton Mills by Night*, c. 1782–83, *plate 1*).[2] The figures who appear in the foreground of Wright's painting – a man escorting a horse-drawn buggy, ghostly human forms around the base of the mill – have little to do with the angry crowds that surged through the Midlands countryside. Wright's figures allude instead to the conventions of picturesque painting: swains, rustics and country labourers living out their Arcadian pleasures in the foreground of Italianate paintings. Wright's figures are pendants to the main action of the canvas. They provide human perspective to an image that finally is not about people at all. Wright is concerned with power, with a new visual technology different from the brass and steel machines that drove Awkwright's mills. Wright refuses the viewer any single focal point and forces the eye to scramble between moon and clouds on the one hand and terrestrial factories on the other. Whether the two collaborate or compete is uncertain. What is clear from the painting is that we, rather than the figures within it, are the only ones granted access to the full drama.

We might conclude from this that Wright has painted history out of the canvas. He has eliminated the fierce contest between workers and manufacturers that marred the Midlands landscape. The women and children who laboured at night through twelve-hour shifts, with one hour off for meals, have been replaced by rows of illuminated windows. They have disappeared into quadrants of light, just as their peers outside the factory have merged indistinguishably with the staffage of picturesque art. In their stead, and in place of the social turmoil that their lives

Detail from Joseph Wright of Derby, *Experiment on a Bird in an Air Pump*, 1768 (*plate 3*).

Art and Technology in Early Modern Europe *Edited by Richard Taws and Genevieve Warwick*.

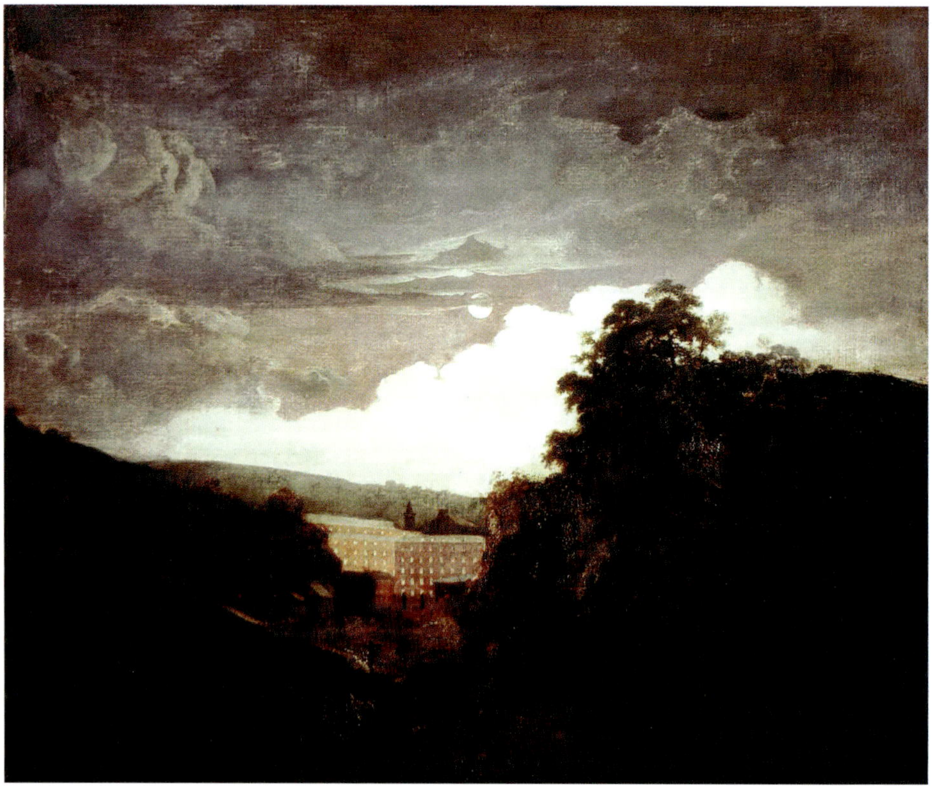

1 Joseph Wright of Derby, *Arkwright's Cotton Mills by Night*, c. 1782–83. Oil on canvas, 99.7 × 125.7 cm. Derby: Derby Art Gallery. Photo: © Eileen Tweedy/ The Art Archive at Art Resource, NY.

chronicle, we have aesthetics: the binary economy of moon and factory, the safe but distant vantage of the viewer, the strangely serene atmosphere, and the emergence of a sublime subject from a picturesque composition. If this is history, then it is a roseate one, a romanticized account of the Industrial Revolution as an interlude – and a not uncurious one – in the history of aesthetics from Burke to Wordsworth.

We might add that Wright's own class position appears to emerge clearly from works like *Arkwright's Cotton Mills by Night*, and the subsequent portrait of Arkwright painted a few years before his death. In *Sir Richard Arkwright*, 1789–90, Wright places on the table to Arkwright's left a model of the cotton-spinning rollers that underwrote Arkwright's achievements (*plate* 2). The model substitutes for the family emblems or elegant possessions that we might expect in a more traditional portrait. Wright painted the image soon after Arkwright had been stripped of his patent rights to the cotton roller by a court decision ruling that Arkwright had stolen rather than invented the machine.[3] The court ruling helps explain the 'history-be-damned' attitude that characterizes not only Arkwright's expression but Wright's awkward collusion in an alternative version of history. Through its innovative realism, as direct and blunt as the man it portrays, Wright's painting reunites Arkwright with his machine. The canvas asserts the power of an industrial elite to ignore or defy legal niceties in its self-portrayals; it also affirms the right of the artist, riding the tailcoats of that power, to claim history as nothing more than a story of his own making. Arkwright emerges from Wright's portrait as the master not only of his machine but of the new commercial power it produced. And Wright stands as an apologist for a new world order, the chief mythographer of the Midlands manufacturing elite.

Such at least is the story that any socially minded art historian might tell. But I am not convinced that this is the right story, let alone Wright's story. For all its efforts to harness art and history together into a sympathetic yoke, this account reduces Wright to the role of spokesman for the industrial classes that bought his work, and interprets his painting as a second-order phenomenon: the legitimation of events understood as outside or prior to the painting itself. If we return to *Arkwright's Cotton Mills by Night* for a second glance, we find hints of a different story. Wright seems to be moving the centre of attention from the content of the canvas – the mills, the night, the landscape – to the process of seeing itself. The binary play between sky and factories expands into triadic terms: moon and mills are triangulated against the viewer, whose comprehending gaze subsumes the entirety of the picture at a glance. The importance of the viewer is highlighted by the contrast he provides with the figures within the painting. They lack perspective of the whole. They function not only as reminders of the picturesque origins of the painting's conventions, but as a testament to the partiality of all viewing positions from within the canvas proper. Only by standing outside the picture frame do we gain the necessary vantage point.[4]

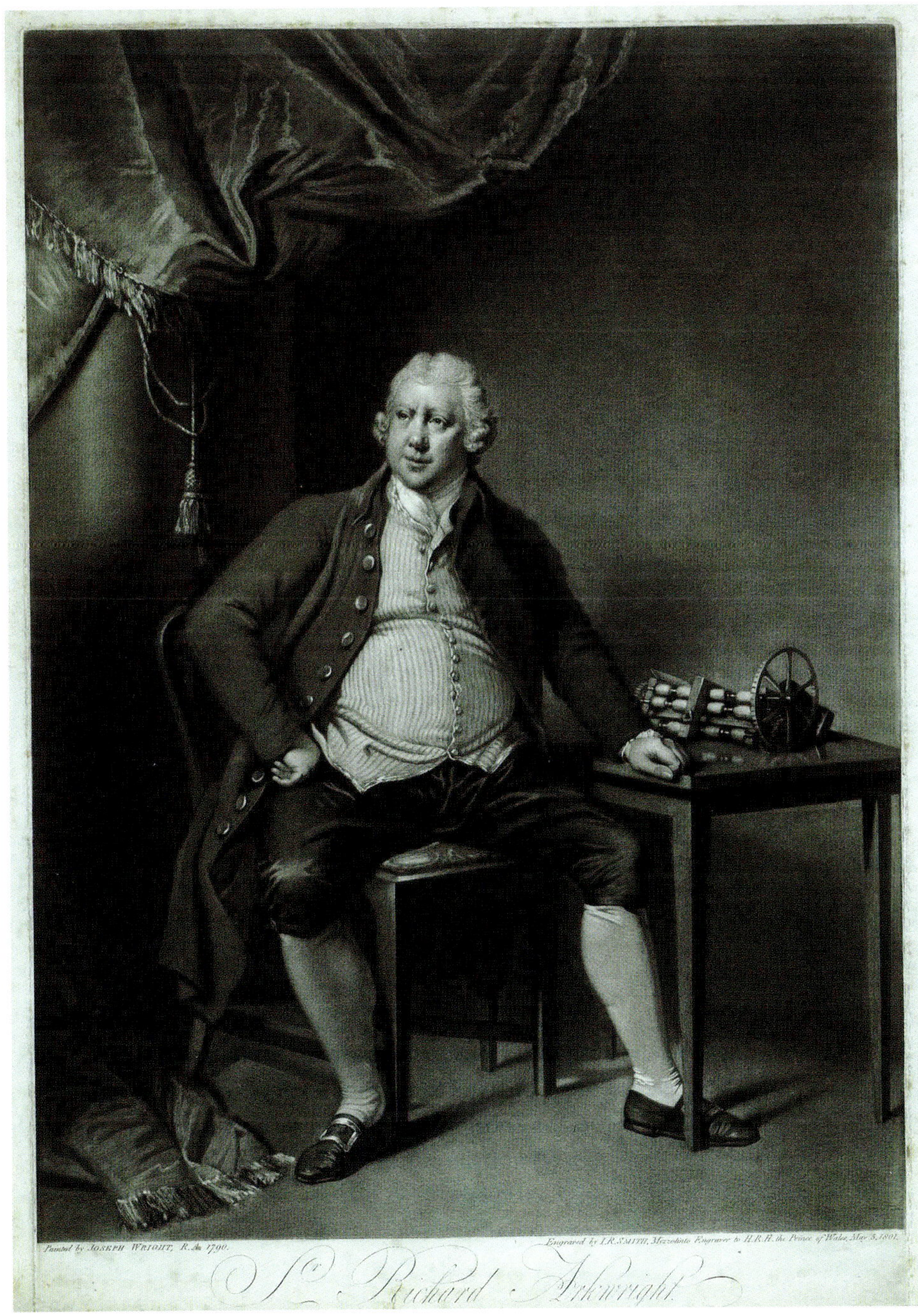

2 Joseph Wright of Derby, *Sir Richard Arkwright*, 1790. Engraving, 68.4 × 52.5 cm. Derby: Derby Art Gallery. Photo: © John Webb/The Art Archive at Art Resource, NY.

The canvas, in other words, is populated with figures who are ciphers of partiality and then centres, beyond the frame, an observer defined by the difference his distance makes. We might understand Wright's construction of the viewer in political terms. Wright's canvas is saturated with civic humanist ideology. It is constructed to flatter the viewer as a disinterested observer whose vision of the whole fits him for tasks of a civic nature. The painting grants a privileged position to an observer whose class position, like his visual one, places him beyond parochial or partisan interests. John Barrell has characterized ideologies of 'civic humanism' as inflating the value of disinterestedness at the expense of manual labour.[5] The labour of the rustic dooms him, like Wright's figures, to partiality and dependence on others. Our ability to see from the other side of the picture plane thus tropes our fitness for cultural or political leadership. We are enfranchised by our perspective, our sense of the whole, a capacity to maintain a larger vision in the face of competing parochial concerns.[6] Those immersed within the world of labour occupy a structurally anomalous position. They are disqualified by their limited point of view from the business of governance. Seeing thus confirms already extant social hierarchies.

Alternatively, we might say that Wright has put the sublime to political work. He has wed a fantasy of control (the viewer's) to a fantasy of power and irrationality (the mill, the moon, the nocturnal landscape). He has turned this relation into a parable of class relations: those outside the picture frame, those with the leisure to contemplate it or money to possess it, belong to a distinctly different rank of people from those visible within the canvas's four sides. These relations in turn mirror the logic of civic humanist ideology, which itself has been transplanted from an agrarian to a commercial context.

But there are really two claims at work here, and they lead in different directions. The first concerns class: it returns us to the role of social historians by turning Wright into a painter of the Midlands industrial elite. It equates the position of the viewer with that of Wright's patrons, and so converts Wright into a not unsubtle champion of the Arkwrights of his day. Those who see are, by sublime formula, those who control and possess. But there is a catch here, and this leads to our second claim. The right of authority granted to the viewer of *Arkwright's Cotton Mills* stems from his position as a spectator. What is most radical about the painting is not the predictable link it establishes between class and authority, its implicit claim that the Arkwrights of the world constitute a new ruling class. Rather it is the construction of *seeing* itself as a new medium of power that interests us. What I am suggesting is that Wright's deck is stacked in a revealing way. Seeing behaves as a wild card in Wright's hand. It provides an access to power that may or may not coincide with traditional definitions of class. In the lamplit pictures of the early phase of Wright's career, as in the later nocturnal images of landscapes, volcanoes and grottoes, Wright plays out a fascination with seeing and the act of representation that complicates questions of class relations. His canvases betray an anxiety about the status of art and a willingness to claim for painting cognitive roles modelled at various moments on science, crafts or industry. What concerns us are not the metaphors by which Wright seeks to understand his own painting, nor the ideological jockeying that he performs on behalf of his Derbyshire patrons, but his willingness to make radical claims on behalf of seeing itself. If I might anticipate where I wish to conclude, I believe that Wright's painting claims for the artist a power of representation – a new 'technology of seeing' – that differs from commercial forms of production and sets him at odds with his own industrial clientele. However much Wright might legitimate the aspirations of a new commercial elite within his art, his own interests lie elsewhere: in ways of seeing

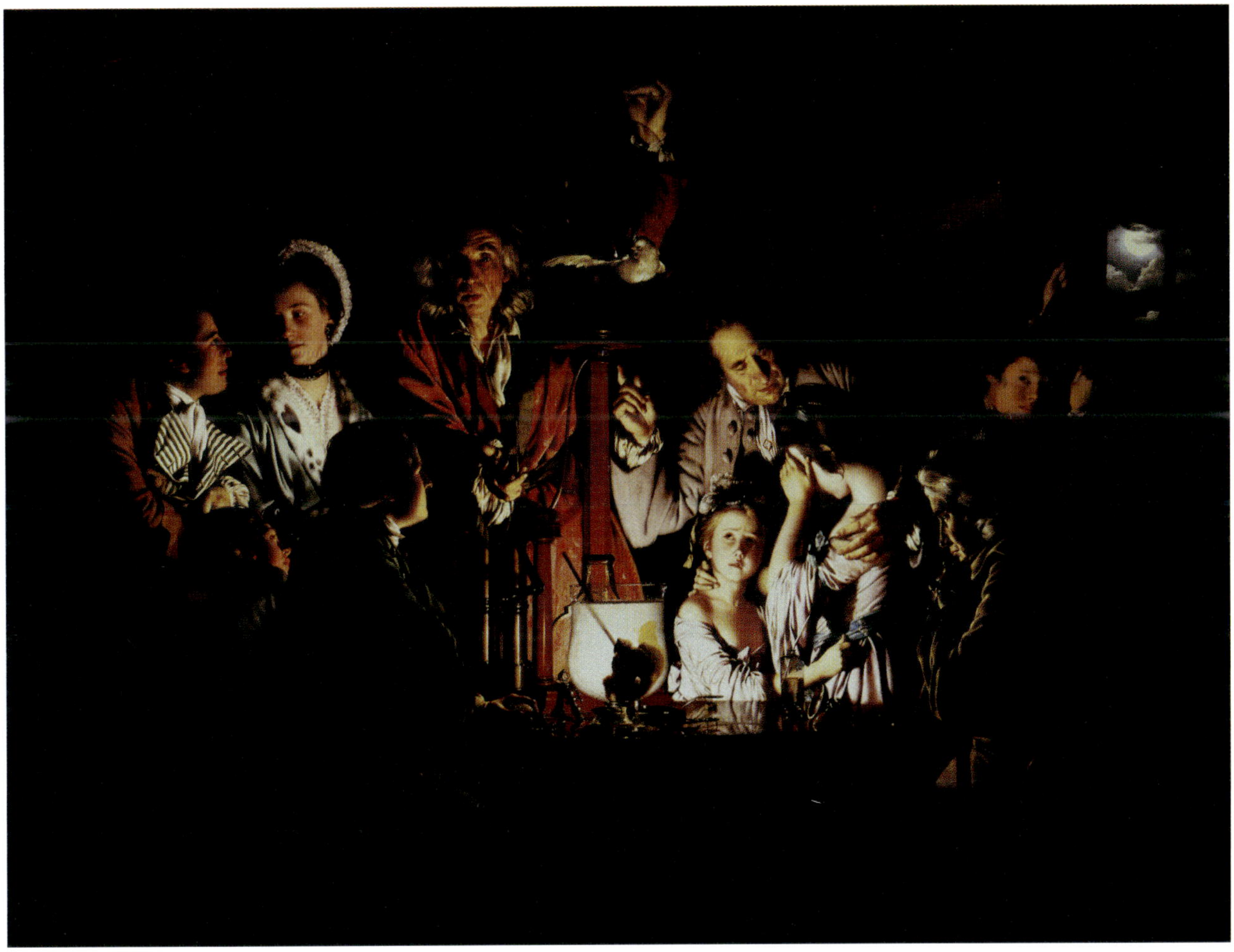

3 Joseph Wright of Derby, *Experiment on a Bird in an Air Pump*, 1768. Oil on canvas, 182.9 × 243.9 cm. London: National Gallery (Presented by Edward Tyrrell, 1863). Photo: © National Gallery, London/Art Resource, NY.

that codify culture in new forms; in the relation that power bears to representation; in culture itself as an instrument of social control.

These concerns are present from the very start of Wright's career. They are explored most profoundly in his early masterpiece of 1768, *Experiment on a Bird in an Air Pump* (*plate* 3). What happens in 1768 – or thereabouts – is that art changes horses in mid-stream. John Barrell has suggested that the central narrative of history painting in eighteenth-century England centres upon the role of civic humanism in the production of artistic ideologies.[7] The thread that links Reynolds to Blake, and both to Hazlitt, revolves around their common effort to position painting within a discourse of disinterested humanism. Joseph Wright's painting suggests a somewhat different story. Though civic humanist tenets surface frequently in his art, the logic that animates the painting stems neither from Country Party ideology nor the concerns or agenda of the Lunar Society, a group with which he is frequently linked.[8] Rather *Experiment on a Bird in an Air Pump* marks a new way of thinking about painting in the eighteenth century. It is a self-reflective image, a meditation on its own status as art, a study, as Michael Fried might term it, about 'representing representation'.[9] The painting questions and ultimately refuses the distinction between public and private that grounds the discourse of civic humanism. It turns instead to the powers of representation as an alternative source of social and cultural authority. Its claims are both epistemic and ontological: that art inaugurates ways of knowing different from those available elsewhere in the culture; that *what* one sees is less important than the *how* of seeing; that power lies in the extension of vision outside traditional class lines.

Wright's art is possible only because it has been preceded by a long prehistory that has less to do with the history of art than it does with the history of property. Wright lived in a post-Lockean world. Though Locke's *Second Treatise of Government* justified property and enclosure from the point of view of the landed estate holder, his terms belied the influence of commercial ideologies on agricultural interests in Restoration England. When Locke proclaimed that 'every Man has a Property in his own Person', and reasoned from this premise that the 'Labour of his Body, and the Work of his Hands ... are properly his', he set himself on the road to a labour theory of value detached from agrarian assumptions.[10] Though Locke thinks of property primarily as land, his successors such as Adam Smith could extend his logic without his political and social alliances to the gentry. The result was a notion of property as fluid, mobile and subject to exchange; property as part of a market economy rather than property as a fixed and unchanging agricultural unit.

Wright's art is unthinkable except in the context of this transformation. He can reconceive the idea of representation because the social and economic matrix that frames it is different from what it had been a century earlier. Wright's task is to rethink the role of painting in a commercial culture defined by new forms of leisure and consumption, altered forms of social relations, property understood as mobile rather than fixed, and ideologies of uplift, education and sentiment. As social life becomes increasingly subject to the laws that govern commodities, so too aesthetic life takes new form under the influence of the market. Wright represents that moment in British painting when the artist takes full stock of this change. Not, I should note, because he commented directly upon the commercial or legal revolutions surrounding him (though his account book reveals a business-like willingness to charge interest on the unpaid portion of purchases), but because he perceived these issues as aesthetic ones. And aesthetics, in turn, had historical work to perform. In reimagining the role of art in a commercial society, Wright was reconceiving the very possibilities of painting itself.

Socialization

Experiment on a Bird in an Air Pump represents Wright's meditation upon new technologies of seeing – new modes of vision – within the modern world, a meditation that touches centrally upon issues of gender and authority. The painting carries us very quickly from the powers of science to the art of painting. I would like to focus not on the activities of the shamanistic presenter at the left centre of the canvas, a figure who presides over an outdated but visually arresting air pump, but on the interlude of girls and the accompanying man to the viewer's right.[11]

Their story is easily told. The younger child, to our left, gazes at the bird in the vacuum chamber with horror and fascination. The bird's fate rests with the presenter, who will either turn the stopcock above the vacuum chamber to permit air to enter the glass globe, or will allow the cockatoo to perish as a signal that a vacuum has been created. The young woman to the right, presumably a sister, averts her eyes. Behind both women – and in a visual hierarchy that carries moral authority, above them – stands a paternal figure who gestures to the bird in the receiver as if in explanation. As his right hand points up to the cockatoo, his left arm reaches around the older girl, the hand resting on her shoulder in a gesture of reassurance. Together the three form a visual subplot within the painting's larger narrative, a transposition of the painting's drama from a major to a minor key. Science gets recast here as sentimentality; cognition yields to emotion; and the collective and public nature of experimentation is rewritten as a domestic drama, private and unspoken. The two girls form a circle in which knowledge of pneumatics turns into knowledge of the heart.[12]

The father figure's presence in turn expands the arc of the girls' relation. As he literally encircles them with his left arm, he visually subsumes their world within his. His body forms the upper half of a greater circle that reaches through his head and shoulder down to the figure of the younger girl and then up again through her right arm and the back of the standing sister. Without him, the girls present the viewer with a circle of femininity, private, gendered and self-contained. With him, they enter into a domestic circle defined by its relation to, and tension with, patriarchal control.[13]

The significance of this drama lies hidden in the patriarch's hands. Each hand tells a different story: the right a narrative of lines, the left a story of circles. The right hand points up, the thumb and index finger leading directly to the bird. Its gesture is male: phallic, erect and unbending, a fleshly counterpart to the vertical column of the receiver. The man's left hand, by contrast, curves with the feminine world that it embraces. It offers comfort that is tactile and warm upon the woman's exposed shoulder. It communicates by touch rather than sight, a striking contrast with the role of the right hand as a visual signifier, gesturing in referential fashion to the floating bird.

Both hands represent different moments in an ideological narrative that defines culture in terms of masculine virtues. To move from the left hand to the right – from the immediate foreground to the centre ground – is to venture from the tactile and bodily to the visual and abstract, from emotion to reason, from circles to lines, from the feminine to the masculine. The power of the patriarch is the power to police the private. He stands at the midpoint between the domestic sphere and public life. His gesturing right hand exists at a tangent to the girls' circle; it interrupts their domestic sphere and re-figures it as a spiral that does not so much turn in upon itself as move out into new realms. His task, as defined by Wright, is not simply to mediate between the private and public, but to perform the work of the latter: to engage private energies in social projects. What started as a parable of domesticity concludes as an allegory of culture: an account of how civilization directs its singular energies to social ends.[14] 'The Man not only protects and advises', David Fordyce wrote in 1754, 'but communicates Vigour, and Resolution to the Woman.'[15]

What the eighteenth century termed 'female education', then, we might call today *socialization*, a process that rewrites bodily and privates states as cultural ones. The girls provide more than a dramatic vignette of emotional – and unscientific – private life. They serve as the vehicle for a narrative whose tenor is acculturation. Their behaviour is governed by forces that we might describe, awkwardly, as already-in-the-process-of-being-regulated. That hyphenated phrase represents an attempt to capture in language the logic of socialization that animates the girls: a logic of both acquiescence and resistance to forces that the painting names as 'science' and 'culture'.

Acculturation

This allegorization of the processes of culture is not unique to Wright. Ten years later, John Singleton Copley – another of England's 'provincial' painters – will transform the story of merchant Brook Watson's escape from a shark in Havana harbour into a parable of social accommodation (*Watson and the Shark*, 1778, *plate* 4). Watson's naked figure shares the foreground plane with the shark: his glazed eyes and open mouth echo the form of the shark and suggest, despite the struggle that divides them, their deeper participation in a realm both creaturely and 'natural', a realm understood as pre-social. Like the half-submerged beast with whom he shares the water, Watson is unclothed and unsocialized. His upward-reaching right arm is suspended equidistantly between the shark to the viewer's right and the inverted triangle of human figures that reaches down to him from the left. Watson is caught in the middle of a modern psychomachia. He is suspended between two realms, figured here as natural and social, and his salvation, translated in the painting into an act of acculturation, depends upon the capacity of the larger middle-ground community to assimilate the isolated individual into its own socially constructed space. The fragility of the enterprise is emphasized by the tension between the heroic and triangulated forms of the centre ground (boat and figures create an isosceles triangle) and the precarious and off-centred position of the figures relative to the boat that supports them. They are threatened not only by the roiled sea, but by their own asymmetric positioning within the craft, whose rear half is empty. Their heroism is defined by Copley as an ability to differentiate themselves from the world of pre-social forces below. Copley's painting construes bourgeois heroism as an enterprise of both social transformation and instinctual repression (notice the force of the boathook poised above the shark).[16]

Though it would seem rather ungallant to equate Copley's sharky foreground with Wright's sentimental heroines, the parallels between the two are worth noting. Where Copley has cast his drama within the oppositional language of nature and culture, shark and mariners, Wright has formulated his antithetical

4 John Singleton Copley, *Watson and the Shark*, 1778. Oil on canvas, 183.51 × 229.55 cm. Boston: Museum of Fine Arts (Gift of Mrs George von Lengerke Meyer). Photo: © Museum of Fine Arts, Boston.

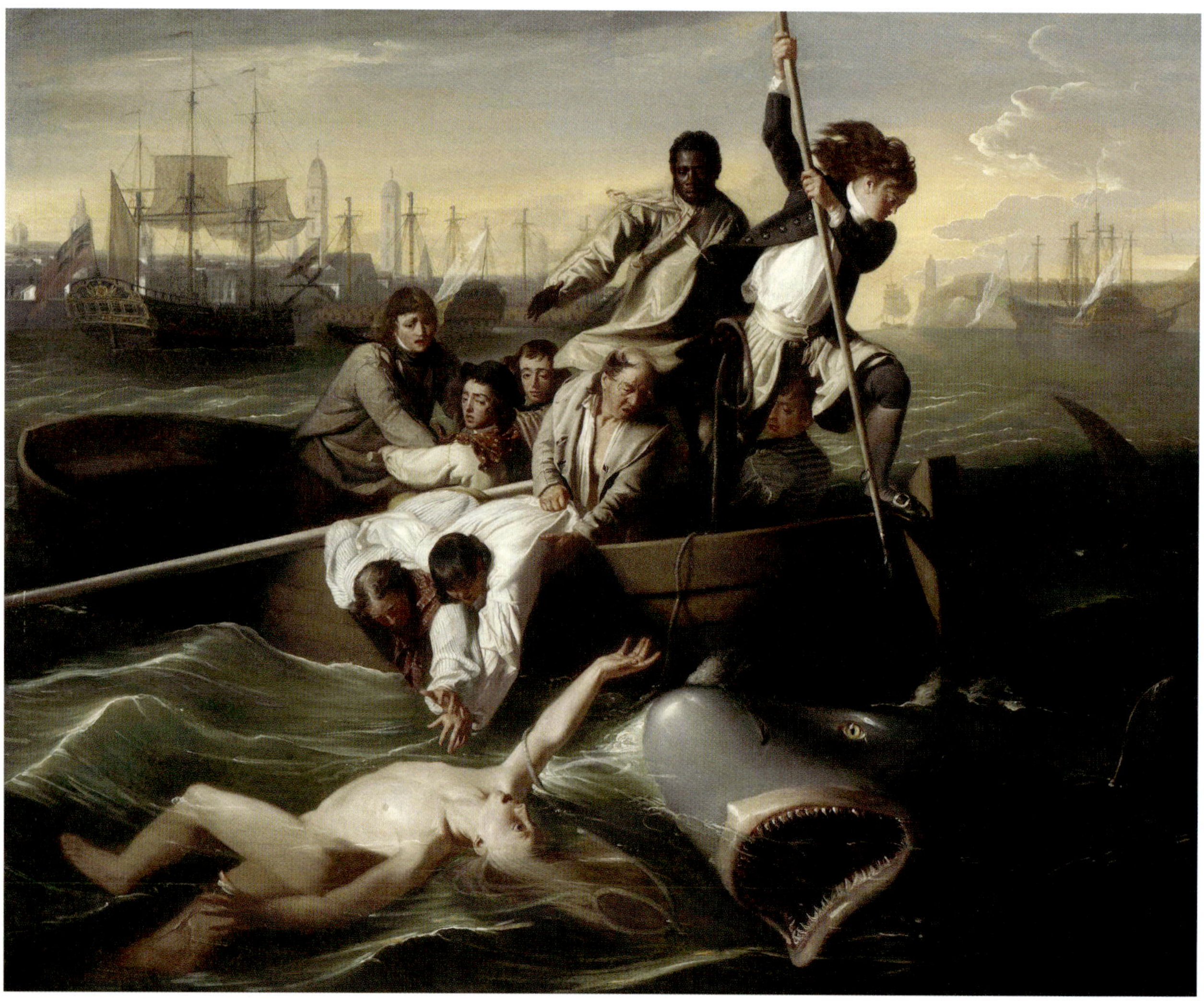

terms as the tension between the private and the public or domesticity and science. Within this binary opposition, Wright then maps out an economy of exchange parallel to Copley's. To move from the girls to the presenter is to undergo a process of translation: to rewrite a set of energies understood in Wright as private and sentimental and in Copley as natural and unsocialized. The differences between Wright's and Copley's categories quickly blur and cross over, each sharing attributes attributable to the other. Copley's masculine world divides into different degrees of masculinity: the closer one is to the foreground, the more 'feminized' the cultural space. Watson's figure is pale and sensual; his long dishevelled hair lends him an androgynous quality. The painting defines its foreground relations by an emotional logic: the pathos of Watson's situation, the horror of the viewer's response. Where Wright names gender as the key to cultural production, and conflates socialization with masculinization, so Copley too moves his drama in the direction of an increasingly male set of attributes: verticality over horizontality, activity over passivity, public definition over private states of emotion.

For both Wright and Copley, culture is understood as an intervention that carries the individual away from private energies to collective purposes.[17] Civilization-making occurs as an activity of rescue or repair. In Copley, the individual self sets its natural energies against its acculturated potential, and in Wright, where natural and feminine tend to coalesce, the gendered and domestic must be directed to socializing forces beyond itself. Though we understand those forces today as masculine, the painting denies its own gendering rhetoric by universalizing its terms as a set of categories that expand concentrically from 'natural history' to science to culture.[18]

And this introduces us to the question of death. In Copley, Watson's survival is pitted against the shark. The salvation of the former, visualized as a passage upwards, contrasts with the suppression of the latter, imaged in the downward thrust of the boathook. If this seems both predictable and unbearably proto-Freudian (civilization achieved through the suppression of libido), then Copley is one step ahead of us. There is a brown void at the centre of the painting, a circle virtually emptied of human content at the prow of the boat. The figures in the boat, together with the drama of the shark and boathook, revolve around this visual void. The (negative) circle forces us to read the action of the painting not in terms of the up-down pneumatics of Watson (up) and shark (down), but as a process of continual revolution, meaning by that term a recycling of energies from the shark to Watson to the rescuers in the boat and then back down again to the shark ad infinitum. To pick up on the Freudian thematics here, we might say that we have left the realm of repression and entered that of sublimation. Copley's painting conceives of culture as a continual act of 'rewriting', a translation of unsocialized energies into culturally useful ones, an endless process of conversion rather than repression. From this point of view not to enter the boat is to die: it is to fail to make the transition from the local and individual to the universal.

But to say this is to re-enter the realm of Whig ideology, and that, perhaps more than anything else, is what holds Wright and Copley together. We might read both paintings at one level as dramas of containment. Each painting is concerned in its separate way with two things: the creation of a sphere of feminized activity segregated from the rest of the canvas, and the ability to reclaim that sphere, to convert or translate it, into a drama about culture that functions at the same time as a narrative of masculinization. The question is *why*, and the answer, I suspect, lies in what J. H. Plumb has called the 'commercialization of leisure' during this period.[19] The public sphere in the eighteenth century, as both John Brewer and Terry Eagleton have shown, came to be defined through the language of culture.[20] This new public culture was reasonable in its means, disinterested in its methods, and universal in its ambitions for the public good. It was also fearful of two factors that could disrupt it: economic greed and sexual passion.

And this is where the commercialization of leisure enters the picture. Culture, as Brewer notes, came increasingly under the sway of the market: its ends were commercial rather than moral. The new cultural entrepreneurs, associated with the Opera, the theatre, Vaux Hall and other scenes of uplift or entertainment, saw culture, like prostitution, as an exercise in salesmanship, an effort to provide the public with what it most wanted.

Copley and Wright take up the gauntlet at this point. Each is allied with a project of re-masculinization, an effort to reclaim culture from its degraded, commercialized and feminine associations and to reconfigure it as moral, instructive and universal. When seen from this point of view, both *Experiment* and *Watson and the Shark* need to be understood as participating within a larger movement in eighteenth-century England that addressed issues of a social or political nature in terms of gender, from Robert Filmer's conservative defence of royal politics through the language of patriarchy to William Cadogan's *Essay Upon Nursing and the Management of Children*, 1748. Cadogan urged men to supervise the upbringing of their children, recommending 'every father to have his Child nursed under his own Eye'. As Ludmilla Jordanova comments, 'Paternal authority … takes the form of active management, for which the sense of sight is the key metaphor.'[21]

Consider again the two girls in Wright's painting in relation to the hovering man. His upper torso draws them into a disciplinary network, converting their individuated anguish into a circle of public, rational instruction. The politics of that circle, for Wright as for Copley, are the politics of Whig socialization.

Patriarchal Pointing

If we ask then what the younger girl gazing upward in *Experiment* actually sees, we might offer two answers from within the painting's complicated perspective. The first, and most obvious, is the bird in its death throes. The child not only sees, but cannot *not* see, the drama of life and death surrounding the cockatoo. Her inability to turn away from the bird's plight, as her older sister does, helps define the differences in the sisters' relation to each other. The older sister suppresses her emotions: she effaces herself, literally, as a gesture of female propriety. She stands psychologically (and spatially) on the other side of the abyss that separates child from young adult. Wright defines growth or maturity as a process of immersion ever more deeply into the socializing discourses of one's gender and class (modesty, propriety, polite behaviour).[22]

Which brings us to the second answer to the question of what the younger girl sees. Though she clearly witnesses the bird, she may also observe the patriarch's gesturing right hand, which stands visually in line with her upward gaze. If so – if her access to the bird is mediated by the man's explanatory gesture – then we suddenly find ourselves at the heart of a conundrum that stands very close to the heart of the painting's politics. And those politics are the politics of seeing. The younger girl has before her two objects: a dying bird and a pointing hand. To look at the bird directly is to bypass the mediating presence of the hand, while to see the bird through the hand is to learn of nature's secrets only as they are explained by cultural authority, an authority that intervenes between the girl's perception and her knowing. The difference between these two options, between direct perception of the bird and the monitory presence of the hand, is the difference for Wright between youth and maturity.

The young girl is situated in this narrative midway between the two. She exists at some pivotal moment in her own history when 'natural' seeing has not yet been reconstituted as a cultural act: when the rawness of seeing has not been 'cooked'. Wright has created a miniature bildungsroman for us, a narrative of female education that exists at the same time as an allegory of seeing. He posits youth as a state of natural appetites and curiosities grounded in the self's direct access to the sensory plenum of the world. Into this world he then inserts a man's hand, an image of mediation that intervenes between the eye and its object. What the girl now sees is not the bird but the bird-as-pointed-to. Her primary visual experience of the death of the cockatoo has been superseded by a new experience: the *explanation* of the death of the cockatoo.

What that hand tells us, in other words, is that we have advanced one step beyond Locke. Locke links seeing to the other senses as the source of the individual's access to nature. He then posits the Understanding as a faculty that rearranges and composes these sense data.[23] Wright takes a different tack. He interposes a hand between the seer and the seen, and in the process, transforms Lockean epistemology in two ways. He suggests, in the first place, that the machinery of knowing, which for Locke is always individual and psychological, should rather be understood as social, a work of culture rather than the operation of private minds. And in the second place, he locates seeing as the space or site of cultural encoding. The patriarch's hand tells us that the world is known through acts of explanation, gestures that describe nature by translating it into culturally legible modes, both visual and verbal. Seeing operates for Wright as a social activity in which the choice between raw perception and cultural intervention collapses. To see is to receive an explanation.

This refusal of Lockean perception as anything other than a form of nostalgia, a type of child's play, helps explain why the patriarch directs his gesture to the older girl rather than the younger child. The apparent discrepancy here – that the man points for the sake of the girl who doesn't see his gesture rather than for the one who does – helps us understand Wright's revision of Locke. Wright has reduced Lockean epistemology to the status of a myth: a naive and originary moment in the history of cognition. If the young girl embodies Lockean perception, an effort to see outside the interventions of culture, then she also gets cast in a socially immature role. Her effort to see directly, to know the fate of the bird with or without the assistance of the gentleman, represents an untenable position with the painting's social and visual economy. The difference between childhood and young adulthood is defined by Wright as the difference between naive sensationalism (Lockeanism) and what replaces it: those explanations that now substitute for direct seeing.

The older girl covers her eyes with her hands because she no longer needs to see as Locke would have us see. She blinds herself to any experience outside the orbit of proper culture. What she knows is a function of what he shows. Her younger sister on the other hand has not yet bitten the apple of cultural knowledge. She is young precisely because she still sees outside or around 'patriarchal pointing'. As her axis of vision comes to coincide with his gesture, she too will see as her older sister does, beyond Locke and within culture.[24]

I should add that the cognitive issues that I am describing here are elaborated by Wright with their own ambivalences. One might read the older girl's covered face as a gesture not only of modesty but of *resistance* to the explanation the man offers. She will not see the world as he describes it. So too with the younger child, who may or may not view the bird directly. Her Lockean position within the painting's narrative may represent not just a superannuated epistemology, a cognitive dead-end, but an opportunity rife with political possibilities. She stands at a crossroads, a moment of slippage, when the net of culture has not yet been drawn fully around a robust human nature. From this point of view, Locke exists not as an epistemological embarrassment but as a reminder that the processes of social mediation are never as total in their effects as they might appear. There will always be moments of slippage, modes of vision outside the network of socially informed perception. If seeing represents at one level the scene of cultural instruction, the agency by which perception gets converted into acculturation, then it also stands in opposition to these processes. Wright is concerned not so much with the enforcement of culturally inflected ways of seeing as mapping their possibilities: how vision works, what social ends it serves, what the limits of its ideological powers might be. Wright is engaged in a twofold effort: he links perception to acculturative tasks at the same time as he also disengages it from such ends by allowing for a seeing that evades the interventions of culture. The central lesson to be learned in the politics of seeing is that vision is a two-way street: it provides the space in which experience gets encoded culturally, and it resists its own semiotic closure. The former transforms seeing into an act of social control, while the latter draws on vision as a faculty never wholly assimilable to ideological imperatives.

Self-fashioning

Let us turn for a moment to an image painted two years after Wright's *Experiment*, Copley's iconic portrait of *Paul Revere*, 1770 (*plate* 5). Susan Rather and Ethan Lasser have provided us with nuanced accounts of the tensions between artist and artisan that animate Copley's painting.[25] I would like to tackle a different dimension of the image, what we might term its play with Lockean metaphors. We should note Locke's presence in the painting in a variety of forms: in the focus on bodily labour, in the concern with the divisions between head and hand, in the attention to light, and in the emphasis placed upon Revere as an autonomous actor within the world. All these forces come to a focus in the figure of the teapot, which exists in the canvas as a version of Locke's tabula rasa, a blank slate awaiting inscription. The tabula rasa represents for Locke the human mind lying fallow before the world; it stands as an alternative to Cartesian notions of innate ideas.

In Copley's painting, the teapot exists not only as a double for Revere's head, a corollary to his thought, but as an image of the New World itself, a smooth globe whose history has yet to be written, a world of untried possibilities not yet inscribed by the maker. Copley, like Wright, is concerned to reimagine Locke through the teapot in social terms, to convert the privacy of the knowing mind into a collective, historical adventure.

The teapot is also the place where light collects in the painting. Indeed, if we stare at the image too intensively, we may find ourselves mesmerized by the convex surface of the teapot and the Escher-like optical distortions that it produces. The result is a form of magical realism that represents the one moment within Copley's otherwise restrained image when the painter abandons traditional mimesis and gives himself over to a space of visual fancy and play.

Let's rephrase that to say that the teapot represents the moment of potential *danger* in Copley's painting. It exists as a feminized form: rounded and nurturing, allied with domestic ceremony and aristocratic social customs. If Boston's colonial elite identified itself in part with the British gentry and upper classes, then it also feared the luxury and 'corruption' that it felt attended the world of non-productive, elite behaviour.[26]

Revere's teapot carries the burden of anxiety within Copley's painting. We need to think of it as we previously thought of the two young girls in Wright's *Experiment*: a circular sub-space within the canvas informed by domestic ideologies. Like Wright's painting, Copley's portrait subordinates and controls its feminized spaces through the extended hand of a male figure, whose grasp of the teapot not only prevents it from puncturing the painting's surface plane and thereby spinning out of control, but also returns it to the torso of Revere, positioning it within his own enfolding form. If we think of the teapot as a reified version of Wright's girls, then we begin to grasp the fear of cultural effeminization and the need to police the Lockean world of naive perception at play in both Copley and Wright.

What historical forces are we talking about? The anxiety that Copley's Boston patrons felt about the new market culture they were creating surfaces in the painting in the teapot, a marker not only of feminized cultural spaces, but of social relations that, precisely because they are driven by market conditions, threaten to destabilize

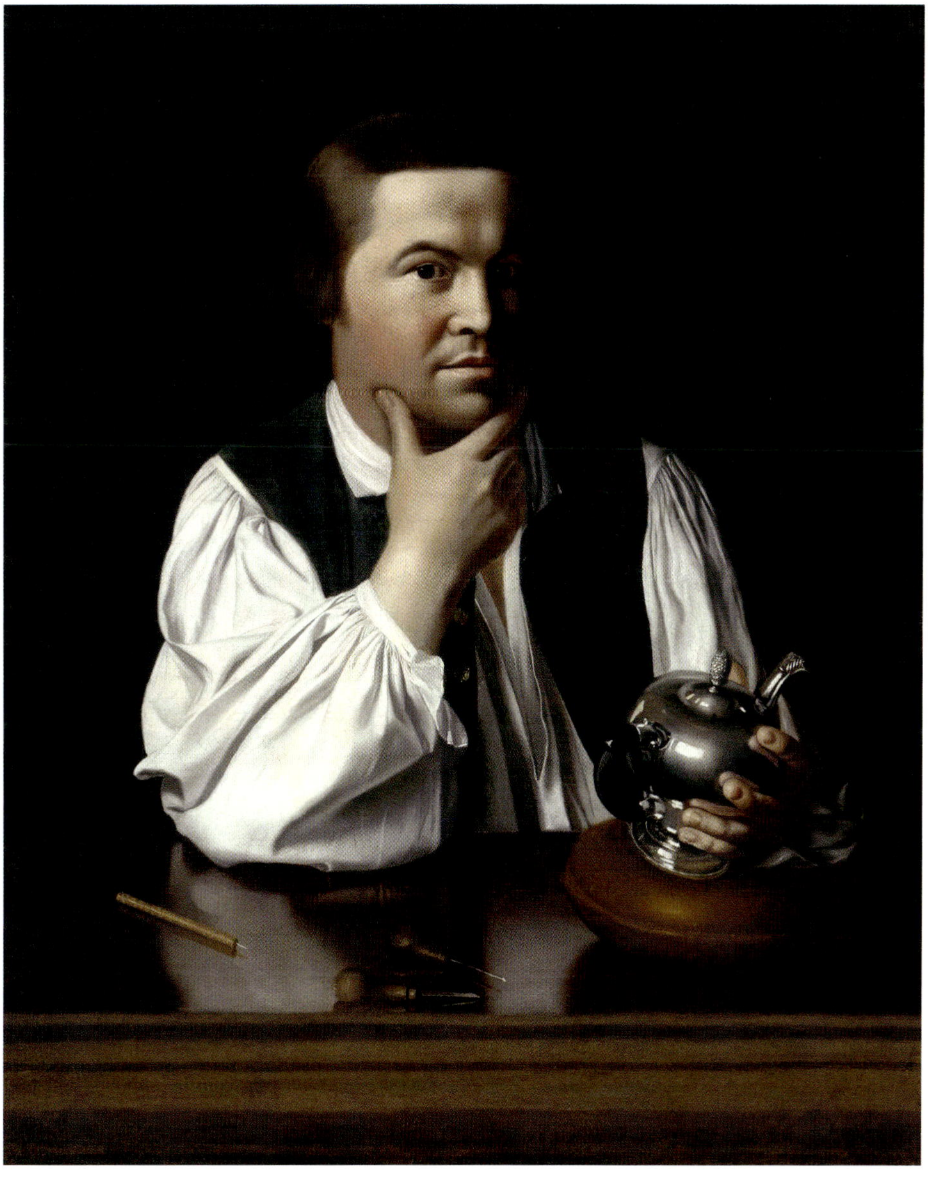

5 John Singleton Copley, *Paul Revere*, 1768. Oil on canvas, 89.22 × 72.39 cm. Boston: Museum of Fine Arts, Boston (Gift of Joseph W. Revere, William B. Revere and Edward H. R. Revere). Photo: © Museum of Fine Arts, Boston.

older and more traditional notions of class and character. We need to remember that the egalitarian dimensions implicit in Copley's image – the eye to eye contact between Revere and viewer, the positioning of Revere as a working man, the refusal of traditional emblems of elite status – are present within the context of *market exchange*. Revere fronts us as equals within the market, partners in a game of exchange that extends from teapot, as an object of consumption, to our position as possible patrons and consumers across the table opposite him.

From this point of view, Revere's control of the teapot has less to do with questions of class or gender than it does anxieties about the market itself. That teapot incarnates property relations in the early entrepreneurial world of Boston: a realm where the penetration of the market into everyday life threatens to disrupt older social relations. In tethering the teapot to Revere's torso, Copley not only assuages his viewers' anxieties about a culture of consumption, but subordinates these new market conditions to an older notion of *character*. He sets the stability of Revere's triangulated form against the magic of the teapot, reassuring his audience that the market itself is but an expression – and subject to the control – of individual human character. It is a form of reassurance repeated in prose by Benjamin Franklin, whose *Autobiography* chronicles the way that young men lacking money can substitute *character* for *capital* and thereby advance in the world.[27]

To understand Copley's use of light, we need to consider the complexity of optical theory in the eighteenth century. Light was characterized by theorists of the era not by its simplicity – its ability to place the observer directly in touch with the external world – but by its many refractions, the complex set of mediations that underlay the process of perception. Edmund Burke, for example, in his *Reflections on the Revolution in France*, compares the rights of man to 'rays of light which pierce into a dense medium'. Such rights are 'by the laws of nature, refracted from their straight line'. They endure 'such a variety of refractions and reflections, that it becomes absurd to talk of them as if they continued in the simplicity of their original direction'.[28] Burke contrasts here the 'simplicity' that appears to govern seeing with the many 'refractions and reflections' that actually define it. Burke uses optical theory, paradoxically, as proof that things are not what they seem, that natural law is perversely refractory, in other words, unnatural.

Copley's teapot, with its marvellous refractions and its 'unnatural' distortions, mirrors Burke's complex trope of light. Copley does so, however, not like Burke to justify the role of custom in shaping natural rights, but to legitimate his own position as artist within the new social order. What we learn from Revere's teapot is that the artist controls the refractions, the mediations, the social codes that underpin his brave new world. Copley's painting is not ultimately about the simplicity and direct address that defines Revere's relation to the viewer, but in more Burkean fashion about the hidden mediations that make social discourse possible. Copley's Revere controls those mediations, that magical and unnatural play of light that defines for the artist the possibilities and limits of his powers. Revere holds more than a teapot, more even than a tabula rasa or New World, in his hands. Like Wright's patriarch, he controls the machinery of acculturation itself, the representational codes by which society constructs, polices and names itself. The power of the artist is the power of civilization-making, a process that has less to do with instruction and uplift than with mediation, masculinization and social control.

There is a term for this sort of power, this new technology of seeing. It is a twentieth-century term. We call it today 'hegemony': the capacity of the artist to convert forces external to his own class or gender, forces imaged as refractory, feminized, and pre-social, into versions of the dominant discourse, doubles of himself. Whether we are dealing with teapots or air pumps, we are witnessing on both sides of the Atlantic the construction of new forms of representation – new technologies of seeing – tied to patriarchal pointing, aesthetic self-fashioning, and the emergence of early market culture.

Power

We have come full circle. We can return at last to seeing as a new technology of power. Michel Foucault long ago outlined what he termed the distinction between 'juridico-political' and more 'technological' regimes.[29] The former operate by the threat of external force: juridical regimes impose their authority from the outside upon those they govern. More 'technological' forms of power, however, disavow physical coercion. They work instead by 'disciplinary' practices that internalize the operations of power: modes of self-restraint and social policing that hinge on language – or in Foucault's account, sexuality – as a *discursive* rather than bodily practice. So too with Wright's technology of seeing. We might think of it as a discursive practice that replaces machines and traditional forms of technology with the more invisible – but perhaps more powerful – operations of the eye, ways of being in the world that work in oxymoronic fashion both to conceal the power that circulates through them and to highlight the centrality of seeing as a social practice. Wright's achievement lies in part in his investment of seeing with newfound imperatives: to reimagine vision's acculturative possibilities and to reinvent sight as a disciplinary force. In the process, Wright simultaneously confirms and disguises his relation to the mercantile-industrial world that informs his practice. He emerges in his art as both a mythographer of the Arkwrights of his world – a participant in their class interests – and as someone who crucially and essentially sidesteps traditional class definitions. In this latter mode, Wright carves out new territory for the artist in the modern world. Wright's artist functions as a broker of representations, an operator hidden behind the scenes, someone who pulls the strings that animate, like so many puppets, the codes and conventions by which his world constructs its version of the Real. For that is Wright's newfound power, not as an owner or master of the machines that drive the industrial age, but as the artist-shaman-showman who presides over the codes (social, ideological, representational) that underpin its operations.

Joseph Priestley, like Wright, associated the air pump and industrial technology with god-like powers. 'The English hierarchy', he once wrote, has 'reason to tremble even at an air-pump, or an electrical machine' because, as Susan Siegfried notes, 'they imitated the operations of nature and, by extension, of God'.[30] Wright carries that possibility one step further. He appropriates the god-like power of the air pump for artists like *himself*. He becomes, in Barbara Stafford's terms, master of the 'visible invisible'.[31] Wright uses the occasion of a scientific lecture to meditate on the conditions of his own art. Realism for Wright is both more and less than what it appears: more in that painting not only mimics the world around it, but virtually invents it, and less in that painting never tells the whole truth about the world that it creates. What it leaves out is its own sleight of hand, that magic by which the artist – like the impresario at the centre of the painting – produces the illusion of a transparent world, one that is readily accessible to the senses and subject to the will of those who would command it.

Wright's achievement, in part, lies in his imagining of an alternative public sphere, one taken out of the taverns and coffee shops central to Jürgen Habermas's later account.[32] Wright instead situates 'rational discourse' within the *domestic* spaces of his painting, and then renders rationality itself only another form of illusion, a fiction designed to hold together those social classes that subscribe to it. Rationality exists as a form of *persuasion* for Wright, and no one speaks its language more cogently than the artist, whose realism persuades its audience that what they attend to is not their invention, but their discovery. Wright's new technology of representation, imaged in the *Experiment* as an almost magical practice, converts the most everyday of processes – seeing itself – into the most powerful. Such seeing is modern because it stands, like Foucault's discursive networks, as a gateway into the newly invented disciplinary habits of the industrial and post-industrial worlds. And it is modern for yet another reason: because Wright transforms seeing from a Lockean and empirical activity into a post-Lockean and discursive one. Such seeing remains hidden in plain sight, a technology, a disciplinary practice, that goes unobserved until, of course, it appears quietly in the fraught gestures of pointing men, the ambivalent responses of young women, and the uncertain fate of exotic birds.

Coda

Let's advance the dial thirteen years, to 1781, the year in which Philippe Jacques de Loutherbourg, a French-born painter who had arrived in London a decade earlier, introduced the crowds at Leicester Square to his new invention, the Eidophusikon (*plate* 6). Part miniature stage, part light show, and all spectacle, the Eidophusikon seated audiences before a faux-proscenium stage that displayed a sequence of moving backdrops behind its painted curtains. The effect, when coordinated with specially orchestrated sounds, resembled 'moving pictures', theatrically framed scenes that came to life before the observer's eyes: images of ships and storms at sea, reproductions of Midlands scenery, and dramatizations of popular literary passages, most notably Satan's consorting with the fallen angels from Milton's *Paradise Lost* – all in motion, all without human actors, all achieved by 'technological illusionism'.[33]

If the Eidophusikon participated, at one level, in the late eighteenth-century culture of spectacle that gave it birth, then it also operated, at another level, as an *antidote* to the spectacle it was promoting. David Kornhaber speculates that the popularity of the Eidophusikon was tied less to its realism – its ability to entertain its viewers with dramatic vignettes of recognizable scenes – than to the sense of mastery that it provided its audience over the spectacle they witnessed.[34] The Eidophusikon gave its audience two things: the thrill of illusionism, and a knowing wink that the illusion they witnessed was all an elaborately staged game. In this way the spectator was both absorbed by the scene in front of him, and admitted behind stage, as it were, to discover how the spectacle was produced. The former fulfilled one's expectations of a dazzling performance, while the latter assured the spectator that even magic could be explained.

What the Eidophusikon achieved, in other words – and where it returns us to Joseph Wright of Derby – is a moment of resistance to its own powers of absorption. It flaunted its sleight of hand even as it continued to trade in illusion, and in this way, it supplied its audiences with the capacity to recognize the very forces that enthralled them. By virtue of its reduction of the theatre to a doll-like set, a miniaturized version of the stages of Drury Lane, the Eidophusikon empowered the viewer with the 'meta' pleasure of witnessing his own amusement.

And that is precisely what *Experiment on a Bird in an Air Pump* achieves: it tempts us with the magic of its own representational power at the same time as it educates us in the means by which that power is achieved. Through the interlude of the patriarch and two younger women, it permits us to examine in mirror-like fashion the terms of our own engagement.

Put differently, Wright's painting toys with the limits of its own acculturative ambitions. It yokes aesthetics to power, seeing to socialization, at the same time as it draws attention to – by virtually diagramming – its own operations. Like the

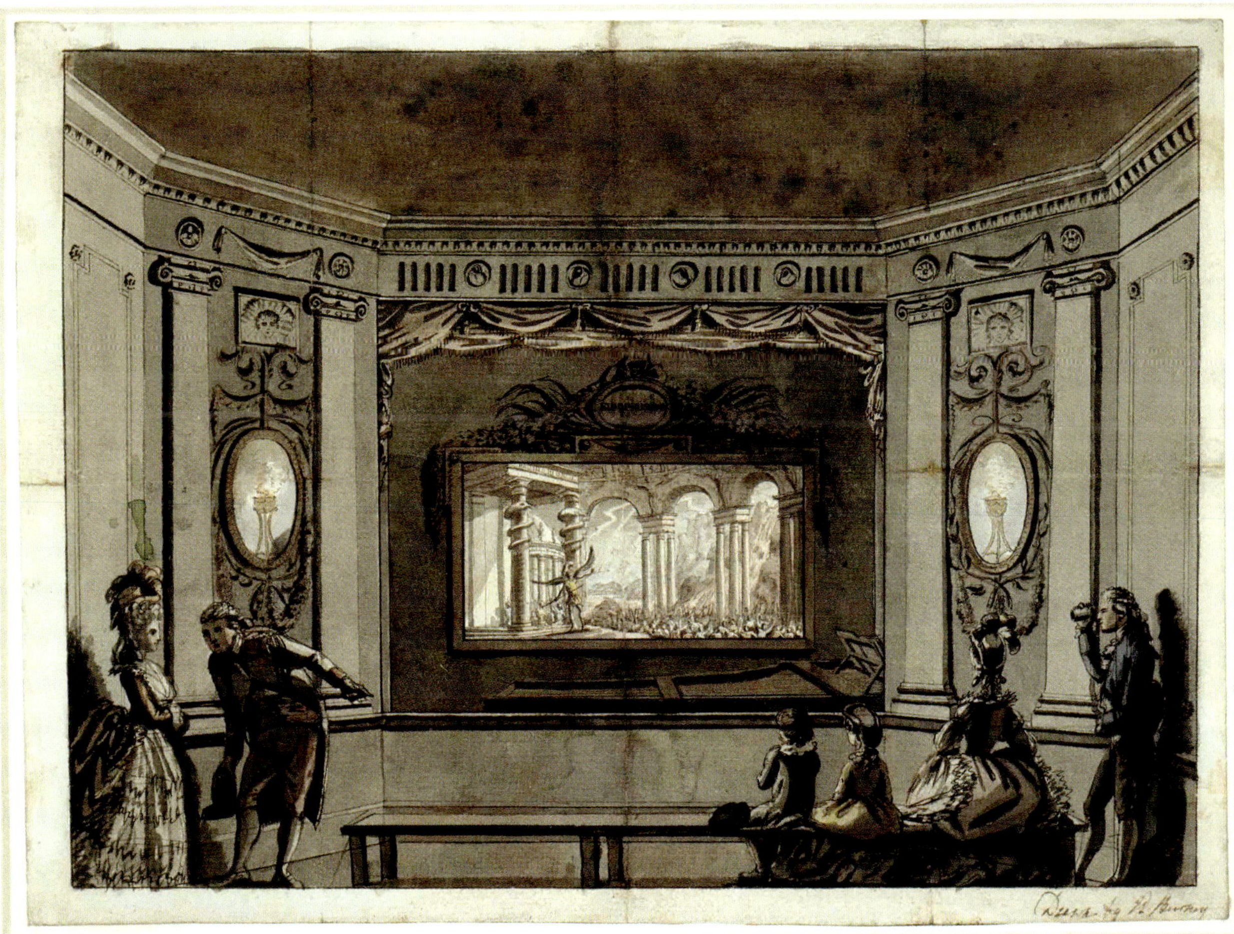

6 Edward Francis Burney, *The Eidophusikon showing Satan arraying his Troops on the Banks of a Fiery Lake with the Rising of the Palace of Pandemonium from Milton*, c. 1782. Watercolour on paper, 19.7 × 27.3 cm. London: British Museum. Photo: © Trustees of the British Museum.

Eidophusikon, the painting both entertains us as spectators and then inoculates us against its own pleasures, not by withdrawing those pleasures, but by framing them figuratively within quotation marks, showing us not only spectacle, but the machinery that supports it.

Neil Harris has described this process of making the audience a party to its own entertainment as an 'operational aesthetic'. Harris develops the notion in relation to P. T. Barnum, the nineteenth-century American showman and entrepreneur, who discovered early in his career that audiences will gladly be deceived if they are offered at the same time an opportunity to unmask the means of their deception.[35] Their pleasure comes not only from the scene or spectacle before them, but from figuring out how the entertainment has been staged. Harris links this pleasure in unmasking the means of deception to mid-nineteenth-century audiences anxious about their fate in a rapidly industrializing world. For two bits, audiences could pit their analytic skills against Barnum's efforts to deceive them, and in the process recover, if only for the duration of their visit to Barnum's 'museum', a fantasy of restored mastery over the world. What Barnum invented, then, was neither entertainment nor humbug, but an antidote to the experience of modernization and urban anonymity. Deskilled workers who could not master the larger social forces of early modernity could, for a small fee, inspect the deceptions of a showman to understand how those deceptions operated. And in the process, they could revive – if only momentarily – the illusion that the world before them was, like Revere's teapot, something that fit into the palm of their hand: comprehensible, malleable, and subject to their agency.

The Eidophusikon worked along similar lines. It generated an 'operational aesthetic', an illusion of mastery over the machinery of spectacle. Kornhaber notes:

> If the desire latent in the attraction towards spectacle is the desire to be overwhelmed, then it seems to come with a corresponding counter-desire for mastery ... If to go to the theatre was to give oneself up to something larger than life, then to see the Eidophusikon was to be in control of that same experience.[36]

The 'operational aesthetic' of the Eidophusikon, then, cues us to a counter-current at play in Wright's *Experiment*. The painting teaches us more than the power of seeing, more even than the coded nature of culture and the artist's privileged relation to those codes. Like the Eidophusikon, Wright's painting instructs us in mastery *over* deception. *Experiment on a Bird in an Air Pump* introduces us to new technologies of seeing – modern modes of power – at the same time as it educates us as viewers in the terms of its own operations: the ways in which seeing functions as both an empirical process and a socializing discourse. The painting reminds us of the gap between what we see and what we get. We see a bird, a vacuum pump, a patriarch and two girls. But what we get is something else: a capacity to understand our own seeing, an ability to advance from seeing as an empirical process to vision as acculturative technology and then – the pivotal moment – to the observation of both operations critically. Wright's *Experiment* thus engages its viewers in the post-Lockean task of refusing innocence, not because we fail to see as children see, but because we see the machinery that transforms us from children into fully acculturated adults. And that seeing, that alternative form of magic, that insistence on a mode of vision that is neither wholly hegemonic, nor simply counter-hegemonic, is what – for Wright – art promises.

Notes

1 Albert Boime, *Art in an Age of Revolution, 1750–1800*, Chicago, IL, 1987, 256.

2 For an overview of Wright's paintings of Awkwright's mills, including the now lost *Awkwright's Cotton Mills by Day*, see Benedict Nicolson, *Joseph Wright of Derby: Painter of Light*, Vol. 1, London and New York, 1968, 164–9, and Judy Egerton, *Joseph Wright of Derby*, London, 1990, 197–200.

3 For an account of Awkwright and his relation to the history of industrialization and the Midlands elite, see R. S. Fitton and A. P. Wadsworth, *Strutts and the Awkwrights, 1758–1830: A Study of the Early Factory System*, Manchester, 1958; and Francis D. Klingender, *Art and the Industrial Revolution*, London, 1968. See also Paul Elliott, 'The birth of public science in the English provinces: Natural philosophy in Derby, c. 1690–1760', *Annals of Science*, 57, 2000, 61–100; and Celina Fox, *The Arts of Industry in the Age of Enlightenment*, New Haven, CT and London, 2009.

4 David Solkin discusses the tensions between partial vision and 'the comprehensive wisdom of the classical citizen' in his excellent account of Wright's earlier painting, *Experiment on a Bird in an Air Pump*, 1768. Solkin interprets Wright through the lens of Shaftesbury and the modern 'division of labour'. 'ReWrighting Shaftesbury: The air pump and the limits of commercial humanism', in John Barrell, ed., *Painting and the Politics of Culture: New Essays on British Art 1700–1850*, Oxford and New York, 1992, 73–99.

5 John Barrell, *The Political Theory of Painting from Reynolds to Hazlitt: The Body of the Public*, New Haven, CT and London, 1986, 48, 53–4.

6 There is a certain parallel between Wright's concentration of power in a mode of vision abstracted from – and elevated above – the partiality of everyday vision and the 'centralization' of the factory system in the eighteenth century with its consolidation of different operations under one roof. For a useful overview of Wright's ties to the history of industrialization, see David Fraser, '"Fields of radiance": The scientific and industrial scenes of Joseph Wright', in Denis Cosgrove and Stephen Daniels, eds, *The Iconography of Landscape: Essays on the Symbolic Representation, Design and Use of Past Environments*, Cambridge and New York, 1988, 119–41.

7 Barrell, *The Political Theory of Painting*, 1–68.

8 David Fraser, 'Joseph Wright of Derby and the Lunar Society', in Egerton, *Joseph Wright of Derby*, 15–24; and Robert E. Schofield, *The Lunar Society of Birmingham: A Social History of Provincial Science and Industry in Eighteenth-Century England*, Oxford, 1963, 130–1.

9 Michael Fried, 'Representing representation: On the central group in Courbet's studio', in Stephen J. Greenblatt, ed., *Allegory and Representation: Selected Papers from the English Institute, 1979–80*, Baltimore, MD, 1981.

10 John Locke, *Second Treatise of Government*, ed. C. B. Macpherson, Indianapolis, IN, 1980, 19. David Solkin discusses Wright in relation to Lockean theories of labour in 'Joseph Wright of Derby and the sublime art of labor', *Representations*, 83: 1, 2003, 167–94.

11 Barbara Stafford, *Artful Science: Enlightenment, Entertainment and the Eclipse of Visual Education*, Cambridge, MA, 1994, 99–102. Stafford contrasts the 'instructive entertainment' of Wright's paintings – in particular the *Experiment on a Bird in an Air Pump* – with the many varieties of charlatanism and 'duplicity' central to eighteenth-century forms of entertainment.

12 For the interweaving of science and spectacle, see Simon Schaffer, 'Natural philosophy and public spectacle in the eighteenth century', *History of Science*, 21, 1983, 1–43. For the relation of gender to early modern science, see Londa Schiebinger, *Nature's Body: Gender in the Making of Modern Science*, Boston, MA, 1993. For a discussion of the role of the visual in eighteenth-century scientific discourse, see Daniela Bleichmar, 'Training the naturalist's eye in the eighteenth century: Perfect global visions and local blind spots', in Andrew Graciano, ed., *Visualizing the Unseen, Imagining the Unknown, Perfecting the Natural: Art and*

Science in the Eighteenth and Nineteenth Centuries, Newcastle, 2008. See also the collection of essays in Pamela H. Smith and Benjamin Schmidt, eds, *Making Knowledge in Early Modern Europe: Practices, Objects, and Texts, 1400–1800*, Chicago, IL, 2007.

13 For an astute analysis of the relation of domesticity to a masculine public sphere, see Elizabeth Dillon, *The Gender of Freedom: Fictions of Liberalism and the Literary Public Sphere*, Stanford, CA, 2004, 1–48.

14 Susan Siegfried argues, in a somewhat different vein, that the women in Wright's paintings enact a diversity of emotions not permitted the gentlemen in his images. In this way Wright's women expand the emotional and psychological range of his paintings by introducing alternative 'economies of vision'. 'Engaging the audience: Sexual economies of vision in Joseph Wright', *Representations*, 68, 1999, 43–58.

15 Quoted in Solkin, 'ReWrighting Shaftesbury', 94.

16 My discussion of *Watson and the Shark* draws on an earlier account of the painting in *Romantic Re-Vision: Culture and Consciousness in Nineteenth-Century American Painting and Literature*, Chicago, 1982, 123–5. See also Jules Prown, *John Singleton Copley: In England 1774–1815*, Washington, DC, 1966, 270–4; Roger Stein, 'Copley's 'Watson and the Shark' and aesthetics in the 1770s', in *Discoveries and Considerations: Papers on Early American Literature and Aesthetics: Presented to Harold Jantz*, Albany, NY, 1976; Albert Boime, 'Triangular trade and triangular composition', in *The Art of Exclusion: Representing Blacks in the Nineteenth Century*, Washington, DC, 1990, 15–46; Margaretta Lovell, *Art in a Season of Revolution*, Philadelphia, PA, 2005, 104, 169–70, 175; Emily Ballew Neff, *John Singleton Copley in England*, exh. cat., Houston, TX and London, 1995, 102–5; and Jennifer L. Roberts, *Transporting Visions: The Movement of Images in Early America*, Berkeley, CA, 2014, 50–67.

17 Steven Shapin and Simon Schaffer, *Leviathan and the Air-Pump: Hobbes, Boyle, and the Experimental Life*, Princeton, NJ, 1985, 225. Shapin and Schaffer discuss 'the processes by which a private sensory experience is transformed into a publicly witnessed and agreed upon fact of nature'. They are concerned with the social construction of knowledge, including what they term 'matters of fact' as being, in fact, cultural conventions.

18 For a useful overview of the role of science in legitimating cultural authority, see Arnold Thackray, ed., *Constructing Knowledge in the History of Science*, Chicago, IL, 1995.

19 J. H. Plumb, 'Commercialization of leisure', in Neil McKendrick, John Brewer and J. H. Plumb, eds, *The Birth of a Consumer Society: The Commercialization of Eighteenth-Century England*, 1982, 265–85.

20 John Brewer, *The Pleasures of the Imagination: English Culture in the Eighteenth-Century*, New York, 1997, and '"The most polite age and the most vicious": Attitudes towards culture as a commodity, 1660–1800', in John Brewer and Ann Bermingham, eds, *The Consumption of Culture, 1600–1800: Image, Object, Text*, London and New York, 1995; and Terry Eagleton, *The Function of Criticism: From the Spectator to Post-Structuralism*, London, 1984, and *Culture and the Death of God*, New Haven, CT and London, 2014.

21 Ludmilla Jordanova, *Sexual Visions: Images of Gender in Science and Medicine between the Eighteenth and Twentieth Centuries*, Madison, WI, 1989, 31.

22 For a different account of the relation of 'sensibility' to science, see Lorraine Daston, 'Preternatural philosophy', in Lorraine Daston, ed., *Biographies of Scientific Objects*, Chicago, IL and London, 2000, 15–41.

23 See Gerald Hanratty, *Philosophers of the Enlightenment: Locke, Hume and Berkeley Revisited*, Blackrock, Co. Dublin, 1995.

24 Solkin provides a parallel, if different, account of Wright's art as moving beyond Locke in its assumptions. For Solkin, Locke's theory of labour – though revolutionary in its break from older and more traditional notions of culture – fails to account adequately for the role of sensibility, contemplation and 'delicacy' in the making of eighteenth-century polite society. 'Joseph Wright of Derby and the sublime art of labor', 174, 181.

25 Susan Rather, 'Carpenter, tailor, shoemaker, artist: Copley and Portrait painting around 1770', *Art Bulletin*, 79, June, 1997, 269–90; and Ethan W. Lasser, 'The business of Copley's Paul Revere', *American Art*, 26, 2012, 26–43.

26 For an account of colonial Boston's Anglophilia and its identification with metropole London, see Paul Staiti, 'Character and class', in Carrie Rebora *et al.*, eds, *John Singleton Copley in America*, New York, 1995, 53–78. For an account of colonial New England's mistrust of elite British society, see Joseph Ellis, *After the Revolution: Profiles of Early American Culture*, New York, 1979.

27 J. A. Leo Lemay, ed., *Benjamin Franklin: Autobiography, Poor Richard, and Later Writings (Library of America)*, New York, 1987. See especially Books I and II of the *Autobiography*.

28 Quoted in Terry Eagleton, *The Ideology of the Aesthetic*, Oxford, UK, and Cambridge, MA, 1990, 57.

29 Michel Foucault, *The History of Sexuality*, Vol 1, trans. Robert Hurley, New York, 1978, 17–35, 87–9.

30 Siegfried, 'Engaging the audience', n. 36, 56.

31 Barbara Stafford, *Artful Science*, 89.

32 Jürgen Habermas, *The Structural Transformation of the Public Sphere: An Inquiry into a Category of Bourgeois Society*, trans. Thomas Burger with the assistance of Frederick Lawrence, Cambridge, MA, 1989.

33 See William Vaughan, 'Magic in the studio', and Iain McCalman, 'Magic, spectacle and the art of de Loutherbourg's Eidophusikon', both in Ann Bermingham, ed., *Sensation and Sensibility: Viewing Gainsborough's Cottage Door*, New Haven, CT and London, 2005, 165–97. See also Ann Bermingham's article on de Loutherbourg's Eidophusikon in the present volume.

34 David Kornhaber, 'Regarding the Eidophusikon: Spectacle, scenography, and culture in eighteenth century England', *Theatre Arts Journal*, 1, Fall, 2009, http://www.taj.tau.ac.il , unpaginated electronic journal.

35 Neil Harris, *Humbug: The Art of P. T. Barnum*, Boston, MA, 1973.

36 Kornhaber, 'Regarding the Eidophusikon', unpaginated.

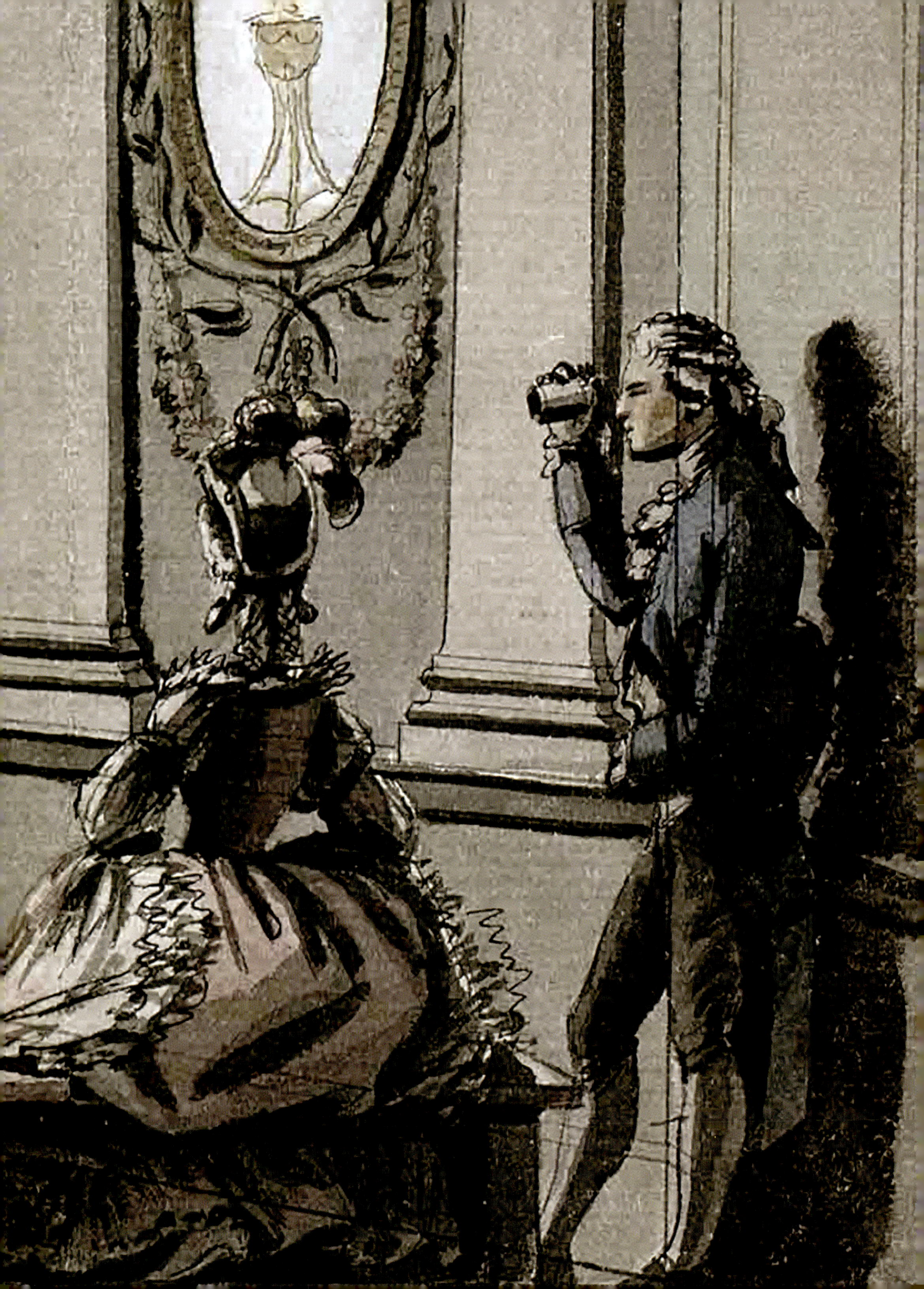

Chapter 10
Technologies of Illusion: De Loutherbourg's Eidophusikon in Eighteenth-Century London

Ann Bermingham

On 26 February 1781 a novel attraction opened in Lisle Street, Leicester Square. Described in a press announcement as 'Moving Pictures representing Phenomena of Nature', the new spectacle was the 'Eidophusikon'. It was the first of several visual attractions with Greek-inspired names – 'Panorama', 'Cosmorama', 'Eidometropolis', and 'Diorama' – that were to appear in the next few years and provide London audiences with immersive experiences of landscape.[1] 'Eidophusikon' derives from *eidoion* ('phantom', 'image' or 'apparition') combined with *phusis* ('nature' or 'natural appearance') and *eikon* ('image' or 'likeness'). The blending of the natural with the supernatural, implied in its name, suited the Eidophusikon's magical simulations of natural scenes. Like so many of the visual attractions of the time, the Eidophusikon hovered between the worlds of art and technology, entertainment and science; moreover, in its illusions, it evoked the magical and uncanny.

The Eidophusikon's mixture of technical wizardry and visual magic touched on a number of issues that would in successive years come to define the boundaries between art and technology, art and entertainment, science and spectacle. As one of the earliest examples of technology-driven visual entertainment, the Eidophusikon clearly prefigures modern cinema. But understanding the attraction's cultural and historical significance as an eighteenth-century object requires consideration of such apparently disparate subjects as theatrical scenography, clockwork automata, theories of the natural world, occult metaphysics, *plein air* oil sketching, landscape painting, and the production of immersive visual entertainments.

The one surviving image of the Eidophusikon is Edward Francis Burney's watercolour drawing from 1782 (*plate 1*). This drawing depicts a theatre interior with some figures entering and others already seated on benches. To the right, separated from the audience by a low enclosure, is a harpsichord. The audience faces a miniature stage framed by a proscenium arch. Two metres wide, one and a quarter metres high, and two and a half metres deep, this stage was used to present a series of scenic illusions designed to mimic natural phenomena. These illusions employed changing light effects to simulate different times of day and various atmospheric conditions. The scenes were accompanied by sound effects and by music by Johann Christian Bach, Michael Arne and Charles Burney, performed on the harpsichord, and occasionally accompanied by singing.

The Eidophusikon's inventor was the Strasbourg-born artist Philippe Jacques de Loutherbourg (1740–1812).[2] A mechanical genius as well as a talented landscape painter and a revolutionary stage designer, de Loutherbourg was a member of

Detail from Edward Francis Burney, *The Eidophusikon showing Satan arraying his Troops on the Banks of a Fiery Lake with the Rising of the Palace of Pandemonium from Milton*, c. 1782 (*plate 1*).

Art and Technology in Early Modern Europe *Edited by Richard Taws and Genevieve Warwick.*

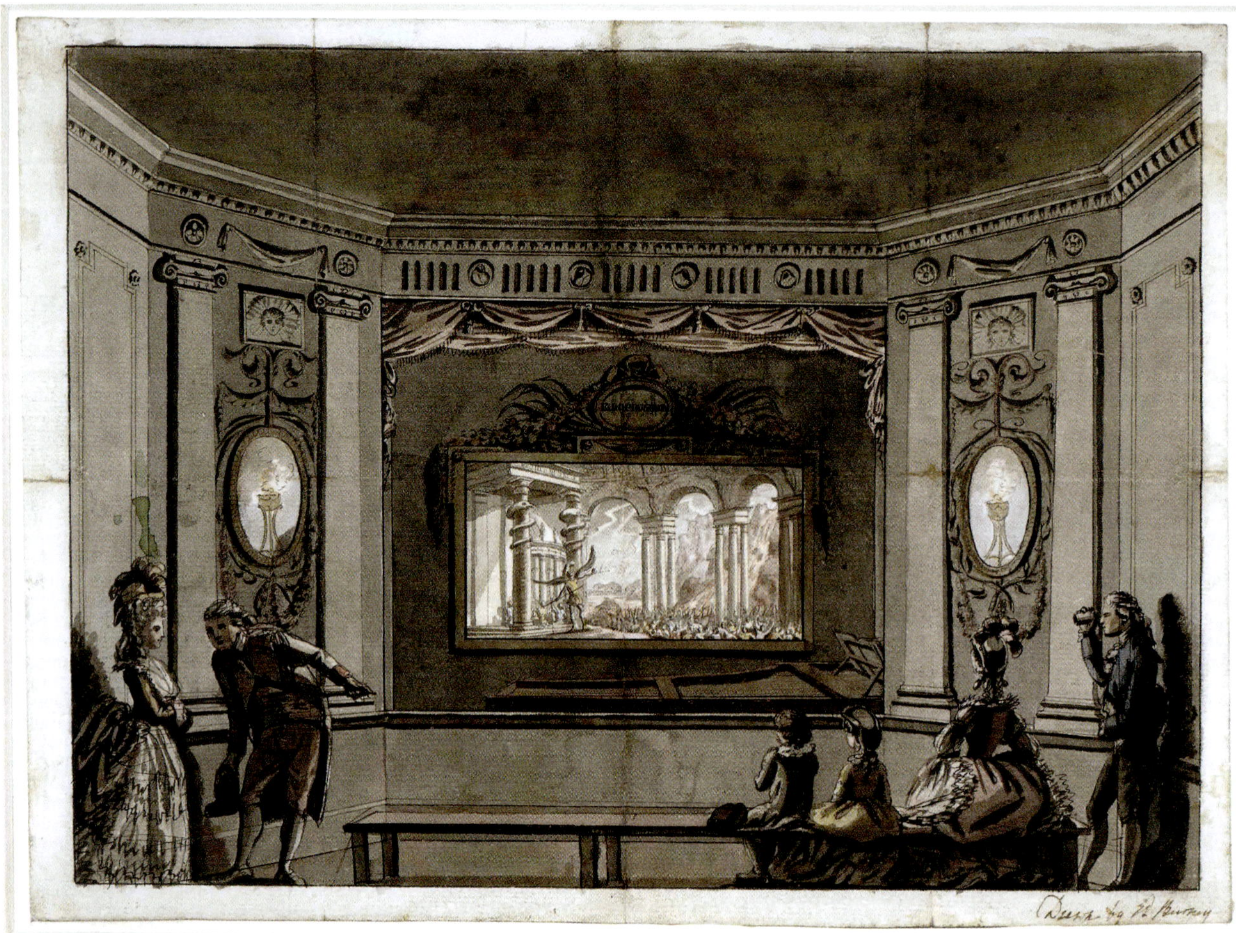

1 Edward Francis Burney, *The Eidophusikon showing Satan arraying his Troops on the Banks of a Fiery Lake with the Rising of the Palace of Pandemonium from Milton*, c. 1782. Watercolour on paper, 19.7 × 27.3 cm. London: British Museum. Photo: © Trustees of the British Museum.

both the French and the British royal academies. His London career oscillated for ten years between the exhibition room and the theatre, and his reputation rested on success in both venues. De Loutherbourg's place in history, however, has been complicated by his career in two countries and by his versatility. Is he a French artist or an English one? Does he belong to the history of painting or theatre? Was he an artist or a mechanic?[3] Also perplexing to anyone attempting to assess de Loutherbourg's significance is his reputation as a 'mystagogue', an occult philosopher, deeply read in alchemy, Behmenist theosophy, Kabbalah, Masonry, Rosicrucianism, and Swedenborgianism.[4] For a short time, he left painting altogether to follow the notorious Count Cagliostro, after which he became, briefly, a mesmeric healer. De Loutherbourg's irregular career can leave biographers bewildered by a remarkable artistic talent apparently wasted on diversions and occult experiments. In this context, the Eidophusikon may appear as one more example of de Loutherbourg's eccentricity.

I would suggest, however, that puzzlement over de Loutherbourg's interests helps to measure the cultural distance between his time and ours. As I hope to show, the Eidophusikon was no mere diversion but central to de Loutherbourg's career and to his conception of himself as an artist and a thinker. As I also hope to demonstrate, the Eidophusikon's aesthetic and philosophical complexity – its seamless incorporation of mechanistic, empirical and spiritualistic theories of nature – reveals something distinctive about the place of art and technology at the end of the eighteenth century. It may be tempting to see the Eidophusikon as looking back to seventeenth-century mechanistic philosophy and forward to nineteenth-century idealism, but my point is that it was the paradoxical mix of the mechanical and empirical with the aesthetic and spiritual that defined the approach to the natural world in this moment.

Artist and Scenographer

The son of an artist, Philippe de Loutherbourg displayed an early love of painting. His father, however, imagined a career for him in engineering, while his mother wanted him to become a minister. The youth was enrolled at the Strasbourg gymnasium, an eminent Lutheran centre, where, perhaps seeking to please both parents, he studied mathematics in order to prepare for engineering, and studied philosophy, languages and theology to prepare to take orders. At fifteen, however, he moved to Paris to become an artist, studying with Carle van Loo and Francesco Giuseppe Casanova. Nonetheless, his early studies in mechanics and what one early biographer calls 'the deepest and most abstruse points' of mystical philosophy were to prove important in his later work, and were crucial to the conception and construction of the Eidophusikon.[5]

In Paris, de Loutherbourg established himself as a prodigy. At twenty-one, he began exhibiting at the Salon where he attracted the interest of Denis Diderot. He was nominated as a *peintre du roi* in 1766 and elected to the Académie in 1767, the youngest man to achieve that recognition. De Loutherbourg was a prolific exhibitor; his submissions to the Paris Salons included battle paintings in the style of his master Casanova, and landscapes derived from the pastoral examples of Nicolaes Berchem or the sublime scenes of Salvator Rosa. In his treatment of light and atmosphere, however, de Loutherbourg followed the example of the great contemporary marine and landscape painter Claude-Joseph Vernet, whose paintings were the product of years of *plein air* oil sketching, beginning when he was a young man in Rome.[6] Vernet's work emphasized varying times of day, seasons, and dramatic aerial effects produced by sunrises, sunsets, storms, and moonlight. Echoing Vernet's example and anticipating the subjects he would later develop in the Eidophusikon, de Loutherbourg debuted at the Salon in 1763 with paintings of the four times of day, followed in 1765 by pendants depicting *A Morning after the Rain* and *The Beginning of a Storm at Sunset*.

In his Salon criticism, Diderot complimented de Loutherbourg's 1765 offerings, speaking of his 'beautiful light, beautiful effects', and 'sublime imitation of nature'. At the same time, Diderot observed in de Loutherbourg's work a tendency toward the formulaic and an occasional heavy-handedness. Diderot concluded his 1765 review by comparing de Loutherbourg to Vernet, noting that the young man's landscapes had less 'tonal refinement' than Vernet's work, but that, nonetheless, their effects were 'fully achieved'.[7] In 1767, however, in commenting on one of de Loutherbourg's submissions, Diderot wrote, 'The sky is one of this artist's heaviest and worst; it is a big chunk of lapis lazuli ready for the stonecutter's blade.' Perhaps more wounding, however, was a sardonic invocation of Vernet: Diderot wrote 'Either Loutherbourg's sky is lumpy and ponderous or comparable skies by Vernet have too much lightness, liquidity and movement. Monsieur de Loutherbourg, go back and look at the sea; you've visited some stables, that's clear, but you have never seen a storm at sea.'[8] Such criticisms of his landscape painting, and in particular his use of colour, would follow de Loutherbourg throughout his career.

In November 1771, for reasons that most likely concerned a disastrous marriage, de Loutherbourg left Paris for London, abandoning his pregnant wife and five children.[9] Travelling with the pyrotechnicist Giovanni Battista Torré, he arrived with a letter of introduction to David Garrick from Jean Monnet, retired director of the Opéra Comique, describing him as one of France's most distinguished painters. Garrick, who was manager of the Theatre Royal in Drury Lane, hired de Loutherbourg to design scenery for entertainments and pantomimes, which, unlike plays, were primarily visual spectacles. Remarkably, Garrick allowed de Loutherbourg complete artistic control over the *mise-en-scène*.[10]

De Loutherbourg's movement into theatre design is not quite as inexplicable as it might first appear. Diderot had already linked the two arts in his writings about theatre, arguing that theatrical scenes should follow the example of dramatic painting with each scene consisting of visually expressive tableaux that would follow each other in a seamless succession As Michael Fried points out, Diderot's attraction to de Loutherbourg's landscapes was due to their extreme effects and their dramatic, climatic subjects.[11] De Loutherbourg's scenography, like the Eidophusikon which was to follow, fulfilled Diderot's desire to make theatrical scenes look like sublime paintings, and to incorporate them into an unfolding visual narrative.

One of de Loutherbourg's greatest theatrical triumphs in this mode was his 1778 production of *Wonders of Derbyshire, or Harlequin on the Peak*, a series of sixteen set designs based on studies of the Derbyshire Peak District that de Loutherbourg had made the previous summer. The sets were fabricated by the Drury Lane carpenters and painters from de Loutherbourg's models, several of which are preserved in the Victoria and Albert Museum. While landscape sets in this period, like many landscape paintings, were typically generic and conventional, these models demonstrate that de Loutherbourg depicted trees, rocks and water in exquisite detail. The sets highlight the sublime Derbyshire scenery, including the cave known as Peak's Hole, where rope makers worked, and the famous mountain Matlock Tor. Henry Angelo, a friend of de Loutherbourg, observed, 'Never were such romantic and picturesque paintings exhibited in that theatre before.' They 'gave you an idea of the mountains and waterfalls, most beautifully executed, exhibiting terrific appearance'.[12]

De Loutherbourg's stage designs revolutionized plays as well as pantomimes. He is credited today with creating the realist stage. As Christopher Baugh has explained, rather than presenting actors in front of a scene, de Loutherbourg's designs integrated them into the setting, heightening the illusion of reality and transforming the relationship between the actor and the audience.[13] Prologues and epilogues disappeared; the stage, set off by a proscenium arch, receded. The auditorium became progressively darker, focusing attention on the spectacle and absorbing spectators into the scenic illusion. Theatre became as much a visual as a verbal medium.

'A new species of painting'

The Eidophusikon should be understood as de Loutherbourg's masterpiece in landscape, the summation of all the innovative stage magic he had wrought at Drury Lane as well as in his earlier exhibition paintings. At the show's opening in February 1781, he displayed five scenes (*plate* 2):

1. 'Aurora, or the effects of the dawn, with a view of London from Greenwich Park'
2. 'Noon, the Port of Tangier in Africa, with the distant view of Gibraltar and Europa Point'
3. 'Sunset, a view over Naples'
4. 'Moonlight, a view of the Mediterranean'
5. 'A Storm and Shipwreck'

Sometime during the run, the second scene – Tangier with a distant view of Gibraltar – was replaced with a subject from the recently declared war with the Dutch, 'The Bringing of the French and Dutch Prizes into the Port of Plymouth'.[14] Between scenes, de Loutherbourg entertained the audience with transparencies and musical performances.

Describing the show's effect, *The Morning Herald* reported that 'the eagerness of curiosity is so great, that as the scenes follow each other in a quick succession, the spectators too frequently rise from their seats, as to destroy the perspective effects of the picture'.[15] Commenting on the illusion, the *London Courant* exclaimed that had de Loutherbourg lived at the time of Galileo, he might have been charged with 'conjuration' as one who 'by the black arts, had captivated the sun, moon, and stars, and collected clouds, thunder, and lightning, by the aid of the Devil'.[16]

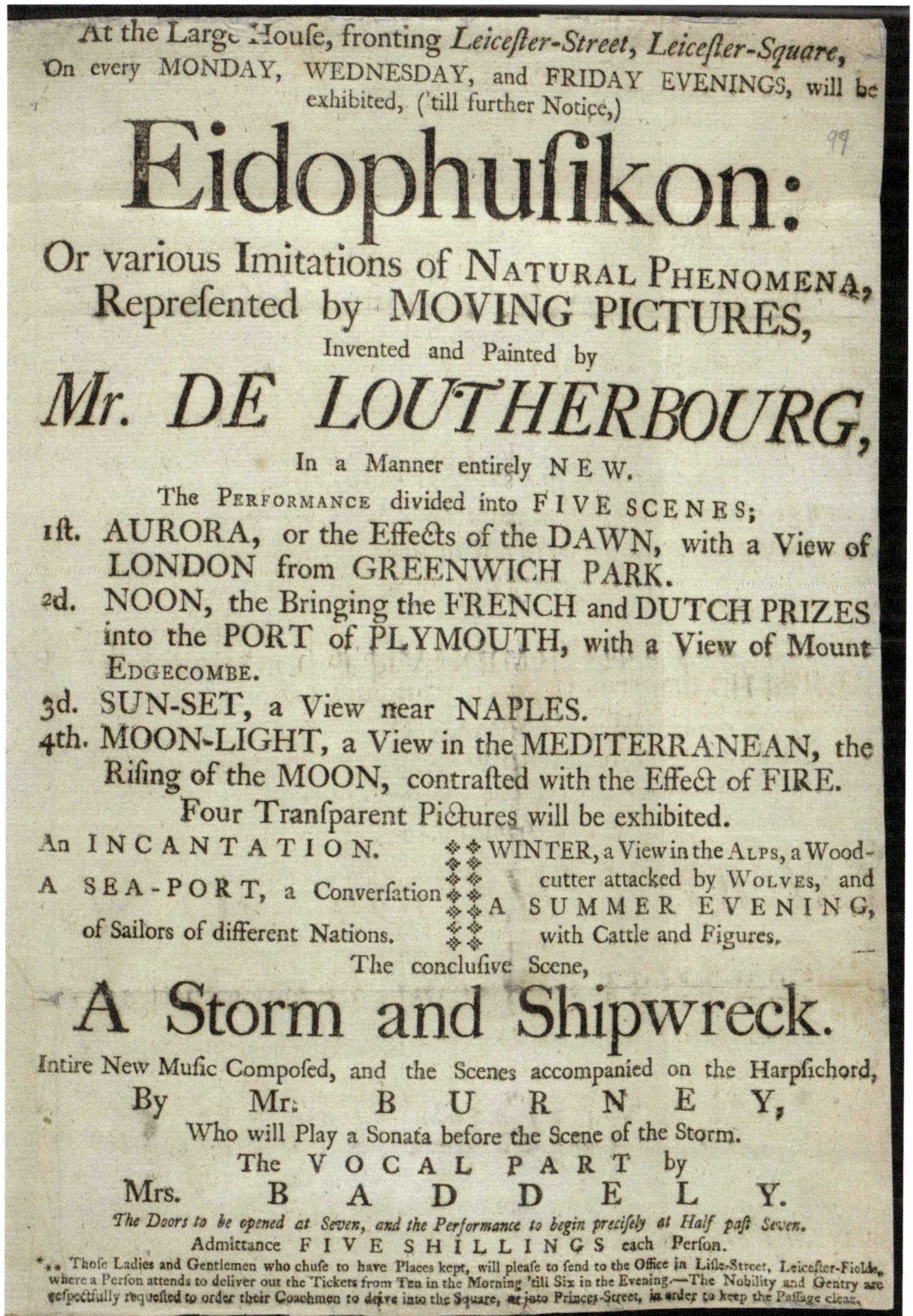

At the Large Houfe, fronting *Leicefter-Street*, *Leicefter-Square*,
On every MONDAY, WEDNESDAY, and FRIDAY EVENINGS, will be exhibited, ('till further Notice,)

Eidophufikon:

Or various Imitations of NATURAL PHENOMENA,
Reprefented by MOVING PICTURES,
Invented and Painted by

Mr. DE LOUTHERBOURG,

In a Manner entirely NEW.
The PERFORMANCE divided into FIVE SCENES;

1ft. AURORA, or the Effects of the DAWN, with a View of LONDON from GREENWICH PARK.
2d. NOON, the Bringing the FRENCH and DUTCH PRIZES into the PORT of PLYMOUTH, with a View of Mount EDGECOMBE.
3d. SUN-SET, a View near NAPLES.
4th. MOON-LIGHT, a View in the MEDITERRANEAN, the Rifing of the MOON, contrafted with the Effect of FIRE.

Four Tranfparent Pictures will be exhibited.

An INCANTATION.
A SEA-PORT, a Converfation of Sailors of different Nations.
WINTER, a View in the ALPS, a Wood-cutter attacked by WOLVES, and
A SUMMER EVENING, with Cattle and Figures.

The conclufive Scene,

A Storm and Shipwreck.

Intire New Mufic Compofed, and the Scenes accompanied on the Harpfichord,
By Mr. B U R N E Y,
Who will Play a Sonata before the Scene of the Storm.
The VOCAL PART by
Mrs. B A D D E L Y.
The Doors to be opened at Seven, and the Performance to begin precifely at Half paft Seven.
Admittance FIVE SHILLINGS each Perfon.

*** Thofe Ladies and Gentlemen who chufe to have Places kept, will pleafe to fend to the Office in Lifle-Street, Leicefter-Fields, where a Perfon attends to deliver out the Tickets from Ten in the Morning 'till Six in the Evening.—The Nobility and Gentry are refpectfully requefted to order their Coachmen to drive into the Square, or into Princes-Street, in order to keep the Paffage clear.

2 Advertisement for the First Season of the Eidophusikon, 1781. Letterpress, dimensions unknown. London: British Library (937.f.2/8(99)). Photo: © The British Library Board.

The success of the spectacle was such that de Loutherbourg produced a sequel that opened in December 1781 and ran until the end of May 1782. Again he displayed five scenes (*plate* 3):

1. 'The Sun rising through fog, an Italian seaport'
2. 'The Cataract of Niagara, in North America'
3. 'The setting of the Sun, after a rainy day, with a view of the Castle, Town and Cliffs of Dover'
4. 'The Rising of the Moon, with a waterspout exhibiting the effect of three lights, with a view of a rocky shore on the coast of Japan'
5. 'Satan arraying his troops on the banks of the fiery lake, with the rising of Pandemonium, from Milton'

The two seasons, we can note, chart a pattern: images of four times of day (dawn, noon, sunset, and night) followed by a sensational finale. In the first offering, the finale features the storm and shipwreck, and in the second, the rising of Pandemonium. This depiction of Pandemonium, the Eidophusikon's most dramatic illusionist triumph, is the scene portrayed in Francis Burney's 1782 watercolour, mentioned above (see *plate* 1).

One of the most complete contemporary descriptions of the Eidophusikon's illusions comes from the artist William Henry Pyne. His account of the first season's opening scene, 'Aurora, or the effects of dawn', conveys a sense of how subtle and dynamic the spectacle was:

> This scene on the rising of the curtain, was enveloped in a mysterious light which is the pre-cursor of day-break, so true to nature, that the imagination of the spectator sniffed the sweet breath of morn. A faint light appeared along the horizon; the scene assumed a vaporish tint of grey; presently a gleam of saffron, changing to the pure varieties that tinge the fleecy clouds that pass away in morning mist; the picture brightened by degrees; the sun appeared, gilding the tops of the trees and the projections of the lofty buildings, and burnishing the vanes on the cupolas; when the whole scene burst upon the eye in gorgeous splendour of a beauteous day.[17]

Pyne's account of the first season's sublime climax – the storm and shipwreck – suggests why that scene in particular thrilled de Loutherbourg's audiences. The storm, Pyne writes, was 'awful and astonishing'. The 'conflict of the raging elements' was 'described with all its characteristic horrors of wind, hail, thunder, lightning, and roaring of the waves, with such marvellous imitation of nature, that mariners have declared, whilst viewing the scene, that it amounted to reality'.[18]

The effect of the scene of the storm and shipwreck on the audience was uncanny in that it erased the distinction between image and reality.[19] In comparing de Loutherbourg's storm to one raging outside, Thomas Gainsborough, who often visited the Eidophusikon, declared that the artificial tempest surpassed the natural: 'De Loutherbourg, our thunder is best'.[20] At another performance a woman in the audience pointed fearfully to a storm gathering outside the theatre. 'The consternation', Pyne writes, 'caused many to retire to the lobby, some of whom, moved by terror or superstition, observed "that the exhibition was presumptuous". We moved to the gallery, and opening a door, stood upon the landing place, where we could compare the real with the artificial storm . . . it was sagely determined that

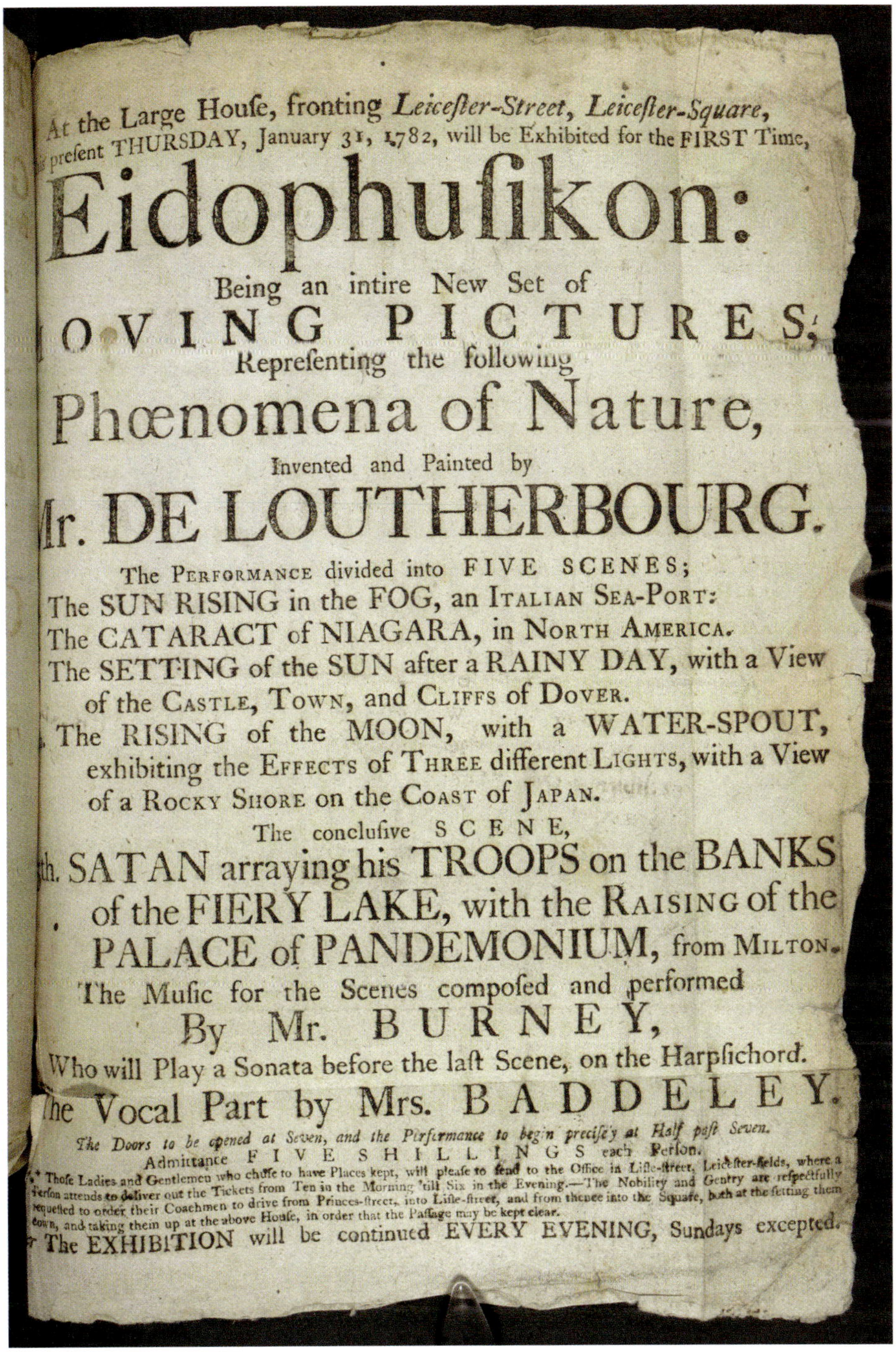

At the Large Houſe, fronting *Leiceſter-Street*, *Leiceſter-Square*,
preſent THURSDAY, January 31, 1782, will be Exhibited for the FIRST Time,

Eidophuſikon:

Being an intire New Set of
OVING PICTURES;
Repreſenting the following
Phœnomena of Nature,
Invented and Painted by
Ir. DE LOUTHERBOURG.

The PERFORMANCE divided into FIVE SCENES;
The SUN RISING in the FOG, an ITALIAN SEA-PORT:
The CATARACT of NIAGARA, in NORTH AMERICA.
The SETTING of the SUN after a RAINY DAY, with a View of the CASTLE, TOWN, and CLIFFS of DOVER.
The RISING of the MOON, with a WATER-SPOUT, exhibiting the EFFECTS of THREE different LIGHTS, with a View of a ROCKY SHORE on the COAST of JAPAN.
The concluſive SCENE,
th. SATAN arraying his TROOPS on the BANKS of the FIERY LAKE, with the RAISING of the PALACE of PANDEMONIUM, from MILTON.
The Muſic for the Scenes compoſed and performed
By Mr. BURNEY,
Who will Play a Sonata before the laſt Scene, on the Harpſichord.
he Vocal Part by Mrs. BADDELEY.

The Doors to be opened at Seven, and the Performance to begin preciſely at Half paſt Seven.
Admittance FIVE SHILLINGS each Perſon.

*** Thoſe Ladies and Gentlemen who chuſe to have Places kept, will pleaſe to ſend to the Office in Liſle-ſtreet, Leiceſter-fields, where a Perſon attends to deliver out the Tickets from Ten in the Morning 'till Six in the Evening.—The Nobility and Gentry are reſpectfully requeſted to order their Coachmen to drive from Princes-ſtreet, into Liſle-ſtreet, and from thence into the Square, both at the ſetting them down, and taking them up at the above Houſe, in order that the Paſſage may be kept clear.

The EXHIBITION will be continued EVERY EVENING, Sundays excepted.

3 Advertisement for the Second Season of the Eidophusikon, 31 January 1782. Letterpress, dimensions unknown. London: British Library (937.b.3). Photo: © The British Library Board.

man was an extraordinary creature, who could create a copy of Nature, to be taken for Nature's self.'[21]

An article that appeared in the *European Magazine* in March 1782 described the Eidophusikon as a 'new species of painting', one that transcended 'common painting' by introducing the element of time. These 'canvases', according to the writer, 'copy the gradual workings of nature in her most important scenes'.[22] In a similar vein, *The Morning Chronicle* called the scenes of the first season 'beautiful beyond description . . . closer imitations of nature than any efforts of art we ever beheld before'.[23] These comments suggest that the Eidophusikon was received, as indeed it was promoted, as a new and improved kind of landscape painting. Moreover, the choice of subjects – dawn, the sun rising through fog, moonrise, a sunset after rain, storms at sea, lightning and fire – reveal how much the 'canvases' owed to Vernet's landscapes and de Loutherbourg's variations on them.

The early accounts of the Eidophusikon emphasize de Loutherbourg's fusion of art and technology. One review, for example, called the scenes of the first season 'the most beautiful representations of nature that were ever effected, by mechanism, and painting'.[24] Another noted that in the show, de Loutherbourg's 'great merit as a painter' was married to 'his skill as a mechanist', and a third observed the happy union of 'the painter and the mechanic'.[25] As Pyne's account makes clear, de Loutherbourg's effects were indeed mechanically-produced. Scenes were animated by changing lights that played over moving parts and coloured scrims. These visual effects were accompanied by sound effects, such as thunder that was created using a suspended copper sheet, rain imitated by turning a box filled with seeds and pebbles, and an eerie whistling wind made by a pair of drum heads rubbing together. Sets were composed of painted flats and three-dimensional models of buildings, trees and rocks. The Greenwich scene, for example, was built from rows of cut-out pasteboard, with a foreground made from pieces of cork that had been cut to suggest a sand-pit covered in moss and lichen, and with individually constructed ships and trees. The moving clouds that played an important part in the effect were images imprinted in semi-transparent colours on a long linen strip some twenty times the span of the stage. The strip was gradually unrolled by a winding machine, with backlighting arranged so that the images cast drifting shadows over the landscape.[26]

As in Drury Lane spectacles, the Eidophusikon's effects relied heavily on dynamic lighting. To vary the quality of light, de Loutherbourg created a machine in which screens of red, purple, blue, and yellow coloured glass rotated in front of lamps hidden in the proscenium arch, thereby projecting changing colours onto the scene. Meanwhile, shifts from front- to backlighting supplied transitions for times of day and atmospheric effects.[27] Remarking on the illusionistic lighting, the *European Magazine* observed that de Loutherbourg achieved 'a harmony in all the movements which completes the deception – There is no harsh, irregular or hasty transition – the progressions are uniform, and have the slowness and constancy of the operations which they imitate.'[28] In the Pandemonium scene, the effect of the coloured lights was particularly dramatic. Pyne reported that in this presentation, the lights 'threw their whole influence upon the scene, as it rapidly changed, now to a sulphurous blue, then to a lurid red, and then again to a pale vivid light, and ultimately to a mysterious combination of the glasses, such as a bright furnace exhibits in fusing metals'.[29] De Loutherbourg's use of lighting effects demonstrated how one could paint with light.

De Loutherbourg charged five shillings admission, a standard fee for a London attraction, though not inconsiderable when compared to the one shilling charged for admission to the annual Royal Academy exhibition. After two seasons, however, he sold the attraction to his assistant, a man named Chapman, who appears to have taken it on tour. Nothing is heard of the Eidophusikon again until spring 1785 when the Pandemonium scene, along with several other unrelated attractions, was exhibited at Exeter Change for one shilling.[30] The following year Chapman reopened the attraction at Exeter Change with several of de Loutherbourg's scenes, including Pandemonium, and the storm and shipwreck, now linked to the recent, tragic loss of the Halsewell off the Dorset Coast.[31] Before closing to refurbish his attraction, Chapman hired a guitar-playing Polish dwarf. The 'Count', as he was called, was described by the *Morning Herald* as, like the Eidophusikon itself, a 'prodigy of nature'.[32] For this amplified programme, Chapman charged three shillings for prime seats, two for others. When Chapman reopened the attraction in Spring Gardens in 1793, again at three and two shillings admission, the show included magic tricks and acrobatic acts. Six years later Chapman advertised a 'new Eidophusikon' in Panton Street, once more priced at three and two shillings. This show included some new scenic illusions, such as a view of Liverpool Harbour, and a variety of other attractions, including a glass harmonica, recitations, popular songs, and a performance by a 'Learned Dog'.[33] On the night of 21 March 1800, a fire started in a nearby brothel and Chapman was burned out.[34] Thus ended the Eidophusikon's nineteen-year history. The downward spiral from elite entertainment to commonplace variety show was typical for eighteenth-century attractions of this sort, and suggests why so few of them survive.

Nature's Clockwork

The Eidophusikon was a hybrid medium that drew both from theatrical lighting technologies and from the clockwork mechanisms of the seventeenth and eighteenth centuries. A revival of interest in clockwork mechanisms began when the famous Swiss artificers Pierre Jaquet-Droz and his son Henri-Louis opened their *Spectacle mécanique* in 1776 at the Great Room in Covent Garden.[35] One of their spectacles was an animated scene called 'The Grotto', a 'mechanical picture' such as those sometimes

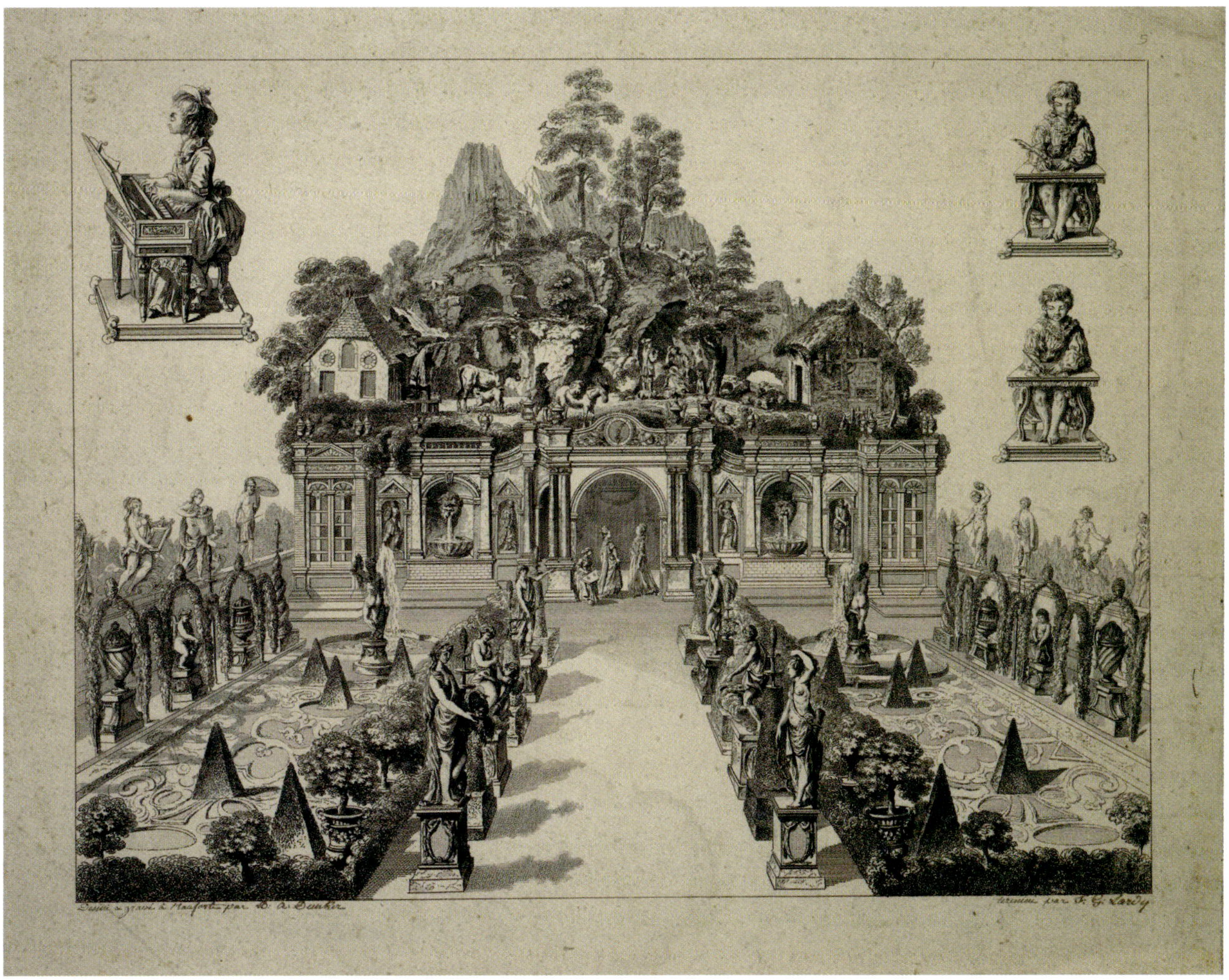

4 B. A. Dunker, finished by Lardy, *Représentation des Ouvrages de Mécanique Inventés et Executés par les Sieurs P. J. Droz et H. L. J. Droz célèbres artistes de Chaudefond*, 1775. Etching and engraving on paper, 40 × 50 cm. London: British Library. Photo: © Trustees of the British Museum.

incorporated in musical clocks but one far more elaborate than the usual clockwork scene (*plate 4*).[36] About 140 cm square, 'The Grotto' was promoted as a 'contrast of art and nature'. In the foreground was a formal garden with parterres and a path lined with topiary trees and classical statues. The path led to an architectural pavilion, behind which rose a Swiss pastoral scene that included a cottage, mill, and rivulet, followed by a distant vista of Alpine peaks. A mechanical sun rose, passed over the meridian and set in exact accord with the seasons of the year. A countryman mounted on an ass crossed from the cottage to the mill, after which a shepherd emerged from a cave and played tunes on a flute. The shepherd's tunes roused a sleeping shepherdess who, taking up a guitar, joined the flautist in a duet. Afterwards the countryman, his ass laden with flour, walked back to his cottage. Meanwhile the trees in the formal garden budded, flowered, and fruited while fountains played, singing birds flew, and animals made appropriate sounds. In the middle of the pavilion, a country girl played minuets on a dulcimer while two young ladies danced, as the prospectus put it, 'with great regularity and grace'.[37]

The promotional description of 'The Grotto' as 'a contrast of art and nature' is curiously ambiguous. Does this formulation refer to the contrast between the formal garden in the foreground of the 'mechanical picture' and the rustic scene in the distance? Or does it refer to the whole mechanical apparatus as an imitation of nature? Numerous contemporary thinkers imagined all of creation as a vast moving machine, populated by smaller machines, a category that included humans as well as animals. The laws of nature as expounded by Isaac Newton were regular, predictable, and mechanical. Such a perspective is represented in Joseph Wright's famous *Philosopher Lecturing on an Orrery* (*plate 5*). In this image, in which Newton's clockwork solar system is demonstrated by the mechanical device of the orrery to a group of attentive laymen, 'natural' and 'mechanical' are synonymous.[38] For those who held to René Descartes' mind/body dualism, only the workings of the mind were exempt from mechanism. Others, however, such as Julien Offray de La Mettrie, argued that mind and body were animated by the same principles. 'Let us then conclude boldly', La Mettrie writes in *L'Homme Machine* (1747), 'that man is a machine, and that in the whole universe there is but a single substance differently modified.'[39] For La Mettrie, the word 'organic' bore no relationship to our ideas of the biological. The only thing differentiating the organic from the inorganic, in La Mettrie's philosophy, was motion. Machines that moved, therefore, were demonstrating the deepest principles of organic nature while also blurring the line between animate nature and the inorganic. Perhaps, then, 'natural' and 'mechanical' were synonymous; the elaborate 'Grotto' is to be understood as something like an orrery, that is, a machine built in imitation of the great machine of nature.

The Jaquet-Droz 'Grotto' has been lost. But three other objects from the *Spectacle* of 1776 survive and can be found at the Musée d'Art et d'Histoire at Neuchâtel, Switzerland, where they still perform. These complex, almost life-size automata are constructed from thousands of individual pieces. Barefoot and dressed in silk robes, two boy figures demonstrate graphic skills. One dips his pen in ink, shakes it, and proceeds to write, leaving appropriate spaces between words, differentiating between lower- and upper-case letters, crossing t's and dotting i's. The second traces the profiles of the kings and queens of France and England along with other images, such as a dog and a cupid in a chariot pulled by a butterfly. The third, a young lady, plays a harpsichord, her fingers striking the proper keys, her foot tapping the measure, her eyes moving, and her bosom swelling emotionally in response to the

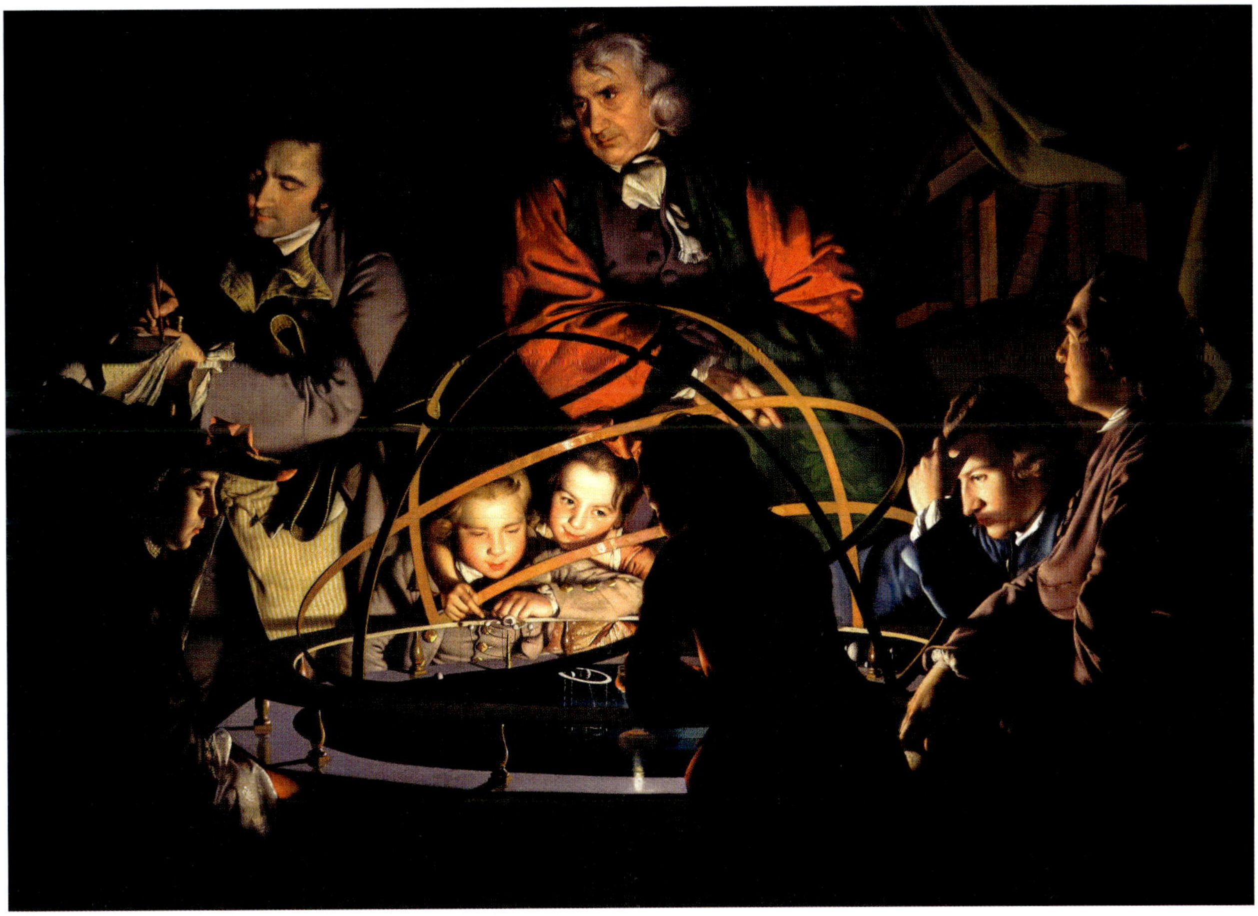

5 Joseph Wright of Derby, *The Orrery*, c. 1766. Oil on canvas, 147 × 203 cm. Derby: Derby Art Gallery. Photo: © Bridgeman Images.

music. The most complex figure, the writing boy, is made up of 6,000 pieces and can write forty lower- and upper-case letters.

The Jaquet-Droz dolls invoke the work of another virtuoso engineer: Jacques de Vaucanson and his renowned artificial duck.[40] Completed in 1739, Vaucanson's duck not only moved like a duck, quacked like a duck, and ate like a duck, but also shat like a duck. Vaucanson claimed to have mechanically reconstituted the biological process of digestion. He insisted that the oats swallowed by his duck were truly digested in a 'chemical laboratory' and excreted through the 'circumvolutions of pipes'.[41] Able to waddle, flap its wings, and wiggle its tail, the duck produced realistic movements due to a multiplicity of small moving parts. One wing alone, Vaucanson claimed, contained no less than 400 parts.[42] Vaucanson's duck was exhibited in London at Spring Gardens in 1742 along with two musical automata: one a flautist, the other a drum and fife player (*plate* 6). These figures actually played their instruments. They had hundreds of tiny levers to operate their tongues, lips and fingers, and angles in their mouths so that the air exhaled through a bellows was properly channelled.

The success of the artificial duck inspired Vaucanson to attempt further feats of mechanized biology. In 1741 he embarked on the creation of a mechanical man that would include 'all the animal operations' of a living human being including respiration, digestion, and the circulation of the blood as well as the movement of muscles, tendons, and nerves.[43] Vaucanson's work on this project was never completed, however, since later in the year he was appointed by Louis XV as Inspector of Silk Manufacture. In this capacity he designed an automatic loom, the first device for programmed patternmaking in textile production. It operated by using long strips of stiff paper punched out with holes to guide the loom's Jacquard weaving mechanisms. Powered by a horse, an ox or an ass, the ingenious loom proved, as Vaucanson provocatively declared, that an animal might make 'cloth more beautiful and more perfect than most silk workers'.[44] The loom was a great success, but by the end of the century, it was powered by steam, not animals.

Because they embodied materialized principles of natural science, the Vaucanson and Jaquet-Droz mechanical wonders were more than mere amusements; like the orrery, they were philosophical machines, objects of enlightenment. We should note, however, that the key impact of the three Jaquet-Droz automata does not seem to have been simply their anatomical or physical realism so much as their psychological credibility. Henry Angelo claims that all who saw the Jaquet-Droz exhibition, including George III, were particularly

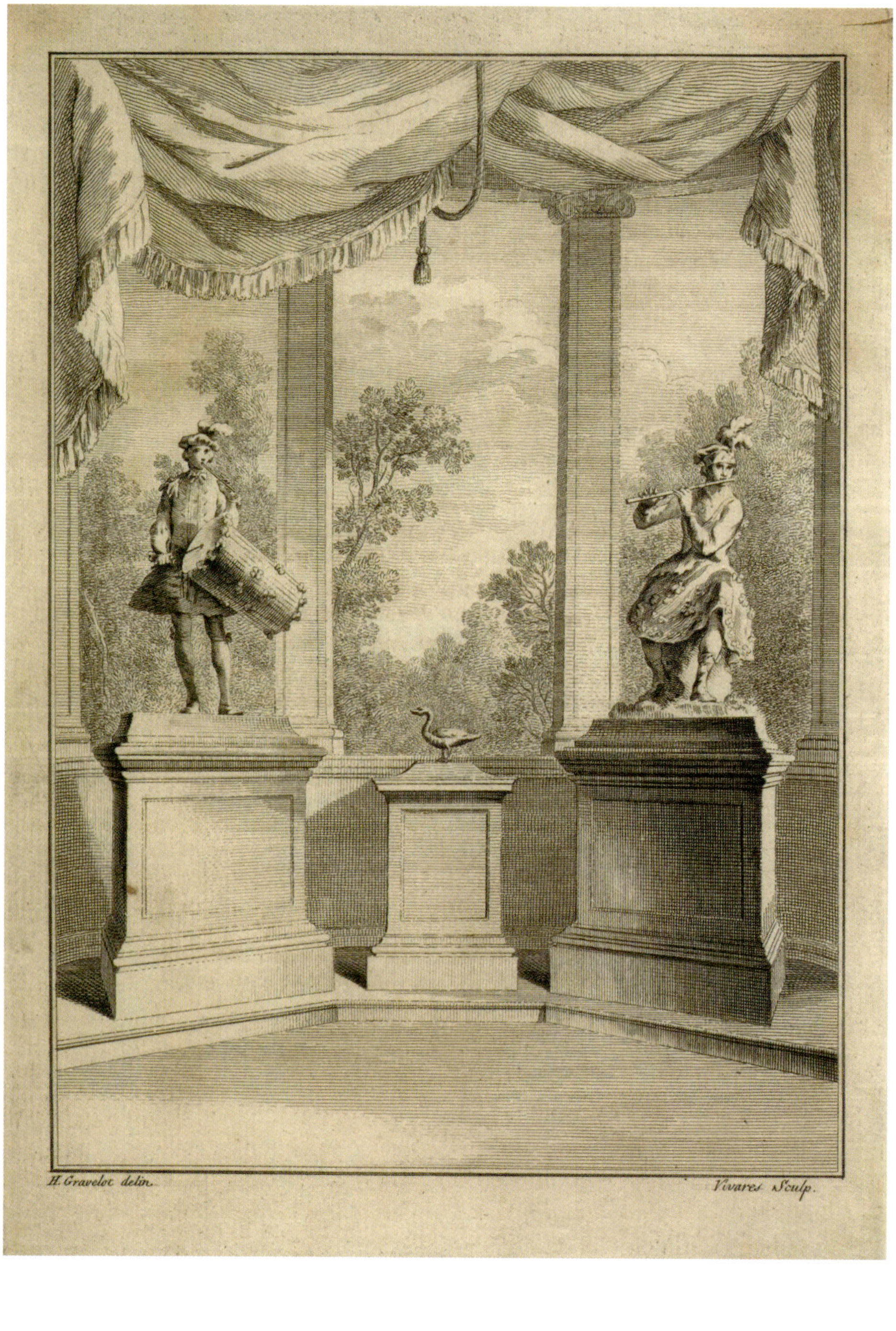

6 François Vivares, after a drawing by Hubert Gravelot, frontispiece for Jacques de Vaucanson, *An Account of the Mechanism of an Automaton or Image Playing on the German Flute*, London: T. Parker, 1742. Cambridge, MA: Harvard University (Hyde Collection 1600–1800, Houghton Library). Photo: © Imaging Services, Widener Library.

fascinated by the writing automaton; as something of an afterthought he also notes there was 'a beautiful Swiss landscape. . . . This was pleasing, but not, like the other [the writing automaton], incomprehensible to everyone.'[45] The writer and draughtsman automata implied to the viewer that the mind might be programmed like a machine. Moreover, the harpsichordist automaton's moving eyes and swelling bosom suggested that emotion, too, might be an automatic neurological reaction.[46] Significantly, all three automata – the penman, the sketcher, the harpsichordist – were conceived as artists of various kinds, suggesting that creativity itself might be demonstrated to be a mechanical function. In short, these philosophical machines not only tested the boundaries between mind and matter but also between mental inspiration and mechanical programming.

From Henry Angelo, we learn that de Loutherbourg knew and socialized with Henri-Louis Jaquet-Droz while the latter was in London exhibiting his automata.[47] Nevertheless, while it may have been inspired by the Grotto, the Eidophusikon's simulations of atmospheric effects were achieved through light and not simply by the mechanical movement of a representation of the sun. The Grotto, synchronized to reveal the sun's position at different times of the year, was an explanatory device. De Loutherbourg's machine, on the other hand, incorporated a ghost, light, and its fluid, ever-changing progressions and transformations seemed magical. As Gainsborough's exclamation – 'De Loutherbourg, our thunder is best' – suggests, the Eidophusikon in some respects had more in common with the uncanny quality of the Jaquet-Droz draughtsman, writer and musician than with clockwork display of 'The Grotto'.[48]

The Eidophusikon was both an empirical demonstration of nature's mechanical laws and an illusionistic replication of nature's appearance. Paradoxically, the more successful any technological simulation of nature is, the more uncanny it becomes. The sensation of the uncanny arises, Freud claims, when one is uncertain if an object or experience is natural or artificial, living or dead, animate or inanimate, familiar or unfamiliar, all of which he saw paradigmatically embodied in automata.[49] The Eidophusikon revealed the uncanny lurking within eighteenth-century rationalism. In its mechanistic simulations of the times of day and natural cataclysms, it threatened to dissolve the rational into the illusory, the material into the metaphysical, and the physical into the psychological. The Eidophusikon demonstrates that the contradictions it produced and held in suspension were not antithetical to eighteenth-century rationalism, but part of its very make-up.

Spiritual Machinery

Light may well have had spiritual significance for de Loutherbourg. Educated at Strasbourg, a centre for mystical theology, de Loutherbourg, like William Blake, Richard Cosway and other contemporaries, seems to have been particularly influenced by the ideas of the German Lutheran mystic Jacob Boehme (1575–1624).[50] For Boehme, light was a manifestation of divine love. This revelation impressed itself on Boehme when he became entranced by a beam of sunlight reflected on a metal bowl. Boehme developed his revelation into a dynamic, dialectical theory of the spiritual structure of the universe that combined features of Kabbalah, astrology, alchemy, and astronomy.[51]

In this theory Boehme identified seven 'properties' of the universe. These properties, generated by a continuing process of divine self-realization, were both objective features of the deity and stages through which the soul needed to pass to

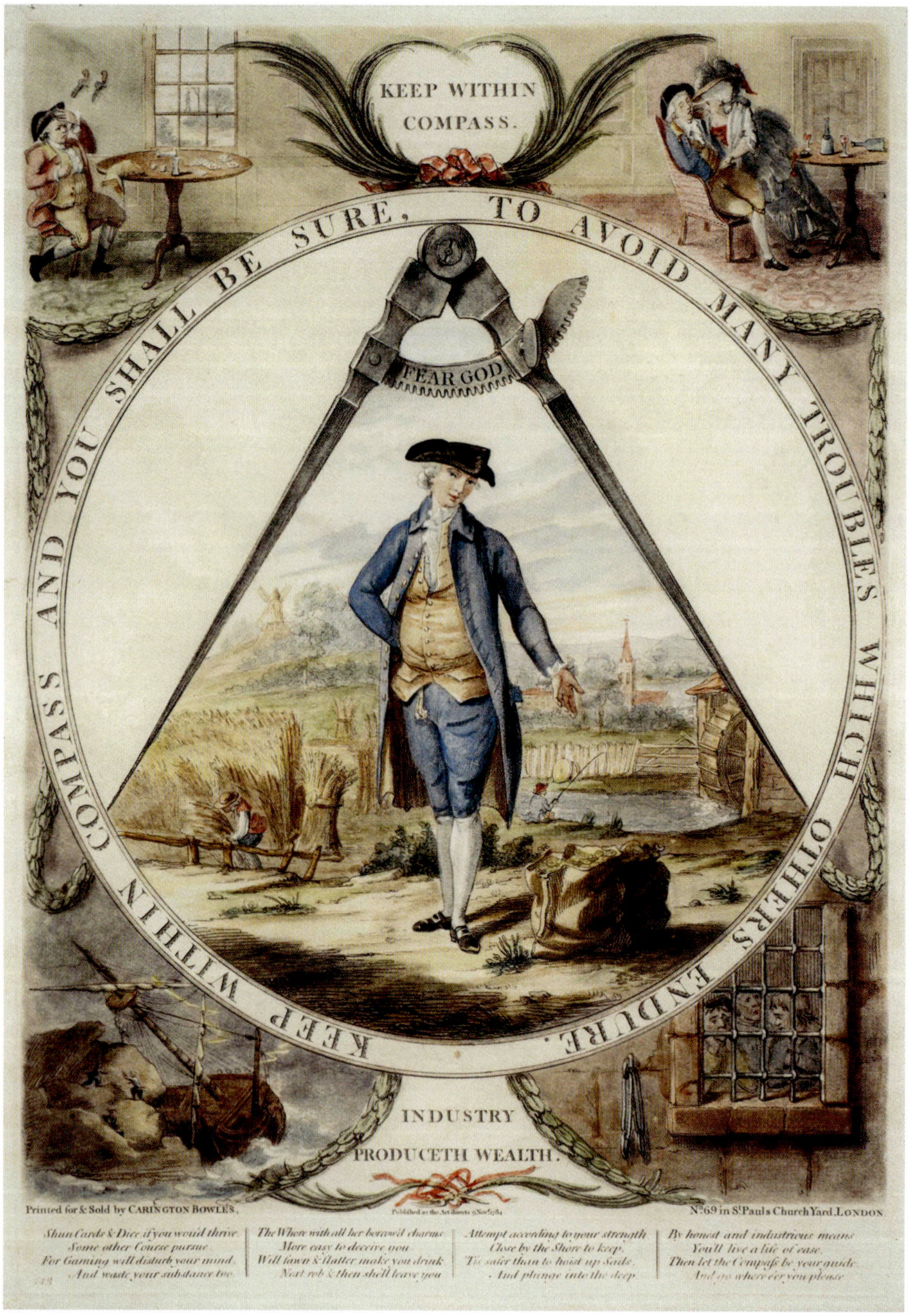

7 Anonymous, *Keep within Compass*, 1784. Hand coloured etching on paper, 45 × 61 cm. Private Collection. Photo: © Tony Mastres.

come to an understanding of divinity. Moving from the first principle – associated with materialism, constriction, and death – to the seventh, 'Sophia' or wisdom, the soul progressed through an alchemical process of spiritual refinement. Boehme identified Sophia with the moon which, possessing the power of reflection, mirrored the six preceding stages. But progress, as Boehme conceived it, was cyclical. Each time the cycle was completed it began again, playing out an eternal progression from darkness to light, birth to rebirth.

The seven stages through which the soul progressed, according to Boehme, corresponded to the three members of the Trinity: Father, Son, and Holy Spirit. The Father, the foundational 'dark world', encompassed the first three stages, while the Holy Spirit or 'light world' was associated with Sophia, the final stage. Between the darkness of the Father and the light of the Holy Spirit was the realm of the Son or earthly life, encompassing the fourth, fifth, and sixth 'properties', associated respectively with fire, water, and sound. This middle realm came into being, according to Boehme, in a flash of spiritual fire, an element he associated with the alchemical transformation of earthly compounds, including metals, stones, gems, and the volatile substances of sulphur, salt, and mercury.[52] The natural embodiment of this positive aspect of fire, identified as an aspect of spiritual desire, was the sun, which had for Boehme the power to liberate the soul from darkness and material bondage. But the middle realm of the Son or earthly life was also the sphere in which the forces of good and evil, spirit and matter, light and dark, and life and death struggled. It was the mid-point in the process of spiritual manifestation, the pivot around which the worlds of the Father and the Holy Spirit revolved, just as the planets revolved around the sun in the Copernican system to which Boehme subscribed.

Boehme's theosophical ideas enjoyed a renaissance in the later eighteenth century when they were taken up and elaborated by mesmerists, alchemists, and spiritualists of every kind including Masons, Rosicrucians, and Swedenborgians. De Loutherbourg was deeply involved with all these movements; he was a member of the London Theosophical Society among other groups, and owned multiple volumes of Boehme's works, as well as those of contemporary followers of Boehme such as the Rev. Richard Clarke.[53] In addition, he had an extensive collection of alchemical, Kabbalistic, Rosicrucian, and Swedenborgian texts.[54] Considering his immersion in Boehme's religion of light, de Loutherbourg may have conceived the Eidophusikon as a metaphysical drama embodied in a series of illusionistic landscape scenes.

The protagonist of de Loutherbourg's spectacle is, of course, the sun. Emerging at dawn to illuminate the darkness, de Loutherbourg's sun triumphs in noon's bright splendour, later suffuses the evening's afterglow, and finally, reflected by the moon, associated by Boehme with wisdom, illuminates the evening. While de Loutherbourg's uninitiated audience might have seen simply a magical simulation

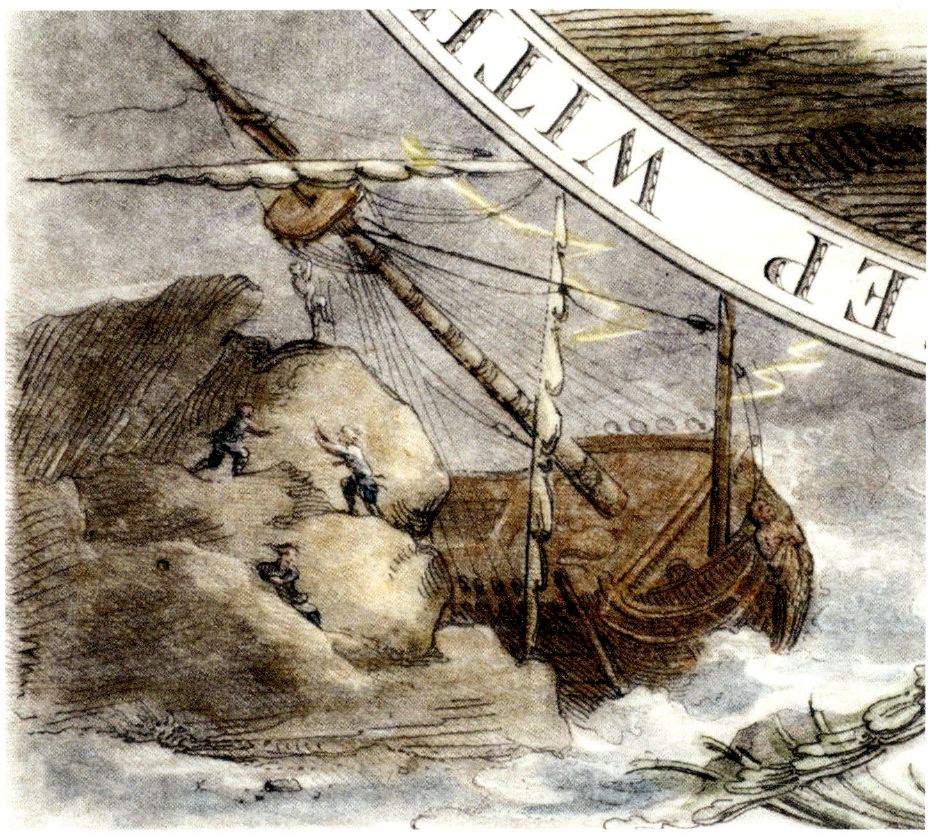

8 Detail of *Keep within Compass*, 1784, showing a shipwreck.

of the times of day, de Loutherbourg himself and others immersed in Boehmenist thought would perhaps have understood the spectacle's first four scenes as a recapitulation of a spiritual progress.

But what about the two seasons' sensational finales: the climactic scenes of shipwreck and the raising of Pandemonium? How might they have been understood in relation to Boehmenist theosophy? The shipwreck does not, so far as I have been able to determine, reference any particular passage in Boehme, but the subject – one treated many times by de Loutherbourg in his paintings – was a traditional emblem of the perils of overreaching ambition or pride, as, for example, in Carrington Bowles's 1784 Masonic and moralizing print *Keep within Compass* (*plate* 7 and *plate* 8).[55] Another common emblem of the perils of pride was the Miltonic scene of Satan constructing his palace, Pandemonium. From early in his theosophical development Boehme was intrigued by the story of Lucifer's fall, which he understood as resulting from the Deity's need to give form to its contrary 'so that the will might find, feel, and behold itself'.[56] Blinded by pride, Lucifer and the fallen angels created a dark, constricted universe in which fires burnt without illumination.[57] Insofar as each version of de Loutherbourg's finale portrayed pride, the Pandemonium scene might be understood as parallel to the storm and shipwreck scene, which it replaced, a warning against the perils of spiritually constructing a self-enclosed world of pride.

The Eidophusikon's drama of movement, its dynamic, ever-changing scenes, may also reflect Boehme's emphasis on the endless process of spiritual enlightenment. Boehme's writings suggest that spiritual enlightenment can reach no final form, no final resolution; to do so would result in a reification, an idol and a falsity. De Loutherbourg's 'new species of painting', one that added 'motion to resemblance' and repeated its demonstration again and again, may well have been conceived as a parallel spectacle of theosophical wisdom, a kind of spiritual perpetual motion machine.[58]

Colour, too, may have held spiritual significance for de Loutherbourg. Changing light effects such as the movement from darkness through the pink light of dawn to the white light of noon in the Greenwich scene, or the modulations in the Pandemonium scene from blue through red to what Pyne described as a 'pale vivid light . . . such as a bright furnace exhibits' could readily be understood as having both theosophical and alchemical meaning.[59] Here perhaps the frequent criticisms of de Loutherbourg's unsubtle colours in easel painting might be reconsidered. The relationship between painting and alchemy was longstanding. Indeed alchemy was revived in the late Middle Ages partly in an effort to understand the properties of minerals in the making of dyes and pigments.[60] Moreover, we know that as a gymnasium student de Loutherbourg pursued alchemical studies, which he used in preparing colours that, as one report has it, were 'more vivid and durable' than those known to others.[61] Diderot's complaint that de Loutherbourg's colours were too bright and unmodulated was repeated in the 1780s by John Wolcott in his satiric 'odes' to the Royal Academicians, which described de Loutherbourg's 'brass skies', 'golden hills', 'marble bullocks', and 'glass pastures'.[62] Nonetheless, de Loutherbourg's distinctive colours tantalized the young J. M. W. Turner, who haunted the older artist's studio in the hope of learning how they were produced.[63] Moreover, like Johann Wolfgang von Goethe, who also studied at Strasbourg and whose *Theory of Colours* was inflected by alchemical symbolism, de Loutherbourg seems to have thought of colour in terms of the alchemical transition from black (*prima materia*) to white. John Gage speculates that de Loutherbourg's curious remark that there were only two primary colours – blue and yellow – can be explained by understanding that he was thinking simply in terms of tone and the alchemical transition from dark to light. Insofar as it is a 'middle tone', red would not be 'primary'.[64]

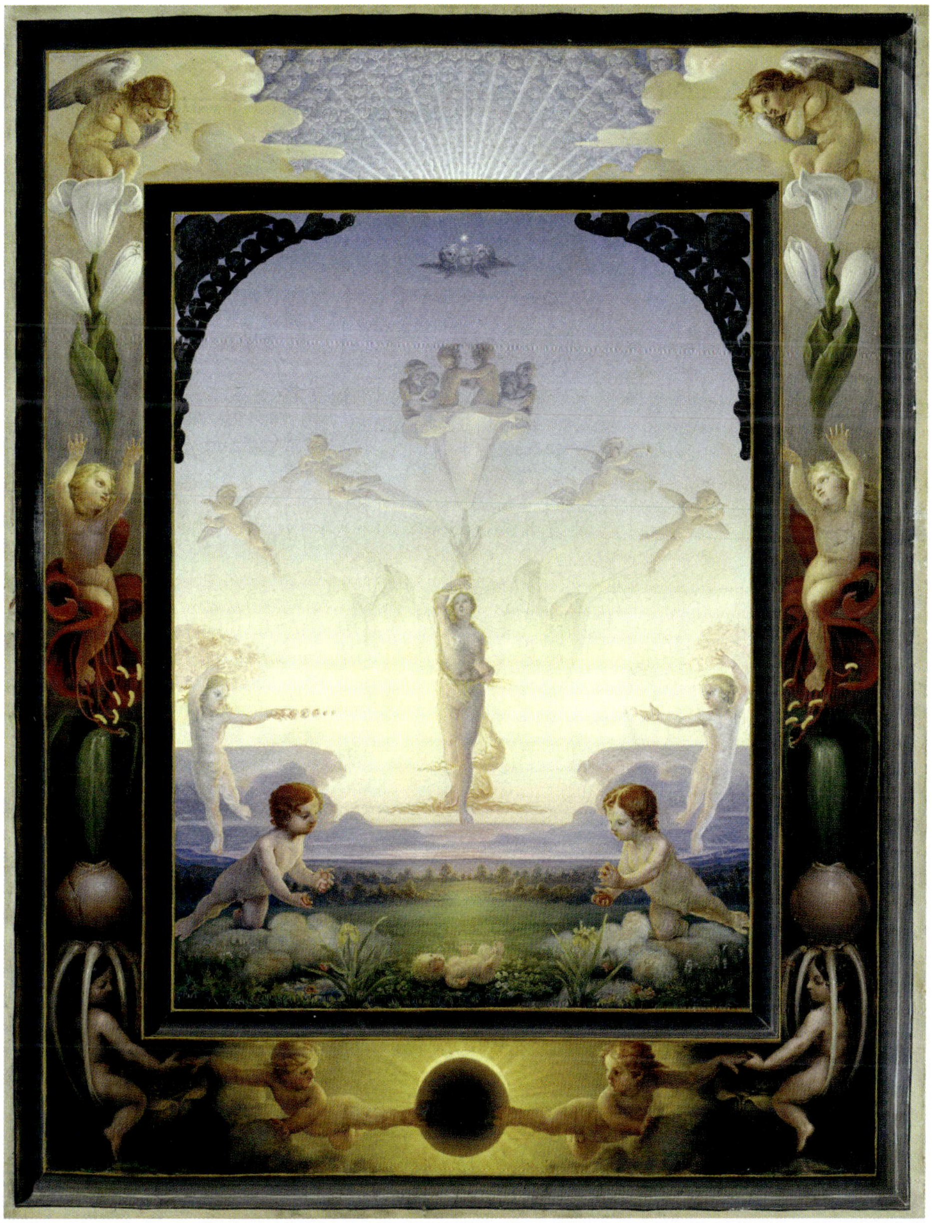

9 Philip Otto Runge, *The 'Small' Morning* [first version], 1808. Oil on canvas, 109 × 85.5 cm. Hamburg: Kunsthalle. Photo: © Elke Walford/Art Resource.

Perhaps, then, precisely the same impulses that led de Loutherbourg to produce what many saw as mineral-like and metallic oil pigments became, when transmuted into effects of changing light in the Eidophusikon, a triumph of colourist art.

Infused with a theosophical dimension, the Eidophusikon would not have been merely a hybrid technology. It would have been, at the deepest level, a contradictory one: an engine at once mechanical and spiritual. Its machinery depended on clockwork that demonstrated the mechanical operations of the natural world, as understood by La Mettrie and others; its illusion, however, depended on immaterial elements like light and colour, infused with spiritual significance. De Loutherbourg's theosophical explorations of light and colour anticipated the Boehmenist projects of Philipp Otto Runge: both Runge's famous colour sphere *Die Farbenkugel* (1810) and his never completed *gesamtkunstwerk*, *Tageszeiten* or *Times of Day* (see *plate* 9).[65] Just as Runge used the times of day, flowers, light, colour and landscape to express Boehme's states of spiritual awakening, so de Loutherbourg, in my understanding, sought to express Boehme's theories by finding their metaphorical equivalents in nature.

Because it reproduced the effects of nature so powerfully, the Eidophusikon passed beyond the mechanical simulations of Vaucanson and the Jaquet-Droz, and into the realm of the phantasmatic explored by the illusionists Philip de Philipstal and Étienne-Gaspard Robertson with their phantasmagoria. The Eidophusikon also anticipates the immersive techniques used by Louis-Jacques-Mandé Daguerre in his diorama paintings. Finally in its combination of movement, light, and sound, the Eidophusikon points, as I have suggested, to modern cinema.

Technologies of Illusion

Contemporary artists received the Eidophusikon with enthusiasm. According to Pyne, Sir Joshua Reynolds 'honoured the talents of the ingenious contriver, by frequent attendance . . . and recommended the ladies in his extensive circle to take their daughters, who cultivated drawing, as the best school to witness the powerful effects of nature'.[66] Thomas Gainsborough, as I have noted, also attended shows frequently. Pyne reports that he was so enthralled 'that for a time he thought of nothing else – he talked of nothing else – and passed his evenings at that exhibition in long succession'.[67] What did painters such as Reynolds and Gainsborough see in the Eidophusikon's illusions that spoke to them as artists? Let us note that in its emphasis on movement and process, the Eidophusikon differed from conventionally static forms of British landscape painting, which tended to dwell on topography or the picturesque, and which generally conceived of pictures either as 'views' or 'prospects'. Like the *plein air* sketches of Vernet that influenced the young de Loutherbourg, the Eidophusikon portrayed atmospheric effects. Moreover, in its power to represent the incidental and the transient through movement, the miniature theatre surpassed the *plein air* sketch.[68] Perhaps it was this ability to capture – or, more precisely, mimic – the evanescent that entranced Reynolds and Gainsborough.

The enthralling dynamism of the Eidophusikon can perhaps be related to a widespread eighteenth-century desire to animate the image. Discussing this impulse, Lynda Nead points to Horace Walpole's gothic fantasy *The Castle of Otranto*, in which the

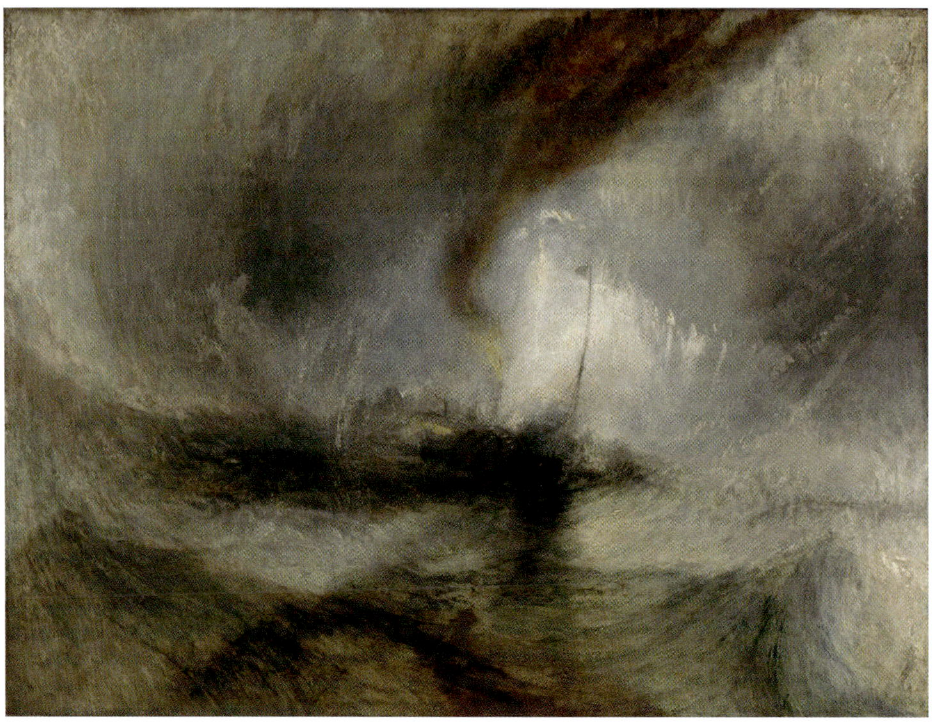

10 J. M. W. Turner, *Snow Storm – Steamboat off a Harbour's Mouth*, 1842. Oil on canvas, 91.5 × 122 cm. London: Tate. Photo: © Tate Images.

subjects of portraits sigh, moan, and finally step out of their frames.[69] We can also recall Diderot's desire for dramatic stage tableaux unfolding into pictorial narrative. In the case of landscape, one can note gardens such as those at Stourhead, designed to guide visitors through a series of constantly changing Claude-like compositions, and to William Gilpin's Wye Valley tour in which boatloads of sightseers observed the passing landscape as a changing series of picturesque views. The Eidophusikon can also be seen as looking forward to such paintings as Turner's *Snowstorm – Steamboat off a Harbour's Mouth* (1842), a dramatic image that strains the conventions of pictorial composition to convey the sensation of wind and turbulent swell of the sea (*plate 10*).[70] We find the fantasy of animation, too, in what John Constable proudly called his 'breezes', those furious brushstrokes that encrust the surfaces of his later landscapes in an effort to suggest movement.

Unlike conventional 'views', the Eidophusikon revealed that the conditions under which a landscape was seen affected not only the object's aspect but also the observer's response. Appearances were relative. Movement, light and colour played central roles in the perception of nature and the depiction of nature's 'moods'. By the early nineteenth century, as Jonathan Crary and others have noted, vision was understood as embodied, subjective, and mobilized by desire.[71] Perspectival painting with its ideal, static viewpoint confined the image within the frame. Paintings such as Turner's *Snowstorm* attempt to erase the frame altogether and project the viewer into the image. The coupling of perception with emotion explains Turner's remark to a visitor when asked about this picture, 'I did not paint it to be understood, but I wished to show what such a scene was like.'[72] Anticipating the fusion of emotion with the perception of nature pursued by artists such as Constable and Turner, the Eidophusikon combined scientific empiricism with the subjective impulses of emotion.

The urge to recreate nature through landscape painting is symptomatic, I believe, of an awareness of humanity's separation from nature. This separation might be taken as an index of human progress and mastery, a triumph over nature's powerful indifference. At the end of the eighteenth century, however, alienation from nature was increasingly felt in art and literature as a deprivation and loss, a development that can be associated, as I have suggested elsewhere, with the transformation of both the human and the natural environment through industrial capital.[73] Large-scale, immersive landscape experiences such as the panorama and diorama exploited the sense of loss and in the process turned subjective states of feeling such as wonder or awe into technologically reproducible commodities.

In part because of its comparatively small size, the Eidophusikon did not fully partake of this commerce in emotion. In fact, like all automata, its success as an illusion depended on the viewer's awareness of artifice. As uncanny as the Eidophusikon's illusions could be, the impulse for audience members was, as Pyne's anecdotes indicate, to compare its mechanical effects with those of nature, and to wonder at the fact that 'man was an extraordinary creature, who could create a copy of Nature, to be taken for Nature's self'.[74] Rather than being fully immersive so that members of the audience truly lost themselves in its illusions, the Eidophusikon ultimately directed viewers' attention to de Loutherbourg's genius as an artist, his deep understanding of nature, and his ability to represent it with unprecedented realism.

Although popular with contemporaries such as Gainsborough and Reynolds, the Eidophusikon and its progeny were not necessarily appreciated by later artists. Upon returning home from Daguerre's diorama in 1823, for example, Constable wrote to his friend John Fisher, 'It is very pleasing & has great illusion', however, 'it is without the pale of Art because its object is deception'.[75] By the time Constable was writing, the thrill that eighteenth-century artists had experienced in the presence of illusionistic technologies had diminished. Such illusions were no longer new. Perhaps more importantly, what had been a source of artistic excitement – the production of the illusion of nature – was now seen as 'deception' and dismissed as outside the territory of 'Art'. As Constable's remark suggests, a distinction had emerged between art and popular entertainment, between art and technology. Such distinctions enabled professional artists to separate their work from popular spectacles such as panoramas and dioramas, and from such technologically grounded procedures of image-making as the daguerreotype and other photographic methods. In short, by the middle of the nineteenth century, high art had come to define itself against the illusionistic machinery of popular visual culture.

The growing divide between art and technology in the nineteenth century points to economic and social changes as well. In the eighteenth century, mechanical simulations like automata had been seen by Vaucanson and La Mettrie as embodying a rational and materialist understanding of the body and nature. By the end of the nineteenth century, however, human beings were being treated by theorists such as Thomas Malthus, Charles Babbage, and Andrew Ure as in effect bodies without spirit, economic units in the ever-expanding universe of industrial capital.[76] Technologically reduced to 'labour', workers such as the silk weavers of Lyon or the cotton spinners of the Midlands could be – and were – seen as machines. As Patricia Fara notes, even Josiah Wedgwood, a notably humane industrialist, wanted to convert his workers into machines that would not err.[77] Duplicable and dispensable, *l'homme machine* had arrived.

In this context, the Jaquet-Droz automata – in which human expression had been recoded and mechanically reproduced – can be taken as emblems that illuminate the historically determined process of proletarianization.[78] Vaucanson's mechanized looms, employing punch cards to create complex Jacquard patterns, offered one example of the ways in which all kinds of artisanal craft were increasingly being displaced by mechanization. In this context, artists and poets such as Constable, Blake, Wordsworth, Haydon, Coleridge, and the Shelleys embraced an aesthetic that prized individuality, originality, and genius. The Eidophusikon's hybrid nature, its fusion of the mechanical and the spiritual, perhaps represents a radically unstable moment that anticipates their rejection of the mechanical. William Blake, himself a Boehmenist, located that rejection most explicitly when in *Jerusalem* he contrasted the spiritual 'wheels' of Eden with the 'Loom of Locke' and the 'Water-wheels of Newton'. Later, critics and painters such as John Ruskin would continue the campaign against the mechanical, looking back to the allegedly more humane Middle Ages for inspiration.

The place where art and mechanism remained closely linked throughout the nineteenth century was in the sphere of popular culture and mass entertainment. Panoramas, dioramas, magic lantern shows, and, by the end of the century, the kinescope, stereoscope and cinema created a mass viewing experience defined by the collective immersion in scenic illusion.[79] Poised between art and technology, art and entertainment, science and spectacle, the Eidophusikon, as de Loutherbourg presented it, was an elite exhibition. But as the Eidophusikon's history shows, the five-shilling audience would eventually be replaced by a three- and two-shilling crowd paying to see a show that included animal acts and popular songs as well as 'Moving Pictures representing Phenomena of Nature'. And the audience for inexpensive, mass-produced popular entertainment would continue to grow, eventually coming to include the very people whose artisanal skills and jobs had been rendered obsolete by technology and whose livelihoods now depended on factory work.

Notes

I have been fortunate in having had opportunities to present my work on the Eidophusikon to a number of scholars whose responses have been invaluable. I would like to thank John Brewer, Iain McCalman, Marcia Pointon, John Bender, Ewa Lajer-Burcharth, James S. Ackerman, Anne Higonnet, Mark Rose, and Emily Zinn.

1 The most complete contemporary accounts of the Eidophusikon are: Anon, *The Whitehall Evening Post*, 1 March 1781; Anon, 'A view of the Eidophusikon', *The European Magazine*, March 1782, 180–1 (which describes the programme of season one in some detail); Ephraim Hardcastle [William Henry Pyne], *Wine and Walnuts; or, After Dinner Chit-Chat*, London, 1823, 2 vols, I: 281–304. Important secondary sources include: Austin Dobson, *At Prior Park and Other Papers*, London, 1912, 111–17, 227–81; Rüdiger Joppien, 'Die Szenenbilder Philippe Jacques de Loutherbourgs Eine Untersuchung zu ihrer Stellung zwischen Malerei und Theater', PhD dissertation, University of Köln, 1972, 342–66; Richard D. Altick, *The Shows of London*, Cambridge, MA and London, 1978, 117–27; and Iain McCalman, 'The art of de Loutherbourg's Eidophusikon', in Ann Bermingham, ed., *Sensation and Sensibility: Viewing Gainsborough's Cottage Door*, New Haven, CT and London, 2005, 181–97. Olivier Lefeuvre's recent indispensable catalogue of de Loutherbourg's work provides a thorough account of the Eidophusikon and includes a collection of eighteenth- and nineteenth-century documents related to it: *Philippe-Jacques de Loutherbourg, 1740–1812*, Paris, 2012, 76–80, 370–3.

2 On de Loutherbourg's biography and career see Lefeuvre, *Philippe-Jacques de Loutherbourg*; Joppien's PhD dissertation and his equally important Kenwood exhibition catalogue *Philippe-Jacques de Loutherbourg, 1740–1812*, London, 1974.

3 See Anne Puetz, 'Foreign exhibitors and the British School at the Royal Academy, 1768–1823', in David Solkin ed., *Art on the Line: Royal Academy Exhibitions at Somerset House, 1780–1836*, New Haven, CT and London, 2001, 229–41.

4 William Beckford was the first to call de Loutherbourg a 'mystagogue' in recounting the special effects the artist produced for Beckford's 1781 Christmas revels. See Guy Chapman, *Beckford*, New York, 1937, 104–5; Beckford's feverish prose has tended to colour our image of de Loutherbourg. While de Loutherbourg did indeed present himself at times as a magus or mystagogue, he saw no contradiction between his scientific and occult interests and his Christianity. His tombstone describes him in italics '*As a Man, and as a Christian*' who associated science with 'Faith, Piety and Liberality'. Quoted in *The European Magazine, and London Review*, 73, 1818, 28.

5 This account is based on an anonymous essay, 'Anecdotes of Mr. de Loutherbourg', *European Magazine*, March 1782, 181, which may have been written by de Loutherbourg himself. See William Whitley, *Artists and Their Friends in England, 1700–1799*, New York and London, 1968, 2 vols, II: 352; and Lefeuvre, *de Loutherbourg*, 371 and 377.

6 Joshua Reynolds recalled seeing Vernet painting from nature in Rome in the 1750s, and a group of oil studies of Roman and Neapolitan scenes were included in the sale of Vernet's effects in 1790. See Philip Conisbee *et al.*, *In the Light of Italy: Corot and Early Open-Air Painting*, New Haven, CT, 1996, 113.

7 Denis Diderot, *Diderot on Art – I: The Salon of 1765 and Notes on Painting*, trans. John Goodman, New Haven, CT and London, 1995, 120–1.

8 Denis Diderot, *Diderot on Art – II: The Salon of 1767*, trans. John Goodman, New Haven, CT and London, 1996, 230 and 238. A year after Diderot's suggestion that he 'go back and look at the sea', de Loutherbourg went to the south of France to sketch coastal scenery. After his arrival in London in 1771, de Loutherbourg also visited coastal ports in England (1784 and 1794) and made pilgrimages to standard picturesque sites such as Derbyshire (1778 and 1783) and the Lake District (1783). The subjects from these trips made up the coloured etchings and aquatints that he published in two volumes: *The Picturesque Scenery of Great Britain* (1801) and *The Picturesque and Romantic Scenery of England and Wales* (1805). Unfortunately no oil sketches, if there were any, survive from these trips.

9 Lefeuvre, *de Loutherbourg*, 28–9.

10 Christopher Baugh, *Garrick and Loutherbourg*, Cambridge and Alexandra, VA, 1990. See too Baugh, 'Philippe de Loutherbourg: Technology-driven entertainment and spectacle in the late eighteenth century', *Huntington Library Quarterly*, 70: 2, 2007, 252–3.

11 Michael Fried, *Absorption and Theatricality: Painting and Beholder in the Age of Diderot*, Berkeley, Los Angeles, CA and London, 1980, 76–80.

12 Henry Angelo, *Reminiscences of Henry Angelo*, London, 1828–30, 2 vols, II: 326.

13 Baugh, 'Philippe de Loutherbourg', 266.

14 Altick, *Shows of London*, 122.

15 Quoted in Altick, *Shows of London*, 212

16 *The London Courant*, 4 May 1781.

17 Pyne, *Wine and Walnuts*, 1: 286. Pyne would have been twelve years old at the first performances of the Eidophusikon. Altick suggests that Pyne's knowledge of the attraction was gained by viewing the third (or even later) season. Altick, *Shows of London*, 521, note 20.

18 Pyne, *Wine and Walnuts*, I: 292. This scene was recreated at the Australian National University's Super Computer Facility in 2006 by Darran Edmundson and Iain McCalman. It can be viewed at http://vimeo.com/32698316

19 In Freud's classic formulation, uncanny objects blur the line between the lifeless and the living, the real and the artificial, which Freud saw exemplified in waxworks and automata. Sigmund Freud, 'The Uncanny', *The Standard Edition of the Complete Psychological Works of Sigmund Freud*, trans. and ed. James Strachey, London, 1955, 24 vols, 17: 217–52.

20 William Whitley, *Thomas Gainsborough*, London, 1915, 354.

21 Pyne, *Wine and Walnuts*, 1: 298.

22 'A view of the Eidophusikon', *The European Magazine*, March 1782, 80–181. If de Loutherbourg is the author of this account and the 'Anecdotes of Mr. de Loutherbourg' as Whitley and Lefeuvre suggest, it means that de Loutherbourg himself imagined his Eidophusikon to be a new species of painting and a development of his work in landscape painting.

23 *Morning Chronicle and London Advertiser*, 28 February 1781.

24 *Morning Herald and Daily Advertiser*, 1 March 1781.

25 *Morning Chronicle and London Advertiser*, 28 February 1781; 'Anecdotes of Mr. de Loutherbourg', 182.

26 Pyne, *Wine and Walnuts*, I: 296–7.

27 Pyne, *Wine and Walnuts*, 1: 289.

28 'A view of the Eidophusikon', 181.

29 Pyne, *Wine and Walnuts*, 1: 303.

30 *Morning Post and Daily Advertiser*, 4 May 1785.

31 Altick, *Shows of London*, 122. Iain McCalman notes that as a showman, de Loutherbourg was in competition with other popular attractions which may have spurred him to sensationalize his programme. See his 'Specters of quackery: The fragile career of Philippe de Loutherbourg', *Cultural and Social History*, 3, 2006, 241–54.

32 *The Morning Herald*, 6 May 1786.

33 Altick, *Shows of London*, 125.

34 For an account of the fire see 'Dreadful Fire', *Morning Post and Gazetteer*, 24 March 1800. The account of the fire says that Chapman claimed a £600 loss, but it does not say that the Eidophusikon was completely destroyed (unlike the nearby pub). On 7 August 1817, the *Caledonian Mercury* reported that 'Loutherbourg's Eidophusikon, or animated scenery' was at the Pantheon. Whether this was the remnant of the original Eidophusikon or a reconstruction is unclear. Seats for this attraction were two shillings and one shilling. In 1854 a 'Royal Eidophusikon' appears to have opened in Liverpool with scenes of the 'Holy Land' and the 'Turkish Empire' (*Liverpool Mercury*, 7 July 1854). In 1857 the Royal Eidophusikon reopened in Manchester (*Manchester Times*, 6 June 1857). Clearly this was not de Loutherbourg's spectacle. Nevertheless, the use of the word 'Eidophusikon' in the title of these attractions demonstrates the long-lasting artistic significance and commercial impact of de Loutherbourg's original. Altick reports that in 1785 the American artist Charles Wilson Peale created a version of the Eidophusikon in Philadelphia (Altick, *Shows of London*, 126).

35 Altick, *Shows of London*, 59–60, 119. On the international success of the Jaquet-Droz exhibition see Altick, 62–6; and Marcia Pointon, *Brilliant Effects: A Cultural History of Gem Stones and Jewellery*, New Haven, CT and London, 2009, 225–43. On the work of the Jaquet-Droz, see Charles

Perregaux and F. Louis Perrot, *Les Jaquet-Droz et Leschot*, Neuchatel, 1916; and Adelheid Voskuhl, *Androids in the Enlightenment: Mechanics, Artisans, and Cultures of the Self*, Chicago, IL, 2013, especially 37–85.

36 On one such tableau méchanique see Hanneke Grootenboer's essay: 'The Clock Picture as a Philosophical Experiment: The Tableau Méchanique in the Physics Cabinet of Bonnier de la Mosson', in this issue of *Art History*.

37 The full description of the Grotto was written by the Jaquet-Droz to be distributed at the performances of their automata. It is reproduced online along with the engraving by the British Museum at: http://www.britishmuseum.org/research/collection_online/collection_object_details.aspx?objectId=1336666&partId=1. Marcia Pointon gives a thorough account of the exhibition of the four automata, suggesting the importance of the Grotto for the Eidophusikon (223–43). I am grateful to her for assisting me in my research on the Jaquet-Droz.

38 E. J. Dijksterhuis, *The Mechanization of the World Picture*, trans. C. Dikshoorn, Oxford, 1961. My understanding of mechanical philosophy is also indebted to D. J. de Solla Price, 'Automata and the origins of mechanism and mechanistic philosophy', *Technology and Culture*, 5, 1964, 9–23.

39 Julien Offray de La Mettrie, *Man a Machine*, trans. G. C. Bussey, Chicago, IL, 1912, 148.

40 Vaucanson's duck and musical automatons were described at length in Diderot and d'Alembert's *Encyclopédie* under 'Automate'. See *Encyclopédie ou Dictionnaire raisonné des sciences, des arts et des métiers, par une Société de Gens de lettres*, 17 volumes of text and 11 volumes of plates, Paris, 1751–72, 1: 896–7.

41 Jacques de Vaucanson, *An Account of the Mechanism of an Automaton or Image Playing on the German Flute*, trans. J. T. Desaguliers, London, 1742, 23. Unlike his musical androids, Vaucanson's duck was a false automaton for it did not truly digest the oats. Nevertheless, Vaucanson's dream of creating a machine that could perform biological functions seems to have started with the duck. The bibliography on Vaucanson is extensive. In addition to his own account, see André Doyon and Lucien Liaigre, *Jacques Vaucanson: Mécanicien de Genie*, Paris, 1967. An important analysis of the duck and its meaning for the eighteenth century and our own is Jessica Riskin's 'The defecating duck, or, the ambiguous origins of artificial life', *Critical Inquiry*, 29: 4, Summer 2003, 599–633. Recent histories consulted on automata and their significance to eighteenth-century art and culture include Myles W. Jackson, *Harmonious Triads: Physicists, Musicians, and Instrument makers in Nineteenth-Century Germany*, London and Cambridge, MA, 2008; Minsoo Kang, *Sublime Dreams of Living Machines: The Automaton in the European Imagination*, Cambridge MA and London, 2011; Barbara Maria Stafford, *Artful Science: Enlightenment Entertainment and the Eclipse of Visual Education*, Cambridge, MA and London, 1994, 190–215; and Adelheid Voskuhl, *Androids in the Enlightenment*.

42 Vaucanson, *Account of the Mechanism of an Automaton*, 21

43 Quoted in Doyon and Liaigre, *Vaucanson*, 110.

44 Quoted in Doyon and Liaigre, *Vaucanson*, 210.

45 Angelo, *Reminiscences*, II: 329.

46 On musical automatons and their relation to early physics, see Jackson's *Harmonious Triads*.

47 'Many an evening', Angelo recalls, 'have we been diverted with three clever men; Loutherbourg, with his fun and tricks, which were in exhaustible; and the two mechanicians, Martin [sic] and Jaquet D'Ross [sic]'. 'Martin' was a Belgian mechanic named John Joseph Merlin who worked at Cox's Museum in Spring Gardens and who in 1783 opened a mechanical museum at Hanover Square filled with scientific toys of his own invention. 'Jaquet D'Ross' was Henri-Louis Jaquet-Droz, the son of Pierre. Angelo, *Reminiscences*, II: 328–9. See also Altick, *Shows of London*, 72–6; and Pointon, *Brilliant Effects*, 203–21.

48 Interestingly, during the Christmas season of 1781 de Loutherbourg transformed Fonthill, the young William Beckford's father's country estate into, in Beckford's words, a 'realm of Fairy' or 'demon temple' by illuminating it with strange 'necromantic' lights. This immersive environment is described in H. A. N. Brockman, *The Caliph of Fonthill*, London, 1966, 38; Jean-Claude Lebensztejn, 'L'Eidophusikon et Vathek', *Furor*, 21, April 1991, 3–20; Simon During, 'William Beckford in hell: An episode in the history of secular enchantment', *Huntington Library Quarterly*, 70: 2, 2007, 269–88; and Iain McCalman, 'The virtual infernal: Phillippe de Loutherbourg, William Beckford and the spectacle of the sublime', *Romanticism on the Net*, 46, May 2007. http://id.erudit.org/iderudit/016129ar.

49 See note 19.

50 The most easily accessible collection of Boehme's work can be found online at: http://www.jacobboehmeonline.com/. On Boehme's thought, see Kevin Fischer, *Converse with the Spirit: William Blake, Jacob Boehme, and the Creative Spirit*, Madison, WI and Teaneck, NJ, 2010; and J. B. Gibbons, *Gender in Mystical and Occult Thought*, Cambridge, 1996, 90–1.

51 In the eighteenth century, particularly in Britain, there were numerous attempts to make concordances among the many esoteric traditions by establishing similarities among them. See Nicolas Goodrick-Clarke, *The Western Esoteric Traditions: A Historical Introduction*, Oxford and New York, 2008, 9, 120–7. The London Theosophical Society, which was founded in 1783 by the Rev. Jacob Duché, and which counted de Loutherbourg as well as William Sharpe and John Flaxman among its founding members, married the ideas of Boehme with those of Swedenborg. See Jocelyn Godwin, *The Theosophical Enlightenment*, Albany, NY, 1994, 103.

52 See for instance Boehme, *Aurora*, XXII: 77–106; and *The Signature of All Things*, IV.

53 Goodrick-Clarke, *Western Esoteric Traditions*, 86ff. On de Loutherbourg's involvement with theosophy, Swedenborgianism, Masonry, Rosicrucianism, alchemy, mesmerism, and his friendship with Cagliostro see Godwin, *Theosophical Enlightenment*, 100–1, 103, 112–14, 132–3, 153, 228. See also Iain McCalman, *Last Alchemist: Count Cagliostro, Master of Magic in the Age of Reason*, New York, 2003, 161ff.; and his 'Mystagogues of revolution: Cagliostro, Loutherbourg and Romantic London', in James Chandler and Kevin Gilmartin, eds, *Romantic Metropolis: The Urban Scene of British Culture, 1780–1840*, Cambridge, 2005, 177–203; and Marsha Keith Manatt Schuchard, 'Freemasonry, secret societies, and the continuity of the occult traditions in English literature', The University of Texas at Austin, PhD dissertation, 1975, 2 vols, University Microfilms, Ann Arbor, MI, in particular I: 340–2, and II: 374–6. On de Loutherbourg's magnetic healing powers see Mary Pratt, 'A list of a few cures performed by Mr. and Mrs. De Loutherbourg...without medicine', London, 1789, 1ff. On mesmerism in London and de Loutherbourg's involvement with it, see Patricia Fara, 'An attractive therapy: Animal magnetism in eighteenth-century England', *History of Science*, 33: 2, 2009, 127–75.

54 On the contents of de Loutherbourg's library, see Coxe, *A Catalogue*. Coxe's catalogue is reprinted in Lefeuvre, *de Loutherbourg*, 354–6. The items related to Boehme are numbers 7, 8, 23, 66, 68, 69, 70, 71, and 126. De Loutherbourg owned William Law's edition of Boehme's works (item 126), illustrated with diagrams by Dionysius Andreas Freher (1649–1728).

55 The Masonic compass teaches men to circumscribe their passions and keep their desires within bounds. The shipwreck scene in the lower left corner of *Keep Within Compass* is accompanied by the motto: 'Attempt according to your strength, Close by the Shore to keep, Tis safer than to hoist up Sails, And Plunge into the Deep'. On the meaning of the shipwreck in Romantic painting see Lorenz Eitner, 'The open window and the storm-tossed boat: An essay in the iconography of Romanticism', *Art Bulletin*, 37: 4, December 1955, 279–90.

56 Boehme, *Aurora*, VIII: 22–6, 82–7; XII: 131–41; XIII: 1–64; 146–64; XIV: 1–135; XV: 1–97; XVI: 88–116.

57 Boehme, *Three Principles of Divine Essence*, V: 30. On the possible influence of Boehme on Milton see Margaret Lewis Bailey, *Milton and Jakob Boehme: A Study of German Mysticism in Seventeenth-century England*, New York, 1914.

58 On the fascination with perpetual motion see Simon Schaffer, 'The show that never ends: Perpetual motion in the early eighteenth-century', *British Journal for the History of Science*, 28: 2, June 1995, 157–89.

59 In saying this I do not wish to rule out entirely the possibility that de Loutherbourg was also drawn to alchemy for pecuniary reasons, having been impressed by Cagliostro's transmutation of copper pennies into pure silver. See Constantin Photiadés, *Count Cagliostro: An Authentic Story of a Mysterious Life*, London, 1932, 225. On alchemical processes and their meanings see Lawrence M. Principe, *The Secrets of Alchemy*, Chicago, IL and London, 2013. On the competitive

relationship between art and alchemy, see William R. Newman, *Promethean Ambitions: Alchemy and the Quest to Perfect Nature*, Chicago, IL and London, 2004.

60 John Gage, 'The peacock's tail', in *Colour and Culture: Practice and Meaning from Antiquity to Abstraction*, London, Boston, MA and Toronto, 1994, 139–53.

61 'Anecdotes of Mr. de Loutherbourg', 181.

62 'Peter Pindar' [John Wolcott], *Lyric Odes to the Royal Academicians for MDCCLXXXIII*, 'Ode IV', *The Works of Peter Pindar, esq.*, London, 1794, 28–9. Similar criticisms were made by 'Anthony Pasquin' (John Williams) in his *Memoirs of the Royal Academicians* of 1796. These criticisms also conform to what K. Dian Kriz has attributed to a growing artistic nationalism in Britain and with it a dismissal of 'French glitter'. See Kriz's *The Idea of the English Landscape Painter: Genius as Alibi in the Early Nineteenth Century*, New Haven, CT and London, 1997, especially 106–10.

63 Walter Thornbury, *The Life of J.M.W. Turner, R.A.: Founded on Letters and Papers furnished by his Friends and fellow Academicians*, London, 1904, 115.

64 John Gage, *Colour in Turner: Poetry and Truth*, New York and Washington, DC, 1969, 181–2.

65 See in particular Karl Möseneder, *Philipp Otto Runge und Jacob Boehme*, Marburg/Lahn, 1981. Runge planned four large canvases depicting the times of day to be installed in a Gothic chapel where, accompanied by music, they would coalesce into an immersive environment. Plate 9 is a small version or study for the large version of *Morning*, the first painting of the never completed series. In its use of sound, spectacle, and music woven into a single artistic fabric, the Eidophusikon is also a *gesamtkunstwerk*.

66 Pyne, *Wine and Walnuts*, 1: 281

67 Pyne, *Wine and Walnuts*, 1: 296. John Gage suggested that J. M. W. Turner also assisted at the Eidophusikon when it was in Fleet Street. See 'Turner and the picturesque', *Burlington Magazine*, 107, 1965, 24–15. Altick investigated this claim and dismissed it (Altick, *Shows of London*, 126).

68 This subject has also been broached by Silvia Bordini, 'Pittura come spettacolo: le "or del giorno" tra pesaggio romantic e nuove technologie', *Richerche di Storia dell'arte*, 25, 30 March 1985, 5–15. See too Bermingham, 'Introduction', in *Sensation and Sensibility*, 22–3.

69 Lynda Nead, *The Haunted Gallery: Painting, Photography, Film, c. 1900*, New Haven, CT and London, 2007, 45ff.

70 Turner seems to have been particularly interested in immersive visual effects. See Bermingham, 'Landscape-o-rama: The exhibition landscape at Somerset House and the rise of popular landscape entertainments', in *Art on the Line*, 127–44.

71 Jonathan Crary, *Techniques of the Observer: On Vision and Modernity in the Nineteenth Century*, Cambridge, MA and London, 1990 and *Suspensions of Perception: Attention, Spectacle, and Modern Culture*, Cambridge, MA and London, 2000.

72 Martin Butlin and Evelyn Joll, *The Paintings of J.M.W. Turner*, New Haven, CT and London, 1984, 2 vols, 247.

73 See Bermingham, *Landscape and Ideology: The English Rustic Tradition, 1740–1840*, Berkeley and Los Angeles, CA, 1986.

74 Pyne, *Wine and Walnuts*, I: 298.

75 John Constable, *John Constable Correspondence*, ed. R. B. Beckett, Ipswich, 1968, 7 vols, IV: 134.

76 On industrialism and automation see Maxine Berg, *The Age of Manufactures 1700–1820: Industry, Innovation and Work in Britain*, London and New York, 1994; Simon Schaffer, 'Babbage's intelligence: Calculating engines and the factory system', *Critical Inquiry*, 21, Autumn 1994, 203–27; Andrew Zimmerman, 'The ideology of the machine and the spirit of the factory: Remarx on Babbage and Ure', *Cultural Critique*, Fall 1997, 5–29.

77 Patricia Fara, review of Minsoo Kang's *Sublime Dreams of Living Machines*, *American Historical Review*, 117: 5, December 2012, 1652.

78 Gaby Wood, *Edison's Eve: A Magical History of the Quest for Mechanical Life*, New York, 2002, passim.

79 See Walter Benjamin, 'The Work of Art in the Age of Mechanical Reproduction', *Illuminations*, ed. and intro. Hannah Arendt, trans. Harry Zohn, New York, 1969, 234–5, 239–42.

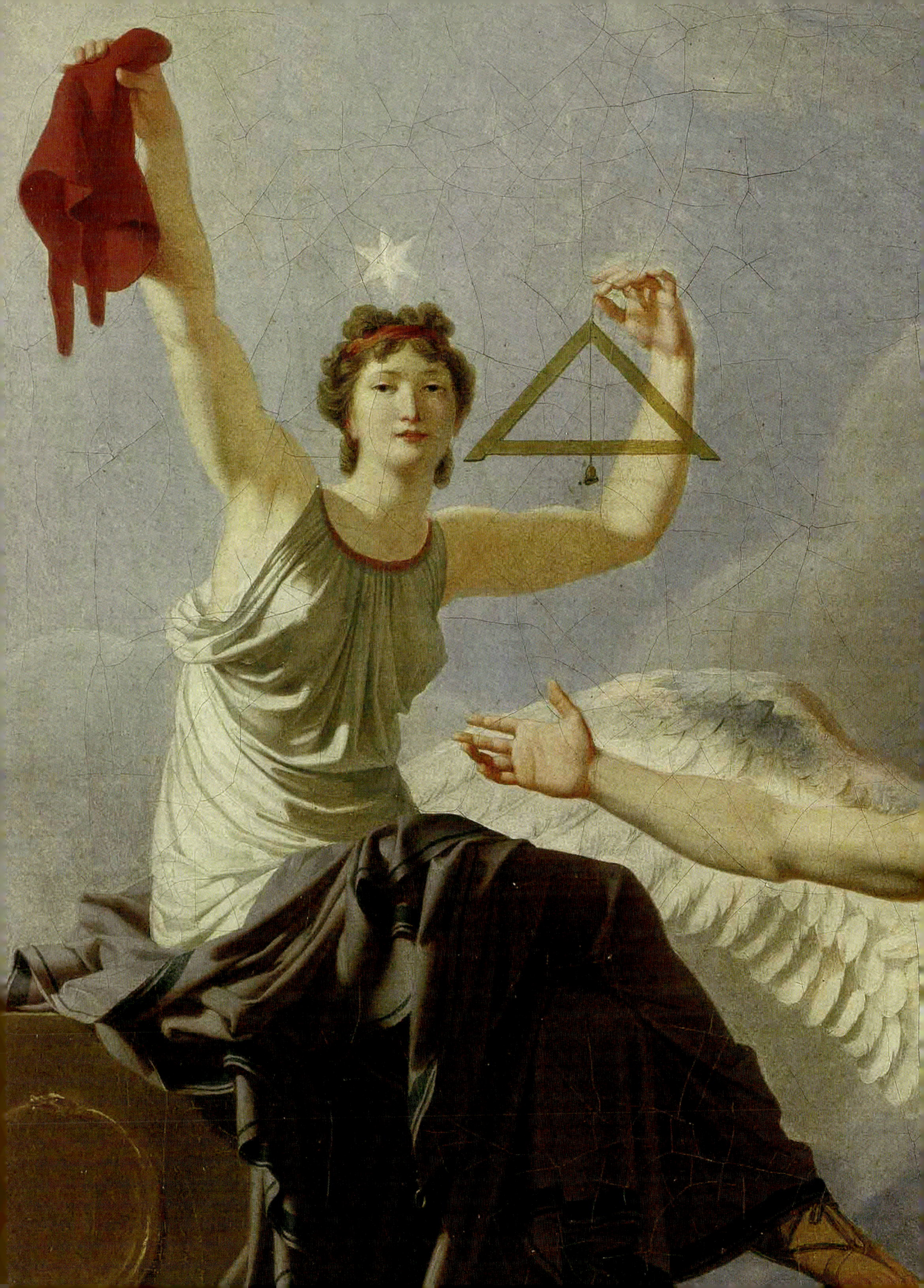

Chapter 11
Telegraphic Images in Post-Revolutionary France
Richard Taws

In *The Paris Sketch Book*, published in 1840, William Makepeace Thackeray offered an idiosyncratic reading of some paintings of the French School. Strangest among these was the novelist's interpretation of Jacques-Louis David's *Oath of the Horatii*, which he illustrated with a rendering of the Horatii as ideographic stick-men:

> Of the great pictures of David, the defunct, we need not, then, say much. Romulus is a mighty fine young fellow, no doubt .... But was there ever anything so absurd as this passion for the nude, which was followed by all the painters of the Davidian epoch? And how are we to suppose yonder straddle to be the true characteristic of the heroic and the sublime? Romulus stretches his legs as far as ever nature will allow; the Horatii, in receiving their swords, think proper to stretch their legs too, and to thrust forward their arms thus,
>
> 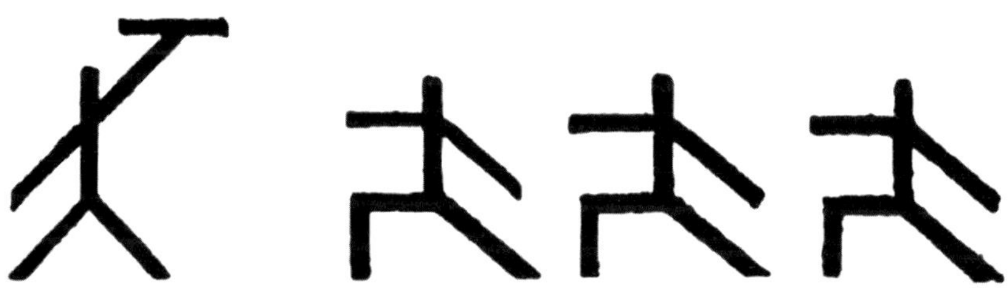
>
> Romulus's is the exact action of the telegraph; and the Horatii are all in the position of the lunge. Is this the sublime? Mr. Angelo, of Bond-street, might admire the attitude; his namesake, Michael, I don't think would.[1]

Thackeray's destabilization of the heroic gestures of late eighteenth-century history painting is intriguing on many levels. Misreading the code – deliberately perhaps – Thackeray reversed the position of the figures in the painting and misrecognized the elder Horatius as Romulus, the subject of David's later *Intervention of the Sabine Women* (1799), hanging close by in the Louvre. Yet if his account of 'David, the defunct' polemicized the artist's obsolescence, Thackeray's image obtained much of its force from a comparison between David's *grande machine* and a more prosaic technology that in 1840 remained a regular sight on the rooftops of Paris: the optical, or semaphoric, telegraph. Imagining painting as telegraphy rendered

Detail from Jean-Baptiste Regnault, *Liberty or Death*, 1794–95 (*plate 5*).

David's painting both declarative and incomprehensible, drawing attention to its seemingly arcane codes and to a discrepancy between its advanced aesthetic claims and an apparently mundane effect. Yet Thackeray's breezy image of the Horatii-as-telegraph was a more complex analogy than he allowed. Accentuating the inherent visuality of telegraphy at this time and connecting it to a wider image world that included academic art, it also exposed the complicated temporal effects of both painting and telegraphy, showing them to be inseparable from the French Revolution's drawn-out afterlives. Thackeray's sketch suggested that optical telegraphy, which operated in this form until the mid-1850s, not only adumbrated a future of smooth technological progress, but provided too a language with which to comprehend images from the recent past whose production preceded its invention.

The machine Thackeray had in mind was the optical telegraph developed by Claude Chappe and his brothers in the early 1790s and brought into official service in 1794, at the height of the revolutionary Terror, where, alongside the cannon, it operated as 'the first of the modern space-annihilating devices', in Lewis Mumford's memorable phrase (*plate 1*).[2] The 1790s and early 1800s witnessed numerous projects aimed at transmitting revolutionary ideas, from printed images of diverse quality and audience to modifications in clothing and speech. The telegraph may, additionally, be considered alongside a number of technologies and protocols invented or modified during the revolutionary decade that aimed to transform the everyday experience of space and time, from the Republican calendar to the standardization of the metre, and from developments in geographic science to the bureaucratic reorganization of the *départements*.

Accorded a unique place among these novel systems, telegraphy promised a universal medium of transparent communication that chimed with revolutionary ideals, giving form to principles that were otherwise abstract. As Joseph Lakanal argued, in a 1794 speech to the National Convention promoting the telegraph's use, the two greatest inventions of the eighteenth century were

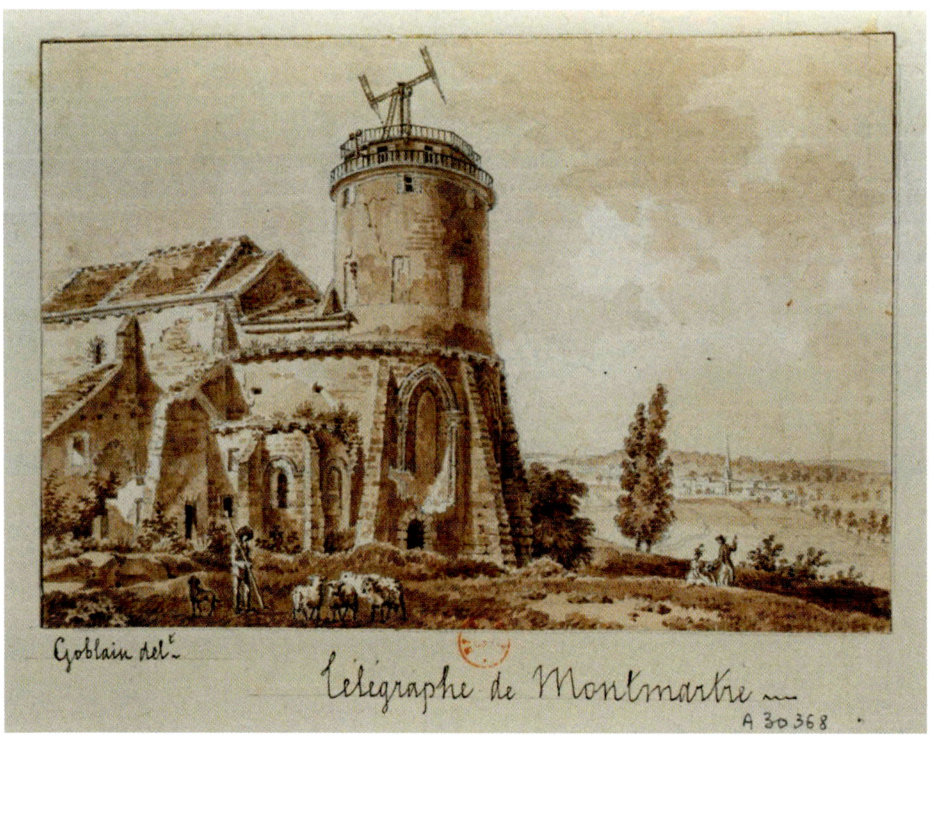

1 Antoine-Louis Goblain, *Télégraphe de Montmartre*, c. 1815. Pencil drawing with brown wash, 9.9 × 14.6 cm. Paris: Bibliothèque Nationale de France. Photo: BnF.

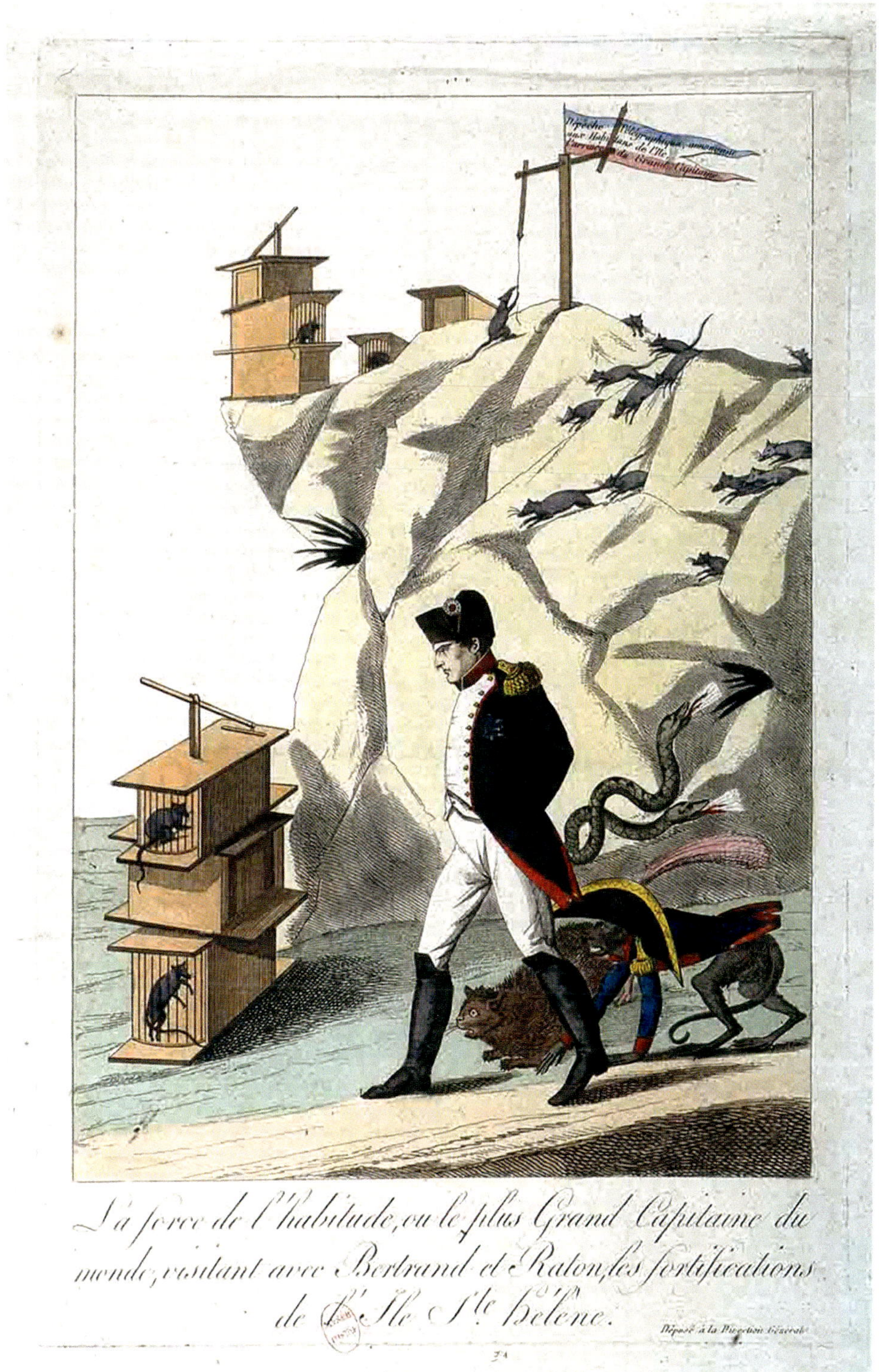

2 Anonymous, *La force d'habitude*, c. 1815. Hand-coloured etching, 43.5 × 27.5 cm. Paris: Musée de la Poste (Inv. 2083). Photo: © L'Adresse Musée de la Poste, Paris/ La Poste/Michel Fischer.

the telegraph and the hot-air balloon.[3] Both were French, or at least appeared to be, and between them they promised to command vertical and horizontal space. Lakanal's evangelism for the telegraph was shared by many of his contemporaries, who considered it a uniquely revolutionary invention whose impact would be comparable to that of printing, the compass, and gunpowder.[4] Yet the 'Frenchness' of the telegraph operated in tension with another set of possibilities: those aligned with the telegraph's potential as a means of universal communication among world citizens. Telegraphy transcended borders in theory, even as it affirmed them in practice. Moreover, by 1798, when writer and politician Pierre-Jean-Baptiste Chaussard referred to the 'winged symbol' of the optical telegraph as a 'bearer of the fates', he did not allude solely to the telegraph's ability to transmit important information, but suggested that it might also, in some way, drive the progress of history.[5]

Emerging from a lengthy period of pan-European experimentation, Chappe's system consisted of a network of relays, usually situated on hills or outcrops, or on tall buildings, on top of which were installed articulated metal 'arms', manipulated via a system of pulleys by an operator below. The telegraph itself comprised a long cross beam, or regulator, set atop a static pole, rotatable through 180 degrees around a central axle.[6] At the end of each arm a smaller, counterbalanced pivoting arm, or indicator, was fixed. This mobile mechanism could be set into discrete positions, each one pre-coded with a letter of the alphabet, or more commonly a word, phrase, or standard instruction (end of message, suspension due to fog, and so on). The signals sent by a telegraph were viewed through a telescope by another operator at the next station, approximately six to eight miles away, who passed messages along the channel until they were decoded and transcribed by an inspector at their final destination.[7] Signals could only be transmitted feasibly during daylight hours and the apparatus was painted black to increase visibility from distance. In its gestural economy Chappe's telegraphic code shared much with the 'reformed' sign language proposed for use by the deaf; indeed, at the time the telegraph took hold, an uncluttered, transparent sign language was being suggested as a potential alternative to speech, for the hearing too, as a revolutionary form of communication between individuals that promised to eliminate contingent meaning (for which read regional accents, patois or slang).[8] As one contemporary commentator put it, 'The discovery of the telegraph isn't just an ingenious speculation; its results leave no uncertainty in the literal transmission of the different characters appropriate to the language of signs.'[9] Telegraphy – 'far writing' – is a revolutionary neologism that evidenced this desire for economical expression.

Chappe's system replaced his earlier experiments with a shutter telegraph based on blocking out sections of light within a sectioned framework, and a telegraph in which a clock face was encoded with the alphabet. By the 1840s the telegraphic system extended for nearly 5000 kilometres within France, its five-fingered spread from Paris to the country's borders mirroring the waving hand from which the telegraph derived its 'natural' authority.[10] Providing a standardized apparatus, and more particularly, a widely accepted code, the telegraph became, especially under Napoleon, a radically far-reaching method to send messages at unheard-of velocities. Although other applications, particularly the transmission of news for print publication, were proposed from the moment of its invention, attempts to turn the telegraph towards commercial communication were resisted, and it was used primarily to send state and military information.[11] Under the Empire, the only other kind of telegraphic message allowed was the dissemination of winning lottery numbers, a use grounded in fears about illicit print; it was hoped that telegraphic transmission would prevent the delay that allowed counterfeit tickets to be produced.

A satirical print from 1815, *La force d'habitude*, represents Napoleon directing a rat-operated telegraph on St Helena, demonstrating the extent to which the association between the Emperor and the telegraph had taken hold (*plate* 2). Mocking the folly of Napoleon's ambition, this print also associates the boredom and madness of exile with the presumed monotony and cruelty of the telegrapher's task. The rats in their traps figure in this image as early instances of what soon became a common trope – the idea that the task of telegraphy had particularly enervating effects on its caged operators, poor exhausted creatures trapped in their personal Bastilles and blind to the messages they transmitted. 'Living chrysalises', Alexandre Dumas termed them in the *Count of Monte Cristo*, 'poor wretch[es]', 'fagged to death with cabals, factions and government intrigues'.[12] Such characterizations had a long shelf-life, continuing into the era of electric telegraphy and providing evidence of widespread suspicion of the telegraph's secret transmissions – they certainly inform writer and photographer Maxime du Camp's 1875 account of the unrelenting labour of telegraphic clerks.[13] Telegraphs produced detrimental physiological and psychological effects in those who operated them. These people, du Camp reported, no longer spoke to one another directly, but conversed by way of coded messages, like the 'intelligent machines' that surrounded them.[14] Yet even from its inception, the telegraph evoked the spectre of a machine-human body imbued with the potential to revolt or subjugate its human masters.

The optical telegraph has been understood by later commentators as contributing to an acceleration of life characteristic of modernity, and as participating in a defining technological transformation; what media theorist Friedrich Kittler, comparing the Chappe telegraph and the Daguerreotype, described as the 'fundamental trend of modern media technologies to replace static values with dynamic values and to replace steadfastness with speed', or what Paul Virilio characterized more locally as 'the increase in the penetration speed of means of transport and communication (such as the Chappe telegraph at Austerlitz)'.[15] For Kittler, the Chappe telegraph marked the point at which this characteristic was allied for the first time to optical image storage, and he noted its importance to the success of the Napoleonic armies.[16] However, the Chappe telegraph did not emerge as an 'inevitable, independent technological object', but as what David Bell calls an 'emblematic artifact created at the conjunction between the technological and the social'.[17] Furthermore, teleological approaches in which technology is the motor of its own forward march obscure the fact that telegraphy looked backwards as well as forwards. Telegraphy enabled new ways of thinking about time, yet these were not all connected to the new speed of communication it permitted, but offered more extensive insights into the relation between technology, time, and revolutionary process. Certainly, we can see both 'steadfastness' and 'speed' applied to collaborative ends in depictions of the telegraph apparatus on a crumbling church tower at Montmartre, a dialectical image if there ever were one (see *plate 1*).

Influencing Machines

In Paris in 1840, a city Ludwig Börne described as the 'telegraph of the past', Thackeray would not have had to walk far outside the Louvre to encounter such a device.[18] In fact, for many years a telegraph had been positioned atop the newly established museum, until its relocation to Saint-Sulpice in 1805. In its early days, the telegraph's presence was felt profoundly, and aroused suspicion; a police document from 1793 reports bemusedly on 'a type of obelisk, raised to the height of a windmill' on the Belleville road, and Chappe's early experimental apparatuses were destroyed by those who believed them to be secret means of counter-revolutionary communication.[19] Yet as with all media, the telegraph's visibility was proportional to its 'newness', and the majority of commentaries such as this one appeared in the interregnum between the introduction of the network and the social acceptance of its hardware and protocols. These processes of naturalization were often laid bare

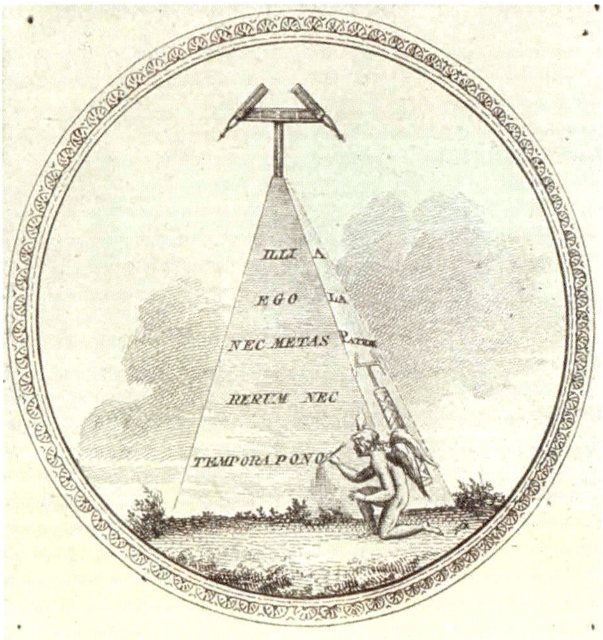

3 Anonymous, *Carte d'admission*, c. 1794. Etching, 7.5 × 7 cm. Paris: Musée de la Poste (Inv. D11038). Photo: © L'Adresse Musée de la Poste, Paris/La Poste.

in artistic responses to the telegraph, many of which attempted to come to terms with the silences and breakdowns that accompanied the telegraphic message – its peripheral, even irritating, presence at the limit of vision – situating them in a wider field of political and visual representation and non-representation. If the telegraph's most obvious order of representation, the shapes thrown by its arms, was paralleled by the storage and translation of this information in the official registers of telegraphic clerks, then its more insidious, accidental, or disruptive appearances were manifest across a range of artistic images, whose variety was congruent with the diverse ways in which the telegraph invaded the public sphere. Rather than posit the optical telegraph as a dehistoricized precursor to the internet, or as a relic of a terminal phase before electromagnetic communications, this article argues that Chappe's network is best understood as an intersection of cultural, technological, sensory, and, crucially, *artistic* phenomena grounded in the particularities of its use in France in the first half of the nineteenth century, and that the meanings of optical telegraphy extend beyond the development of the technology or its immediate application. Representations of telegraphy, or even images that foreshadowed telegraphy's implementation, regularly made sense of telegraphy itself, rather than the other way around.

Indeed, as Jennifer Roberts has argued, in assessing the visual impact of telegraphy 'the best way ... is not to locate some direct causal impact of telegraphy on picture making, and not to limit analysis to the iconography of telegraphs as they were depicted in paintings, but rather to delineate concerns about coding, replication, and transmission shared by both visual media'.[20] I take Roberts' position as exemplary, although it is worth noting that optical telegraphy in early nineteenth-century France depended on visual transmission to a far greater extent than the electromagnetic methods she discusses, and was connected to diverse forms of cultural expression in ways that began with, but were not limited to, the iconography of the telegraph's angular choreography. Chappe himself was more than aware of the striking symbolic effects of the telegraph, over and above its operational function. A printed identity card issued to a telegraph operator exploits a formal slippage between the telegraph's arms at rest and the masonic triangles, pyramids, and scales of justice featured in Republican symbolism (*plate* 3). Meanwhile, the masthead for a schematic map of the inaugural Paris to Lille telegraph line combines telegraphic arms with an eye of surveillance and a *bonnet rouge* to create a compound vignette that is at once innovative and immediately recognizable (*plate* 4). The porosity of such images carries over to more self-consciously artistic works, such as Jean-Baptiste Regnault's 1795 *Liberty or Death* (*plate* 5), which recalls in turn vignettes on official paperwork, and is, while not a representation of the telegraph, nonetheless a 'telegraphic image' par excellence: reducing a complex political scene to its simplest formulation as binary choice, a *génie* of France semaphores the Revolution's global message.[21]

In the half-century of their operation, optical telegraphs became a ubiquitous sight, altering significantly skylines across France, as well as the buildings and landscapes that accommodated them. Telegraphs were placed on purpose-built turrets, but also on existing structures, from the roofs of prominent buildings to church spires, where, taking over the traditional role of church bells, melted down during the Revolution, they figured as secular, mobile crucifixes. The visual register of the telegraph is particularly pronounced in a work such as Charles Norry's *Vue de l'intérieur du Louvre d'après nature l'an 4ème* of 1799, which represents the optical telegraph on the roof of the Louvre (*plate* 6), presenting the encrypted signs transmitted by the modern telegraph in proximity to the artist-made signs in the building below.[22] Made in the period from 1796 to 1801 when the Louvre was closed, Norry's image focuses attention instead on the exterior of the building, making the telegraph the main subject of the work, hovering implacably over the ill-defined figures transporting a sculpture across the courtyard beneath it. Yet this work, and others representing this subject from multiple angles, is no straightforward glorification of Consular power. Rather, oscillating between dispassionate observations and allegories of representation itself, such works address the movement of images across time and media, and as with representations of the telegraph at Montmartre, they accentuate the telegraph's modernity by highlighting its juxtaposition with an obsolete or transformed architectural milieu, the traces of which remain necessary to its operation.

Some of the telegraph's associations were less readily locatable in space and time. As Alix's print after Jean-Frédéric Schall's remarkable image shows, the telegraph invaded somewhat retro allegories of love, where, reworking Rococo aesthetics for an age of total war, it is deployed as both semaphore and improvised missile launcher (*plate* 7).[23] As Mary Favret has proposed, 'with the advent of mass media ... and in an increasingly popular visual culture of prints, panoramas, and theatrical performances, wartime stepped forth as a mediated relationship to distant violence'.[24] Given its primarily military use, the Chappe telegraph and its representations might well be included among the technologies that mediated conflict in this period. However, as Favret makes clear, it is one thing to document instances of military

4 Claude Chappe, *Plan de la ligne télégraphique de Paris à Lille*, 1794. Drawing and watercolour on paper, 170 × 36 cm. Paris: Musée de la Poste (Inv. 22411). Photo: © L'Adresse Musée de la Poste, Paris/La Poste.

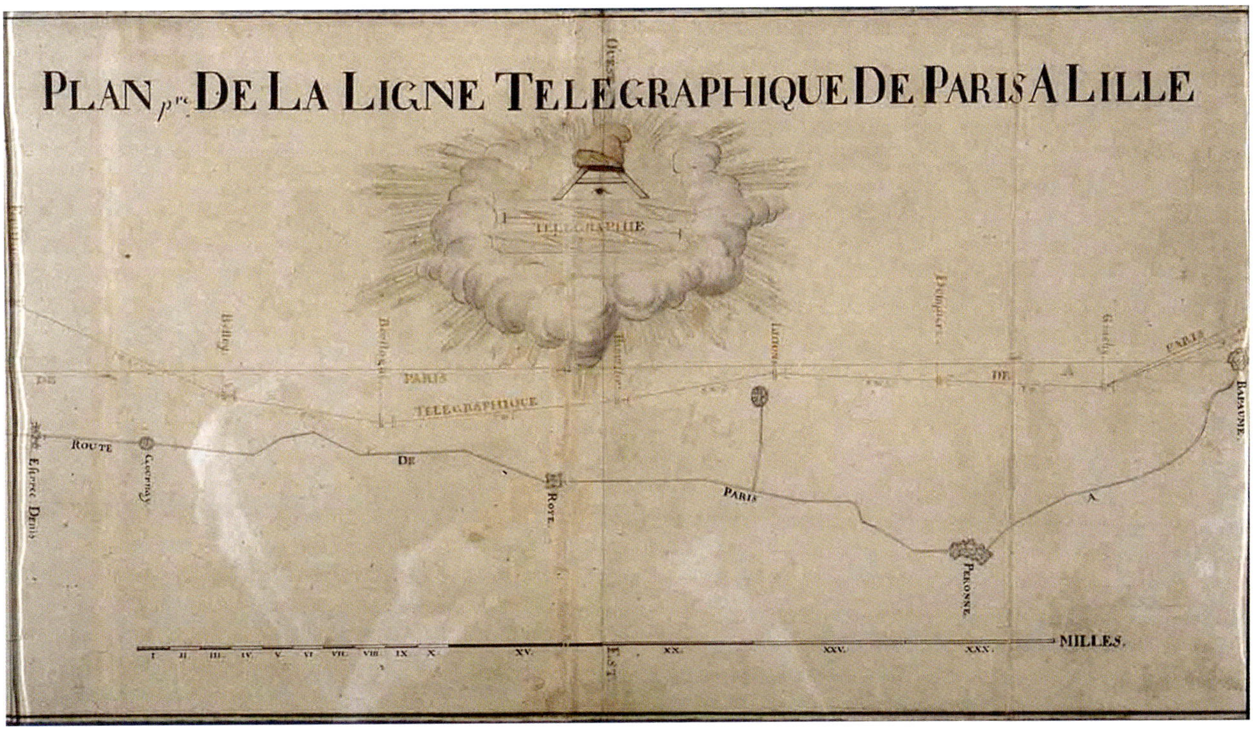

communication, but quite another to account for the vagaries of wartime affect. Viewing a telegraph in action rendered distant conflict visible, collapsing centre and periphery, forcing attention to the temporality of war and rendering it an object of potential knowledge. However, this visibility was perplexing, for it actually revealed none of the information being transmitted. Consequently, the knowledge gained from the telegraph, for those on the ground at least, was less a documentation of the actual events of the battlefield than a manifestation of the visual experience of those who lived through a time of war. Mediations of mediations, images of telegraphs in action provided a means to think about transmission more generally, to consider the historical dimensions of the revolutionary everyday, and to trace connections between different representative media. The power of such images exceeded the life of the telegraph itself. Sending the image of telegraphy back to the front, in 1942 Emile Louis Macé's statue of Claude Chappe and his telegraph, erected on the Boulevard Saint-Germain in 1893, was melted down to make guns for the war effort.[25]

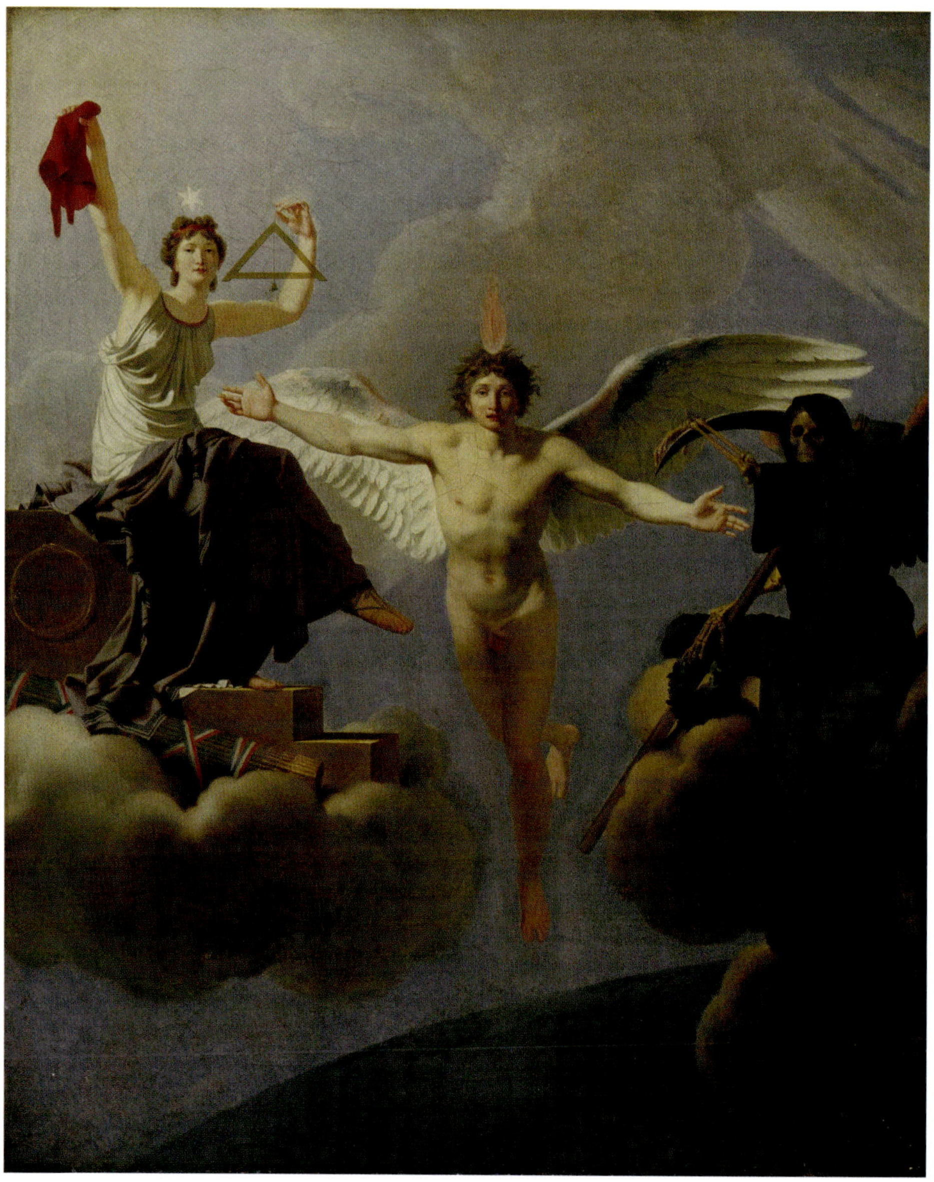

5 Jean-Baptiste Regnault, *Liberty or Death*, 1794–95. Oil on canvas, 60 × 49 cm. Hamburg: Hamburger Kunsthalle. Photo: bpk/ Hamburger Kunsthalle/ Hanne Moschkowitz.

6 Charles Norry, *Vue de L'intérieur du Louvre d'après nature l'an 4eme [sic]*, 1799. Pen and India ink drawing with watercolour, diameter 35.5 cm. Paris : Bibliothèque Nationale de France. Photo: BnF.

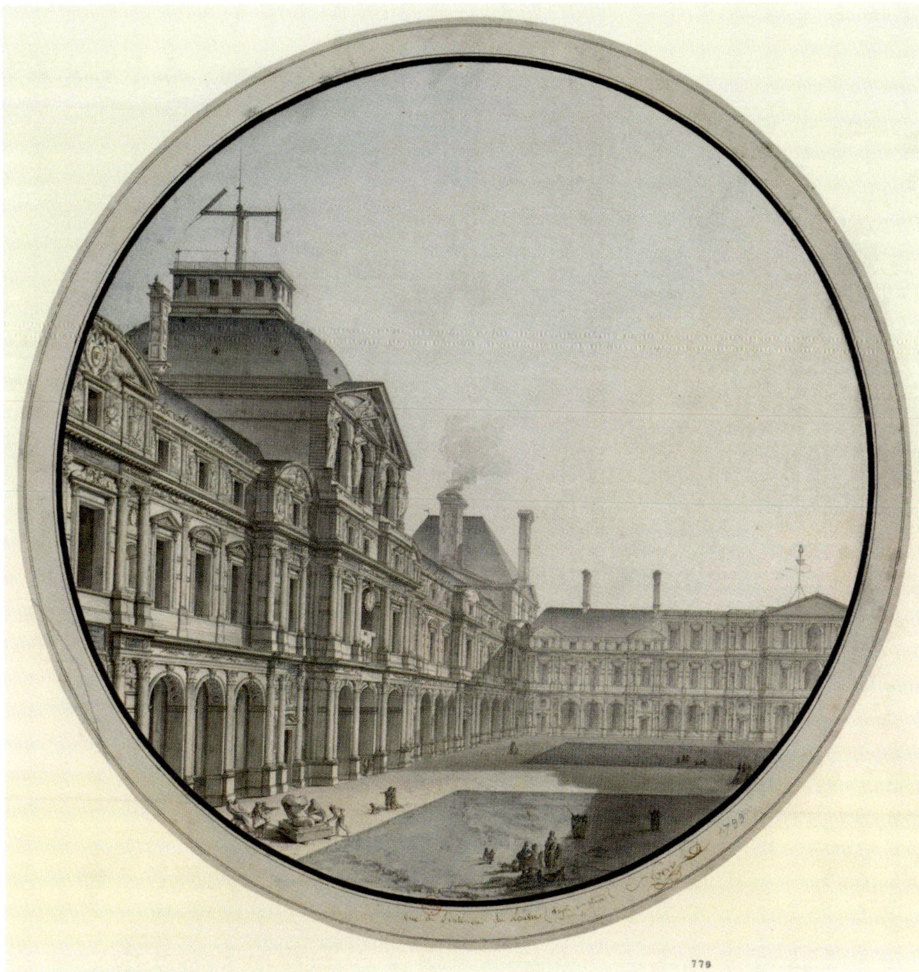

Telepresences

Nearly forty years after its introduction, in the aftermath of the 1830 Revolution, the optical telegraph reappeared on the horizon of revolutionary politics. In Honoré Daumier's famous caricature, the citizen-king Louis-Philippe in the guise of Gargantua squats gross and immobile on his throne, his body a machine that, fed with a stream of taxpayers' money, defecates sinecures and privilege to the greedy politicians who scurry around like portly dung beetles beneath him (*plate* 8). The caricature, which sent Daumier to jail, represents Louis-Philippe as a technology of transformation, turning the taxed coins of the starving French people into the printed paper of state. *Gargantua*'s success as a political satire is usually attributed to Daumier's radical simplification of form, Louis-Philippe beginning his mutation into the pear-like shape that would dominate his caricatural image during the July Monarchy. It is worth pausing, however, to shift the focus, to reorientate this image towards a barely visible speck in the distance. In the nebulous Parisian cityscape to the right-hand side of the print, a telegraphic apparatus cranks silently atop a far-away building whose asymmetrical twin towers identify it as Saint-Sulpice.

Collapsing the space of the print, the telegraph invokes a form of invisible communication that troubles the insistent visibility of the lithograph. The materiality and immediacy of lithographic technique – the gestural effects of the artist drawing on stone – contrasts with the intangibility and airborne inscrutability of the telegraph. At the same time, the smeary impression of this planographic

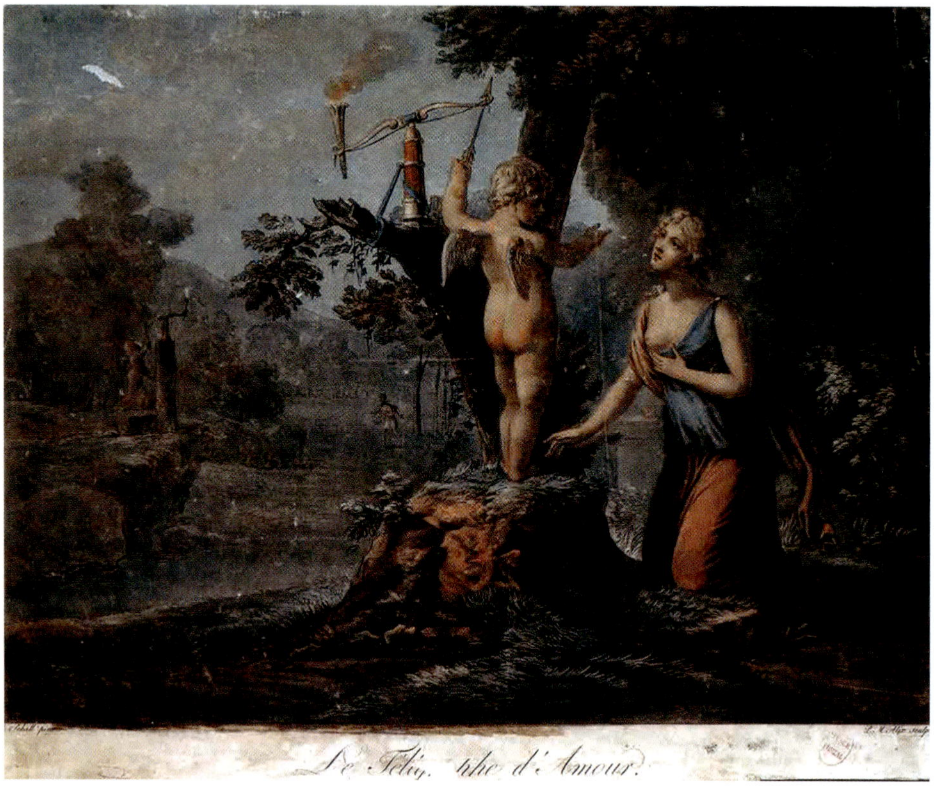

7 Pierre-Michel Alix after Jean-Frédéric Schall, *Le Télégraphe d'amour*, 1804–15. Colour engraving, 33.5 × 27.5 cm. Paris: Musée de la Poste (Inv. 7491). Photo: © L'Adresse Musée de la Poste, Paris/La Poste/Michel Fischer.

process seems suited to the representation of telegraphy's visual 'noise'. Happening above heads, in the corner of the eye, at the periphery of vision, it evokes a kind of pictorial interference that structures the whole scene. Daumier recognized that while telegraphs made a substantial visual impact in the first half of the nineteenth century, yet more important was what they left to the imagination. Who, or what, was this telegraph trying to contact, and what news was it hoping to convey? Daumier provided something by way of response in a print made three years later, one of several lithographs that reference the telegraph directly, which bears comparison with his design for *Gargantua* (*plate* 9). In this image, the telegraph stands in for the body of Louis-Philippe as an engine for the production of ministerial and clerical corruption, making explicit what was hinted at in the earlier image: that *Gargantua* is, in part at least, an image about telegraphy, and that telegraphy, finally withdrawn from control of the Chappe family in 1830, was an agent of the state – the silent, sinister means of transmission that enabled the July Monarchy to stay one step ahead of the people.

Gargantua marks the point from which, due to imprisonment and censorship, Daumier's attacks on *juste milieu* repression became ever more encoded. Telegraphy, as represented in print at this time, was constructed as its antithesis, its communications recast as blind enigmas and its cryptic secret-state modalities contrasting vividly with the oppositional potential of graphic satire. By 1831 the revolutionary potential of universal communication implied by Lakanal and Chappe in the 1790s had, in these works, become something distinctly anti-revolutionary, and any possibility that it might be harnessed to speak to power, as well as for it, was extinguished with an 1837 law establishing prison sentences for transmission of 'signals, from one place to another'.[26] The imagined community to which the telegraph spoke was enabled, as in Benedict Anderson's classic analysis, by print capitalism, but it increasingly involved a return to arcane codes and privileged access, rather than the democratic, universally available vernacular that Anderson viewed as key to print's transformative role in modernity.[27]

That said, Daumier's print reanimates the telegraph as a revolutionary technology, or at least, a formerly revolutionary technology against which contemporary politics might now be judged. Part of the force of these images comes from their situation of the telegraph in a historical network and in relation to other forms of political representation. Daumier's fellow caricaturist Grandville alluded to this in a sketch for a print representing Louis-Philippe as a wizard, casting the telegraph into darkness, where, rather than embodying the practices of a particular repressive regime, the telegraph functioned more abstractly as a cipher for communication itself.[28] In both cases, the telegraph provided a means to think Revolution beyond the bounds of its immediate manifestations, to theorize its corruptions, but also its possibilities. Daumier's image looks back to the enlightened possibilities of telegraphy presented by its advocates in the early 1790s, but also to the visual landscape within which it evolved. Faced with the impasse presented by Daumier's repressive telegraph, we might do well to return to the revolutionary origins of the telegraph, to a point in time when it seemed to presage an enlightened future of uninhibited contact between peoples, to track another hidden message.

Lisa Gitelman has observed recently how 'even media that seem less involved with representation than with transmission, like telegraphs, offer keenly persuasive representations of text, space/time, and human presence, in the form of code, connection, and what critics today call "telepresence," that feeling that there's someone else out there on the other end of the line'.[29] 'That feeling', of the coexistence of, and perhaps even identification with, an unseen other, certainly informs the meanings the telegraph was loaded with, and those it produced, in the final decade of the eighteenth century. At the inception of the telegraph in the early 1790s, the potential for communication across distance raised the possibility, not only of military advantage and future commercial success, but the transmission of revolutionary ideas. That sense

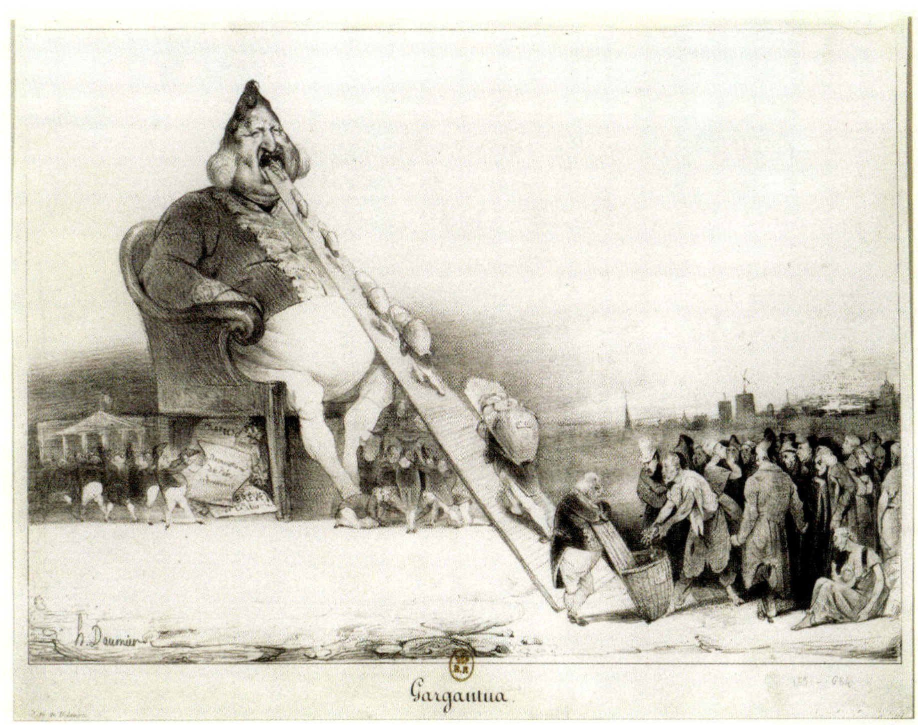

8 Honoré Daumier, *Gargantua*, published in *La Caricature*, 15 December 1831. Lithograph, 30.5 × 20.4 cm. Paris: Bibliothèque Nationale de France. Photo: BnF.

of feeling someone else on the 'end of the line' could incite fear or suspicion, or provide concrete evidence of government surveillance, but it might equally, given the right conditions, suggest an empathetic community of like-minded citizens.

During the French Revolution, attempts to centralize authority and stabilize the boundaries of France operated in an ambivalent and at times contradictory relationship with the Revolution's global potential. While some revolutionaries imagined the Revolution as a model of emulation for other countries, others argued against the transformation of liberty into an export, and suggested that France should concentrate on internal politics before attempting to revolutionize the world. The outbreak of the Haitian Revolution in 1791 demonstrated that this debate could not be contained within the confines of France. Political philosophers were no longer ruminating on the problems for formal government of political relations formed between strangers within the metropole.[30] Rather, legislators were forced to take the practical implications

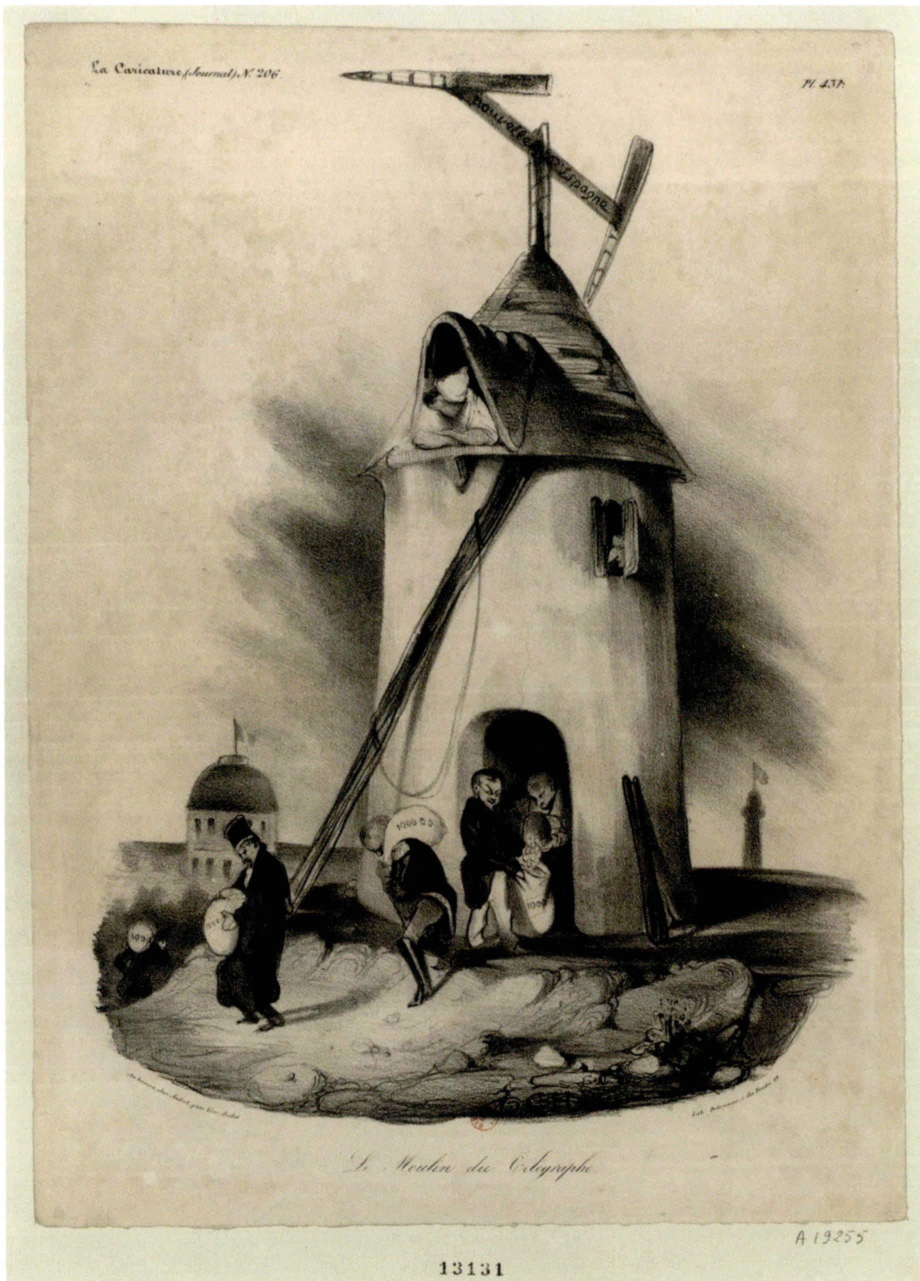

9 Honoré Daumier, *Le moulin du télégraphe*, published in *La Caricature*, 16 October 1834. Lithograph, 31.8 × 24.5 cm. Paris: Bibliothèque Nationale de France. Photo: BnF.

of the Rights of Man into account, considering claims for equality and citizenship from those they had previously believed to be less than human. As Emma Rothschild has framed it, this 'new diversity of political subjects' unsettled an antique republican model of citizenship, whereby 'all citizens should be the same sort of person'.[31] Instead, communication between strangers was incorporated, in ways that were frequently unresolved and contradictory, into everyday revolutionary practice. To be free was to communicate. The utopian rhetoric of telegraphy at this moment, as enabling both free trade *and* communication between equal citizens, can be read in the light of the conundrum that the Haitian Revolution presented to the French: how to justify the legitimacy of a supposedly universal and equal civil society while profiting from the sale of sugar produced by far-away slaves?

Du Camp's dystopian vision of modern wage-slavery, of telegraphic workers dictated to by the whims of their machines, is clearly of a different order to the horrific brutalities of slavery in the French colonies. However, telegraphy had, by rationalizing and collapsing abstract principles of 'commerce' and 'communication', already provided a language with which to speak, in a distanced way, about such matters. Although it finally came into official use six months after the abolition of slavery in France's colonies in February 1794, telegraphy – more as an idea than as an apparatus – encapsulated efficiently the twin, contradictory, desires that underpinned the initial revolutionary approach to the colonies; the irreconcilable aims of universalist freedom and commercial advantage. Arguably, telegraphy's colonial associations were, almost immediately, sublimated, first in terms of Eurocentric resistance to 'antique' or political slavery (freedom from despotism), or, later, via repeated reference to the cruel conditions of those who operated the telegraphic apparatus itself: the rats in cages on St Helena, the 'poor wretches' described by Dumas, and so on, whose misery bore the traces of another, more distant servitude. Such associations were not solely a consequence of the increasingly parlous work conditions that accompanied industrialization, but were present in some form from the beginning. Conveying to the National Convention the news of the first telegraphic message in 1794, Lazare Carnot reported that 'Condé is restored to the Republic; its surrender took place at 6 o'clock this morning'. The actual words of this original message, sent from Lille to the telegraph on the Louvre on 28 thermidor an II and translated by the inspector at the end of the line, are slightly different, however: 'Large Austrian garrison of 3000 slaves has put down arms and surrendered to the mercy of the conqueror'.[32] That telegraphy, optical and electric, played a crucial role in colonial expansion through the course of the nineteenth and twentieth centuries is only part of the story. Telegraphy was imbricated, in unresolved and often unrecognized ways, in the colonial imaginary of post-revolutionary France, its mute disturbance of the visual field chiming with the silences, stoppages and disconnections that characterize the colonial archive.

Coding Commerce

Returning to this early phase of telegraphy means, however, that we must, like Thackeray, acknowledge that the 'telegraphic' aspects of an image might not be coincident with the implementation of the system, but that its creation might even, as with David's 1784 *Oath of the Horatii*, precede it. An engraving by Jean-Charles Levasseur of Anicet-Charles-Gabriel Lemonnier's enormous painting *Le Commerce* offers one entrypoint for an examination of the intersection of communication and commerce at this crucial juncture, and it is with a discussion of this image that I will conclude (*plate 10*). Commissioned in 1786 to hang in the Chamber of Commerce of Lemonnier's native Rouen, *Le Commerce* is a work that points in intriguing ways to the global exchange of people and commodities in the late eighteenth century, and to the manner in which reproductive media provided a conduit for the transmission of revolutionary politics across space and time.[33] It is also an image that condenses in visual form several of the debates taking shape around telegraphy in the early 1790s.

Although the large-scale sketch for the now-destroyed painting was exhibited at the 1787 Salon (*plate 11*), the finished version was not presented to the Rouen Chamber of Commerce until 1791.[34] This was a crucial delay, for in the intervening time the artist made several changes to the canvas, based upon recent revolutionary events in France, but also in Saint-Domingue, modern-day Haïti, where the first throes of the slave Revolution had radically reconfigured the insular prerogatives of revolutionary liberty. The final painting was exhibited at the Salon of 1791, while the painted sketch was shown again in 1793, and was considered a great success.[35] The print represents the final version of the painting, in reverse of course, and it was also issued in 1791. Even putting aside the mediation integral to all intaglio printmaking, Levasseur's print was especially conditioned by the images that preceded it, a system of relays that enabled it to speak to the historical character of more than one revolution.[36]

Levasseur's engraving of *Le Commerce* represents the four continents as female allegories, against the background of Rouen's teeming port. On the right-hand side, 'Europe', assisted by Minerva, sits in judgment over the scene. 'Asia', her hand on the shoulder of a servant bearing a platter of perfume and jewels, is represented in antique garb, observing the prospect in front of her, where a plumed, bare-breasted 'America' is unveiled by a *génie* of Commerce carrying a compass. *Le Commerce* reiterates a stock image of the birth of America that had become a common subject in print culture since the American Revolution, and is explicit in its articulation of France's happy role

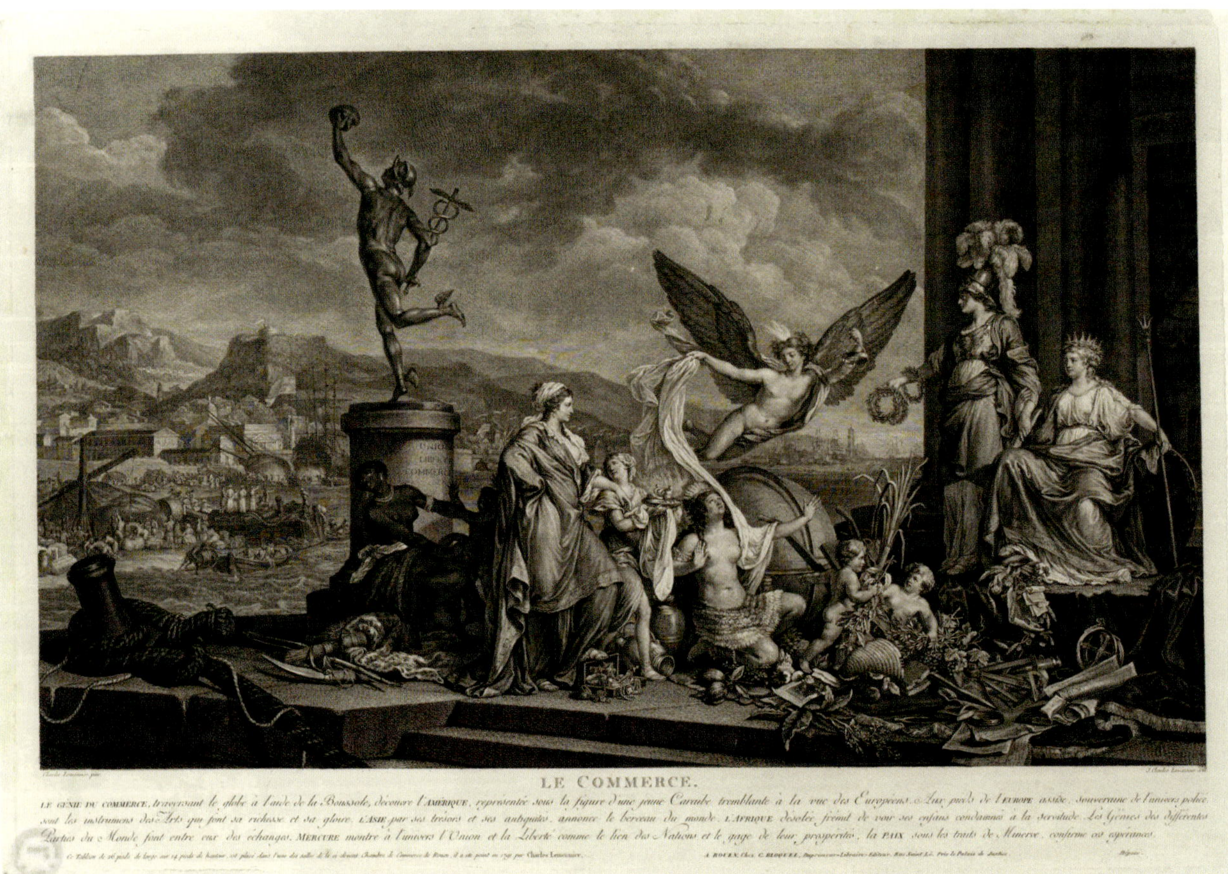

10 Jean-Charles Levasseur after Anicet-Charles-Gabriel Lemonnier, *Le Commerce*, 1791. Etching and engraving, 43.2 × 71.6 cm. Vizille: Musée de la Révolution française. Photo: © Collection Musée de la Révolution française/ Domaine de Vizille.

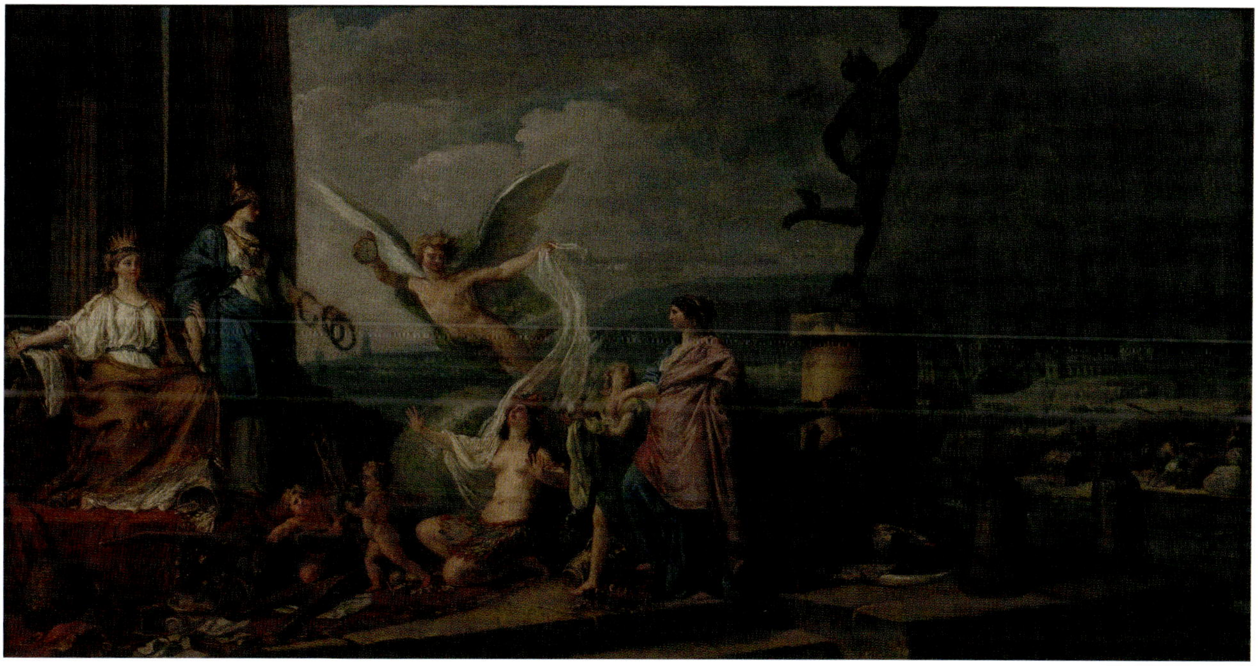

11 Anicet-Charles-Gabriel Lemonnier, *Le génie du commerce découvrant l'Amérique aux yeux des Nations*, 1787. Oil on canvas, 70 × 130 cm. Rouen: Musée des Beaux-Arts. Photo: © Musées de la Ville de Rouen/ Agence Albatros.

in assisting American independence from British rule.[37] However, a dissonant note is struck by the presence of the figure group representing Africa, cowering at the foot of a burnished statue of a flying Mercury with liberty cap held aloft. The Africa group was moved from the centre of the painting after the outbreak of the Revolution in Saint-Domingue, and displaced to this marginal space, likely to neutralize its impact. Two *putti* exchanging French and American products, including telescopes, maps, and a print of Christopher Columbus, take its place in the foreground.

This image is in keeping with the abolitionist sentiments of the merchants and Masonic lodges who had commissioned Lemonnier's painting, expressing many of the ambiguities of their liberal position. Although Rouen was no longer a slave port to compete with the likes of Nantes or Lorient, the commodities that sustained one leg of the Atlantic slave trade, textiles in particular, departed from its harbour in vast quantities, as can be seen in the left-hand background of the print. The artist was overt in his own anti-slavery position. In Toulon in 1784, on his way back to France from his second visit to Rome, where he had spent four years, Lemonnier had met the Abbé Raynal, the *philosophe* author of the influential *Histoire philisophique et politique des établissements et du commerce des Européens dans les Deux Indes*, who was also re-entering France after three years in exile.[38] Raynal had famously predicted the emergence of a 'Black Spartacus' who would free the enslaved people of France's colonies in an outburst of revolutionary violence – the realization of this prediction was often understood to be the revolutionary leader Toussaint Louverture. Raynal was full of praise for Lemonnier's painting, which he saw in the studio on the rue de Soubise, and which expressed visually many of the key themes of his own work (although Raynal had by this point reneged on his anti-slavery position in the light of events in Saint-Domingue). On viewing the painting in 1791 an anonymous critic writing in the *Mercure de France* suggested that 'If Africa ever raises its oppressed head to the level of the organized nations, it will owe much to the brush of M. Lemonnier.'[39] Although Lemonnier's painting aimed to express the mother's inability to face seeing her children born into servitude, this image of a dependent and despondent Africa rejecting her children and violating nature, perpetuated a gendered discourse of servility and ineptitude that underscored much abolitionist sentiment at this time.[40]

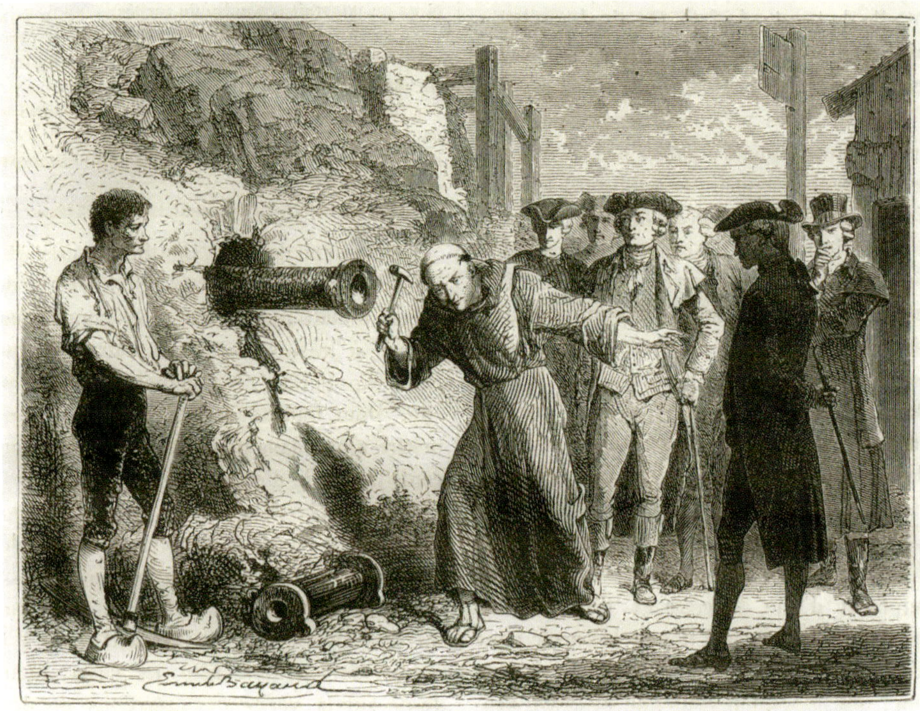

12 Émile Bayard, *Expérience de télégraphie acoustique faite à Paris, par Dom Gauthey, en 1782*, from Louis Figuier, *Les Merveilles de la science ou description populaire des inventions modernes*. Paris: Furne, Jouvet, 1868, volume 2, page 13. Wood engraving, 18.5 × 13.5 cm. Photo: Author.

Despite the presence of these tensions in his work, Lemonnier, himself from a Rouennais merchant family, attempted to obscure them, stating that: 'Commerce, so useful, so great, in that it binds together the men of a single empire, gets an even brighter gloss from communication and reconciliation between different nations. I have considered my homeland (*patrie*) in its commerce with France and in its trade with foreigners ... in order to deal with the history or the spirit of commerce in general' [41] Overt here are Lemonnier's location of his work in time and his sensitivity to the communication of revolutionary values, allied to a tacit understanding of how the reduction in space between people and things is embraced by capital. In his turn to a world-historical perspective, Lemonnier not only reaffirmed his status as a painter of important allegorical themes in the grand manner, but suggested an attempt to come to terms with broader currents of historical change: 'the history or the spirit of commerce in general'. While Levasseur's print of Lemonnier's painting clearly reproduces this original message, it circulated in a revolutionary context in which communication and commerce could not be so easily reconciled.

The image is dominated by a statue of Mercury, messenger and god of commerce. It emulates Renaissance archetypes such as that by Giambologna, the owner of which, the Duc de Brissac, was executed for his anti-revolutionary sympathies in 1792. Indeed, the use of Mercury in this context was well established by this point, not least by the celebrity of Pigalle's *Mercury fastening his sandal* (1736–39), regarded as a sculptural ideal well into the nineteenth century, and it appears in a wide variety of contemporary printed images with diverse political agendas. This sculptural motif finds a metallic echo in Lemonnier's image in the strangely prominent bollard in the left-hand foreground, possibly a cannon captured from a British ship, a relic from the American Revolutionary wars. This cannon is not present in the painted sketch, where more conventional posts suffice; a notable departure, even in the context of the image's crazed excess of objects, inserting a militaristic motif where none had previously been present, but figuring its obsolescence as a sign of the peace conducive to trade. In 1782, four years before Lemonnier received his commission for *Le Commerce*, a proto-telegraphic apparatus designed by a priest, Dom Gauthey, was demonstrated in Paris (*plate 12*). Dom Gauthey's device involved metal tubes embedded in the ground, which would convey sounds from one place to another.[42] Anticipating telegraphy's subterranean future and telephony's sensory effects, Gauthey's project was nonetheless out of step with its time, and was not taken up; forgotten within six months, he emigrated to America. Lemonnier's painting, unsurprisingly, makes no reference to Gauthey's device, yet the introduction of the buried cannon is an image that transparently evokes themes of projection, transmission, and appropriation echoed elsewhere in the work, as well as in the scientific culture of the 1780s. Its appearance in the work in the aftermath of the Revolution in Saint-Domingue is even more striking in this regard.

Captured cannon seldom retained their material integrity at this time, for they were readily transformable into rare metal coinage. Another form of reuse was the monument made from melted-down enemy armaments, which carried a particularly powerful symbolic charge. Indeed, Lemonnier was certainly aware of the contingent nature of materials and the potential for their remaking as revolutionary signs, having inventoried Notre Dame in November 1790.[43] The presence of the buried weapon in Levasseur's engraving, in alignment with the statue of Mercury, suggests that statue and cannon are composed, at least in the frame of representation, of the same matter, pointing to the possibility that the figure represented in the statue may be far from peaceable, and not unequivocally friendly to French interests. In this light, the hefty rope knotted loosely around the cannon (restraining it, even) seems to serve as an analogue for the bound rods – the fasces carried by the lictors of Republican Rome – that were a symbol of revolutionary unity, justice, and in some versions, the immediacy and brutality of state-administered violence. This was a culturally central motif in 1791, following the re-exhibition that year of David's 1789 *The Lictors Returning to Brutus the Bodies of his Sons*, a subject also represented in a sketch by Lemonnier. Furthermore, in its binding of chromatically distinct elements, the rope – led by a white guide rope – inevitably brings to mind the complex coding of colour in this context, what Darcy Grimaldo Grigsby has described as a kind of bodily materialization of the French tricolour, mapped out in racial terms.[44] Does Lemonnier represent such a system, abstracted onto the rope, as a bulwark against colonial violence or as a constraint on liberty, or both? Moreover, in line with this uncertainty, the position of the hawser, and the diagonal approach to the harbour edge, suggest that the viewer is positioned as if aboard a ship, tied to the shore, but at sea. Destabilizing the perspective of a putative viewer, in more ways than one, this conceit compels the question of who exactly the 'we' on the boat is meant to be, conjuring the presence of a potential beholder whose transportation and forced labour ensured France's prosperity.

In the print the statue of Mercury occupies an indeterminate space in the racial politics of an image so clearly coded in terms of difference mediated through

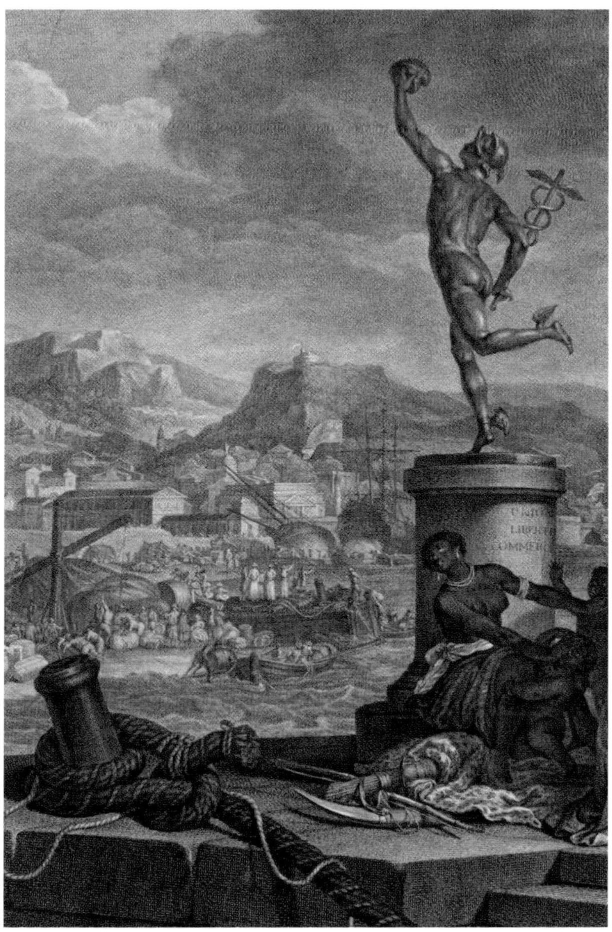

13 Detail of Levasseur after Lemonnier, *Le Commerce*, 1791, showing monument of Mercury on left-hand side of image.

skin colour. If this difference is obscured in the print, whose monochrome tonal gradations make more subtle use of such distinctions, it is explicit in the painted sketch. Nonetheless, it is perhaps Levasseur's engraving that expresses most accurately the nuance with which race as indicated by skin colour was understood in the late eighteenth century. In pre-revolutionary Saint-Domingue, slavery was justified by the imposition of increasingly precise hierarchies of skin colour, between whites, blacks and *gens de couleur*. The representation of Mercury in this manner could not have failed to evoke the complexities of this system to contemporary viewers.

Squarely aligned with the mother and child group representing Africa, and, by extension, Saint-Domingue, the figure of Mercury, read vertically at least, implies a positive futurity altogether lacking in the figure group beneath it (*plate 13*). Whereas the first sketch submitted by Lemonnier had shown Mercury announcing to Neptune that thanks to the Revolution all nations could trade freely with one another, the final image represents a patinated, 'black' Mercury, caduceus in his right hand, revolutionary cap of liberty aloft in his left. This section of the print marks a coming into revolutionary consciousness, a historical narrative of the trajectory from enslaved Africa to revolutionary Saint-Domingue, but also a recognition, and likely, a fear, of how this transmission might 'feed back', for it foregrounds both the communication of Revolution to France's colonies, and the returned message, as Mercury is facing inland, back towards Rouen. Is this 'Black Mercury' an iteration of the 'Black Spartacus' prophesied by Raynal, inflected by abolitionist anxiety about the real consequences of the attainment of liberty in the colonies? Might this Trojan Horse explain, in part at least, and ambiguously so, the fact that Laocoön is invoked here, recast to radical effect as a black woman?

What is more, while the visual homology between the slanted beam of the crane on the left-hand side of the image and a Chappe telegraph apparatus is coincidental, as is the collapsing of distance between the Mercury figure and the fortified outcrop in the distance which from this perspective so resembles a telegraph tower, they raise the prospect of communication between Mercury and a point in space and time that is at once near and far away. In fact, considered from this perspective, this picturesque rendering of Rouen's port is not actually much like Rouen at all, or at

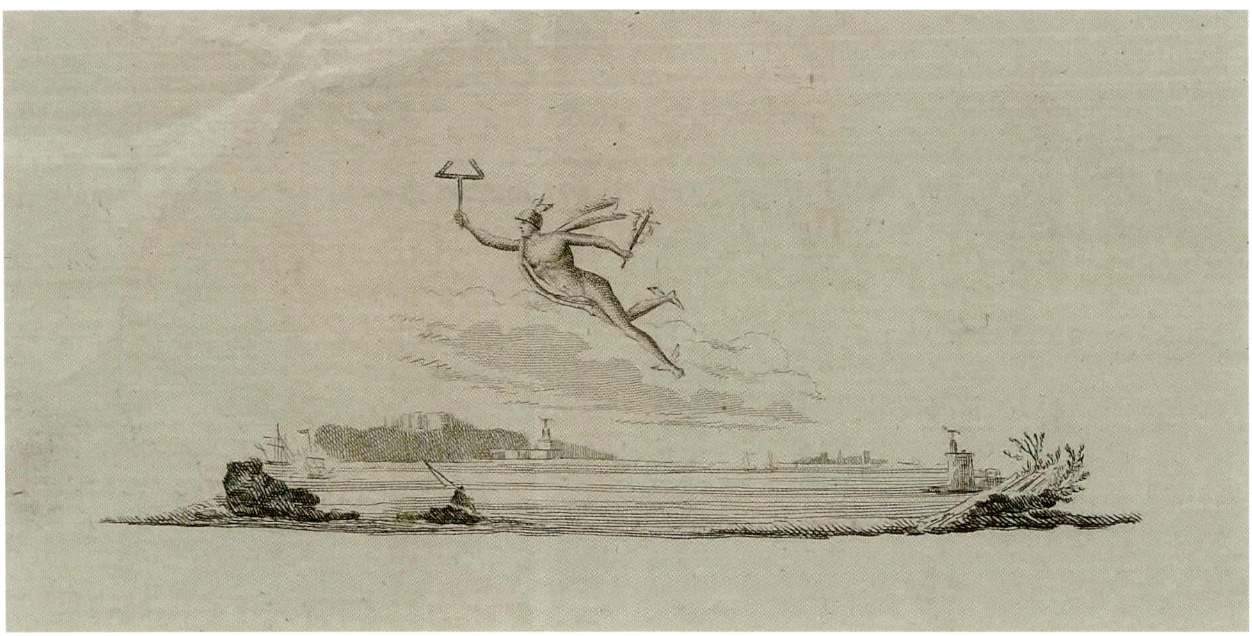

14 Detail of telegraphic dispatch, signed Ignace Urbain Jean Chappe, 19 March 1809, showing letterhead for Chappe telegraphic network. Engraving with manuscript. Washington, DC: Dibner Library of the History of Science and Technology, Smithsonian Institution (MSS 000332A). Photo: Courtesy of the Smithsonian Libraries, Washington, DC.

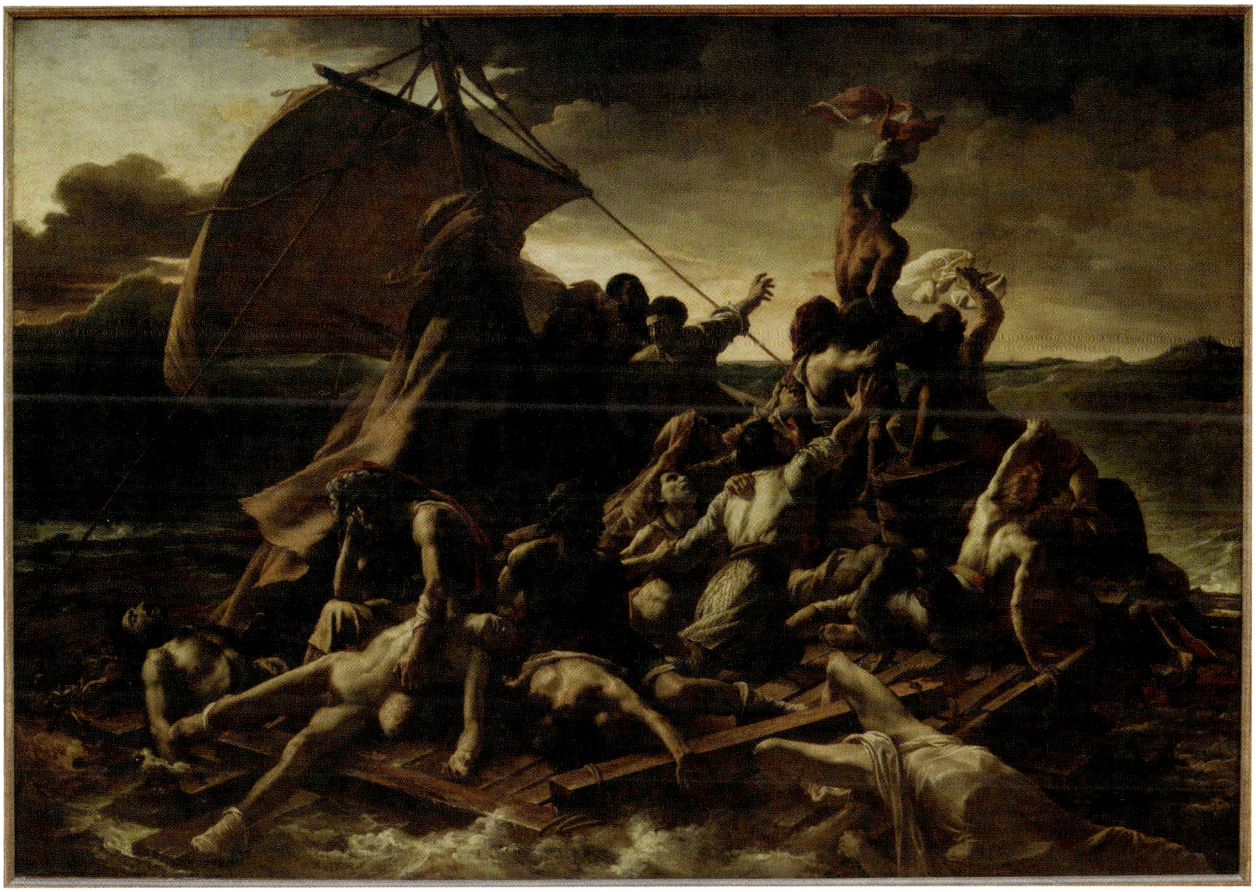

15 Théodore Géricault, *Le Radeau de la Méduse*, 1819. Oil on canvas, 491 × 716 cm. Paris: Musée du Louvre. Photo: © RMN-Grand Palais (Musée du Louvre)/Michel Urtado.

least not *just* Rouen (the proximity of the ruins of Richard I's Chateau Gaillard at Les Audeleys and the Barfleur lighthouse to the right of the print, means that it does not cohere as a representation of an actual place). It seems as close, in fact, with its mountainous landscape and stormy skies, to the topography of Saint-Domingue – can we read it, like the rope, as a binding together of disparate elements, an intervention in pictorial and geographical space that parallels the grafting and hybridization that characterized both the flora of the colonial landscape and the population that inhabited it?

The image of Mercury becomes, in this context, a kind of telegraphic apparatus in its own right, while the print as a whole anticipates, and was later comprehensible in relation to, the telegraph's concrete realization of abstract principles of communication. 1791, the year that Lemonnier's finished painting made its debut at the Salon, and Levasseur's print appeared on the market, also saw Claude Chappe's first live forays into telegraphy. Soon, the figure of Mercury traversing the seas was reimagined as a symbol of communication, rather than commerce, although the commercial possibilities of telegraphy were not disregarded, but were built into early premonitions of the future course the technology would take. An image of Mercury flying from port to port was the natural choice for the letterhead used by the Chappe brothers for correspondence relating to the administration of their network (*plate 14*).

In Lemonnier's image, telegraphy is evoked *avant la lettre* in an allegorized fantasy of communication benefiting commerce that was, as yet, more a concept than a fully realized technology. By foregrounding relations of space and time, telegraphy provided a powerful means to think about history, yet these possibilities did not emerge from a vacuum, and were not, crucially, produced by the technology itself.

When, in 1824, Ignace Chappe published a history of telegraphy that focused on the role he and his brothers had played in the introduction of the optical system in France, he stressed the historically determined nature of telegraphy, which, after decades of similar proposals, only emerged to fruition in congruence with unique historical circumstances.[45] Telegraphy was, in a sense, always already there.

In its early days, the Chappe telegraph was considered inseparable from the commercial opportunities it might offer, and it was impossible to talk of commerce in the 1790s without invoking the Atlantic trade, even if the words themselves were not spoken. However, telegraphy, by opening up a space for thinking about communication between equal citizens in (almost) real time, also provided a model for imagining resistance to a dominant message. 'Our great discoveries will perhaps not make us better', Ignace writes, 'but they help make us as strong as those who would oppress us; and oppression is the worst of all the evils to which the human race can be prey.'[46] Telegraphy, for Chappe, was a specifically anti-despotic technology, and as such, for a brief period in its utopian early 1790s mode, it raised the possibility of confronting the real, not merely conceptual or ideal, conditions of slavery. However, an Enlightenment object par excellence, the telegraph was correspondingly paradoxical, and its use was accompanied by anxieties about the possible effects of the challenge it posed – universalism tempered by a fear of outsiders gaining access to its messages. The wish-image of telegraphy, Levasseur's print of Lemonnier's painting is similarly ambiguous, representing overlapping understandings of empire and global trade marked as much by continuities from previous political regimes, and versions of the image, as by ruptures in this stable order. Purporting to speak in favour of Liberty, *Le Commerce* did so, like the telegraph, at the edge of vision, its message hidden in plain sight.

Twenty-eight years later, a work by another Rouennais artist, Théodore Géricault's *Radeau de la Méduse* (*plate 15*), offered a striking, negative postscript to both Lemonnier's 'telegraphic image' and to the neo-classical aesthetic with which Thackeray later identified the technology.[47] A tangle of limbs substitutes for the clarion legibility of the telegraph's arms, and the frantic semaphore of Géricault's multi-ethnic crew – these people of no state – exposes the horror of the colonial encounter repressed in Lemonnier's image, figuring telegraphic communication, once again, as a means of historical comprehension, but one that is failed, illusory and exploitative, in which, this time, the return message is left unanswered.

Notes

I am very grateful to those who have given me opportunities to present aspects of this work, as well as to participants and audiences on those occasions, for their invaluable comments. I would particularly like to thank Colin Jones for his generous suggestions, as well as Daniel Abramson, Ann Jensen Adams, Joshua Bauchner, Hanneke Grootenboer, Lynn Hunt, Erika Naginski, Sina Najafi, and Katie Scott, whose responses enriched my thinking on this topic enormously. Many thanks are also due to my co-editor Genevieve Warwick, Sam Bibby, and the anonymous readers for *Art History*. The completion of this article was enabled by the award of a Philip Leverhulme Prize from the Leverhulme Trust.

1 William Makepeace Thackeray, *The Paris Sketch Book: by Mr Titmarsh. With Numerous Designs by the Author, on Copper and Wood*, vol. 1, London, 1840, 105–6. Henry Angelo (1756–1835) was the proprietor of a fencing academy on Bond Street, frequented by well-known political figures.

2 Lewis Mumford, *Technics and Civilization*, Chicago, IL, 2010, 89.

3 Joseph Lakanal, *Rapport sur le télégraphe fait au nom du Comité d'instruction publique*, Paris, 1794, 2.

4 *Détail exact du Télégraphe, au moyen duquel la reprise du Quesnoy, de Valenciennes et de Condé a été connue à Paris long-temps avant l'arrivée des couriers*, Paris, 1794, 6.

5 Pierre-Jean-Baptiste Chaussard, *L'Industrie ou les arts: ode publiée à l'occasion de la fête du 1er vendémiaire an VII*, Paris, 1798–99, 16.

6 See Alexander J. Field, 'French optical telegraphy, 1793–1855: Hardware, software, administration', *Technology and Culture*, 35: 2, April 1994, 315–47. Also *Le Télégraphie Chappe*, Nancy, 1993.

7 Telescope lenses were later made of high-quality crystal ground down from captured English luxury goods. See *Lettre de l'ingénieur Fréminville au comité de salut public a.s. de la fabrication des lunettes*. 15 October 1794. Musée de la Poste, Chappe archives, 621.1 carton 7 (dossier central télégraphiques Paris I), D7544 and *Arrêté ouverture atelier de fabrication lunettes*. 25 vendemiaire year III. Musée de la Poste, Chappe archives, 621.1 carton 7 (dossier central télégraphiques Paris I), D6269.

8 See Sophia Rosenfeld, *A Revolution in Language: The Problem of Signs in Late Eighteenth-Century France*, Stanford, CA, 2001. Telegraphy was even relevant to studies of blindness; Auguste-Guillaume Schwenger's

Mémoires sur les aveugles, sur la vue et la vision, suivis d'une déscription d'un télégraphe très simple, Paris, 1800, contains a proposal for a new telegraph, a variation on the Chappe model.

9 *Détail exact du Télégraphe*, 2.

10 See Ernest Jacquez and A. Kermabon, *Atlas des lignes télégraphiques aériennes, construites en France de 1793 à 1853*, Paris, 1892.

11 The naming of newspapers after the telegraph began early – see *Le Télégraphe Politique, ou, Journal des fondateurs de la République*, 3, 10 brumaire an III (1794). On telegraphy's commercial potential see Jean-Claude Bastian, 'L'argent et la naissance des télécommunications françaises', *Les Cahiers de la FNARH*, 107, 2008, 5–15.

12 Alexandre Dumas, *The Count of Monte Cristo*, New York, 1894, 78. On the telegraph in different European literary contexts, see David F. Bell, *Real Time: Accelerating Narrative from Balzac to Zola*, Champaign, IL, 2004, 76–130; Mary Fairclough, '*The Telegraph*: Radical transmission in the 1790s', *Eighteenth-Century Life*, 37: 2, Spring 2013, 26–52; and Bernhard Siegert, *Relays: Literature as an Epoch of the Postal System*, trans. Kevin Repp, Stanford, CA, 1999, 165–85.

13 Maxime du Camp, *Paris, ses organes, ses fonctions et sa vie dans la seconde moitié du XIXe siècle*, vol. 1, Paris, 1875. See Richard Taws, 'When I was a telegrapher', *nonsite.org*, 14, December 2014. At: http://nonsite.org/article/when-i-was-a-telegrapher

14 Du Camp, *Paris*, 140.

15 Friedrich Kittler, *Optical Media: Berlin Lectures 1999*, trans. Anthony Elms, Cambridge, 2010, 129; Paul Virilio, *Bunker Archaeology*, Princeton, NJ, 2009, 17.

16 Friedrich Kittler, 'The city is a medium', trans. Matthew Griffin, *New Literary History*, 27: 4, Autumn 1996, 717–29.

17 Bell, *Real Time*, 86.

18 Inge Rippmann and Wolgang Labuhn, eds, *'Die Kunst – eine Tochter der Zeit': Neue Studien zu Ludwig Börne*, Bielefeld, 1988, 140.

19 Archives Nationales, Paris, DLII/1.

20 Jennifer Roberts, *Transporting Visions: The Movement of Images in Early America*, Los Angeles and Berkeley, CA, 2014, 142.

21 The association between Regnault's painting and vignettes is made in Philippe Bordes, 'David's contemporaries: A generation of artists against the revolution?', in George Levitine, ed., *Culture and Revolution: Cultural Ramifications of the French Revolution*, College Park, MD, 1989, 67.

22 Seventy-one artists signed a petition complaining about the telegraph's location on the dome above the Louvre's *grand escalier*. The Louvre telegraph was demolished in 1806.

23 Schall's image was the pendant to *La Lanterne Magique d'Amour*. Similar themes were explored in J. Servière and C. Henrion's *Le Télégraphe d'amour, comédie en un acte*, Paris, 1801. In this play, the 'mute eloquence' offered by signals transmitted by windmill between the lovers Valsain and Rose allows them to avoid detection by the prudish widow Madame Robert. See also Bruno Bentz, '*Le* télégraphe d'amour', *Les Cahiers de la FNARH*, 93, 2004, 28–31.

24 Mary Favret, *War at a Distance: Romanticism and the Making of Modern Wartime*, Princeton, NJ, 2010, 13.

25 On the commissioning and construction of this statue, Archives Nationales, Paris, F/90/19103.

26 Bell, *Real Time*, 89.

27 Benedict Anderson, *Imagined Communities: Reflections on the Origin and Spread of Nationalism*, London, 1991.

28 Jean-Jacques Grandville [Jean Ignace Isidore Gérard], *L'enchanteur Merlin fesant en plein midi descendre la nuit sur le Télégraphe*, 1835, pen and wash on paper, 20.1 × 30 cm. Musée des Beaux-Arts, Nancy, inv. 877.624

29 Lisa Gitelman, *Always Already New: Media, History, and the Data of Culture*, Cambridge, MA, 2006, 4.

30 As Susan Buck-Morss writes, 'By the eighteenth century, slavery had become the root metaphor of Western political philosophy, connoting everything that was evil about power relations.' Susan Buck-Morss, *Hegel, Haiti, and Universal History*, Pittsburgh, PA, 2009, 21.

31 Emma Rothschild, 'Globalization and the return of history', *Foreign Policy*, 115, Summer 1999, 112–13.

32 Quoted in François Gautier, *L'œuvre de Claude Chappe*, Paris, 1893, 3.

33 The most complete analysis of Lemonnier's career is Christine Le Bozec, *Lemonnier: un peintre en révolution*, Rouen, 2000.

34 The painting was destroyed on 31 May 1944.

35 *Musée de la Révolution française: Premières collections*, exh. cat., Vizille, 1985, 39.

36 On Levasseur's contribution, see E. Delignières, *Catalogue raisonné de l'oeuvre de J. C. Le Vasseur, d'Abbeville*, Abbeville, 1865, 55; also Claudette Hould, *L'Image de la Révolution française*, Québec, 1989, 151.

37 Le Bozec, *Lemonnier*, 51.

38 Le Bozec, *Lemonnier*, 53.

39 *Mercure de France*, 26, 25 June 1791, 152. Le Bozec suggests the author was probably Lecouteulx de Canteleu, a Rouennais lawyer who likely played a key role in the commission. Le Bozec, *Lemonnier*, 51.

40 Joan Dayan, *Haiti, History, and the Gods*, Berkeley and Los Angeles, CA, 1995, 46.

41 Anicet-Charles-Gabriel Lemonnier, *Séance du 1er juillet 1791, Délibérations de la Chambre du Commerce*, cited in Le Bozec, *Lemonnier*, 47.

42 Louis Figuier, *Les Merveilles de la science ou description populaire des inventions modernes*, Paris, 1868, 13–16. See also, Dom Gauthey, *Expérience sur la propagation du son et de la voix dans les tuyaux prolongés à une grande distance*, Paris, 1783.

43 Archives Nationales, Paris, F^{17} 1261, dossier 2, no. 18. See Richard Clay, *Iconoclasm in Revolutionary Paris: The Transformation of Signs*, Oxford, 2013, 66.

44 Darcy Grimaldo Grigsby, *Extremities: Painting Empire in Post-Revolutionary France*, New Haven, CT and London, 2002, 15.

45 Ignace Urbain Jean Chappe, *Histoire de la Télégraphie*, Paris, 1824, 8–9.

46 Chappe, *Histoire de la Télégraphie*, 11.

47 On Géricault's painting in the context of contemporary communication technologies, see Jordan Bear, 'Adrift: The time and space of the news in Géricault's *Le Radeau de la Méduse*', in Jason E. Hill and Vanessa R. Schwartz, eds, *Getting the Picture: The Visual Culture of the News*, London, 2015, 182–9; and Tom Gretton, 'From *La Méduse* to the *Titanic*: Géricault's *Raft* in journalistic illustration up to 1912', *19: Interdisciplinary Studies in the Long Nineteenth Century*, 16, 2013, 1–38.

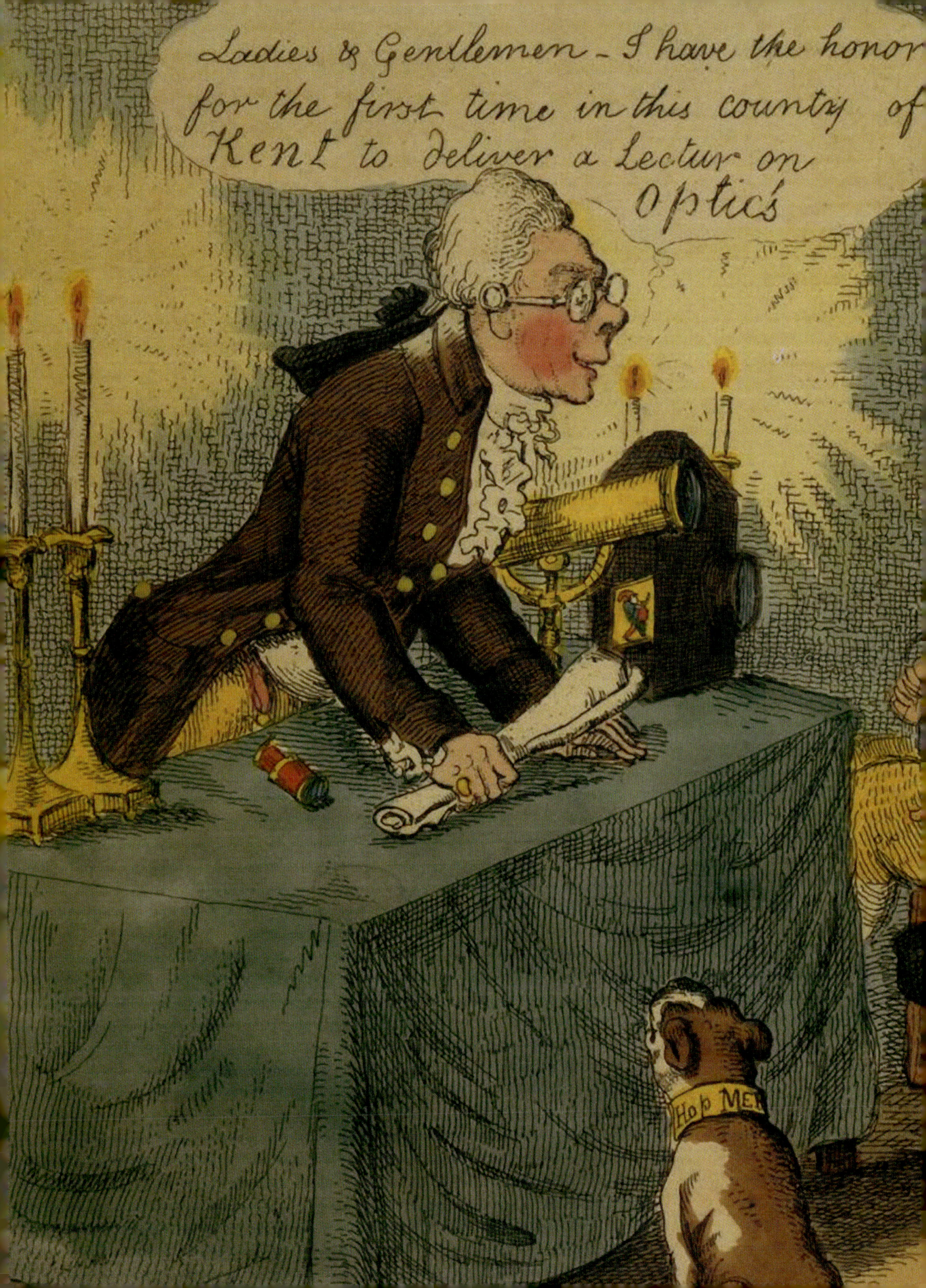
Ladies & Gentlemen - I have the honor for the first time in this county of Kent to deliver a Lectur on Optic's
Hop Me

Chapter 12
Seizing Attention: Devices and Desires

Barbara Maria Stafford

The message we listen for so carefully must be mediated through our own experience, our personality, our heredity, our inner needs. Can we ever break free of the devices and desires of our own hearts?[1]
(P. D. James, 1989)

Whatever else early modern technological devices did, they artfully vied for our undivided attention. Mirrors, cameras obscura, microscopes, telescopes, Claude glasses, kaleidoscopes, clockwork automata, as well as the optical telegraph and the pinhole camera, perceptually and cognitively focused our selective attention.[2] Today, conscious attention – that narrow band of brain function enabling people to be more than fleetingly aware of what they are holding in mind – is becoming scarce. Not surprisingly, then, this precious frontal-lobe real estate (estimated to be only 10 per cent of our total cognitive horsepower) has become the kind worth selling.[3]

Swift electrical patterns of information, quickly vanishing into distractions and inattention, comprise the divided terrain competed for by a dizzying array of software and hardware. Digital apps and optical apparatus abound as the revolutionary tools of targeted marketing. Significantly, this induced noticing is paid for by 'the attention we offer to providers who track our habits and preferences'.[4] Not unlike the use of cognitive-enhancing drugs, these unregulated practices raise issues involving personal agency and the undermining of the value placed on human effort by relying on an external locus of control.[5]

Neurological marketing services depend either on bypassing the pre-frontal cortex entirely by inciting impulsive behaviour, or shielding the instinctive-targeting data in working memory even while a person is engaged in some other task or goal. Recent additions to this addictive armamentarium tailored to our covert desires are emotion-sensing technologies addressing the thalamus, amygdala, hippocampus, and more generally the primal limbic system. Not surprisingly, this affect-seizing apparatus is frequently tied to advertising and 'the monetizing of the moment'. As one of Samsung's researcher's declared: in addition to the tastes, 'if we know the emotion of each user, we can provide more personalized services'.[6]

In the late-nineteenth century, William James drew an important and still valid distinction between two types of attention. Certain kinds of objects, events, and situations effortlessly engage our involuntary attention: 'strange things, moving things, wild animals, bright things' strike us without our consciously wanting to be struck. There are, however, phenomena that do not effortlessly slip into the

Detail from Woodward (artist) and Cruikshank (etcher), *A Kentish Hop Merchant and the Lecture on Optics*, c. 1809 (*plate 1*).

mind: these require our voluntary, directed or selective attention instead. This executive attention – at the heart of intense practice, expertise, and sustained thinking – is fuelled by working memory and lies at the heart of effortful activity, process rather than ends, concentration.[7]

These opposing kinds of attention translate into two different modes of embodied thought. Consider the stylistic difference, say, between Jean-Jacques Rousseau's easeful, arabesqued reveries as he drifted across the rippled mirror of Lac de Bienne. Sailing away from the clamouring duties of ordinary life, the *philosophe*'s unfocused somatosensory system is impressionistically at one with the reflective ambient. Much like a parade of magic lantern slides or unfurling kaleidoscopic images, his daydreams proliferated, crossbred, evolved.

Contrast Rousseau's *plein air* recovery from his prior manoeuvrings within the hectic environment of Paris to the hard demands of sharp thinking. Instead of solitary, soft immersion within the natural world,[8] the harshly abutting political and social patchwork typical, say, of Cruikshank's pointed caricatures roughly focuses the viewer's attention, obliging her to buckle down and piece the incongruent parts together (*plate* 1). The audience to this provincial lecture on optics demonstrates by their exaggerated facial expressions the so-called 'peak shift' effect. This is a behavioural response bias arising from our powers of discrimination whereby primates (but not exclusively) display a directional preference or avoidance. Cruikshank's extreme upturned mouths and downturned noses as well as his squashed or buoyant wigs play with our tendency to respond to really strong stimuli rather than to more straightforward perceptual cues.

As is true for those aperture-ruled optical devices (from the telescope to the pinhole camera), executive neural areas are dramatically exerted when demanding and powerful things oblige us to pay close attention, to create sense out of confusing or overwhelming environmental noise. A leading function of such devices is, in fact, the strengthening of the magnitude of the otherwise simple perceptual stimulus provided by ordinary reality. This saliency also ensures that the viewer's response is directly proportional to the exaggeration.

Recently there has been an international surge of interest in building large brain models. The European Union's Human Brain Project (HBP) has received one billion euros worth of funding, and President Obama announced the Brain Initiative along with a similar level of funding.[9] However, the large-scale models affiliated with both projects do not demonstrate how their generated complex neural activity relates to observable behaviour – a central challenge for neuroscience to be sure but a challenge and opportunity for the visual humanities as well.

We need to be more than archivists, more even than the chroniclers of the viral intermixing contagiousness of imagery across cultures and societies. In books such as *Visual Analogy*, *Echo Objects* and, most recently, the *Field Guide to a New Metafield*, I have argued that we need to collaborate with the neurological community in the study of art history's most enduring problems – from the underlying logic of forms, to the mirroring dynamic of ritualized gesture and performance, to the continuing force of symbols. Failure to do so will only lead to the decline or irrelevance of the field. If anyone doubts this, then Google the growing art-related research produced by the scientific community – good and bad, influential results with or without our help.[10]

A major contribution to an art/science/technology focus area within art history (or better, a cross-sensory experimental history area) would be the formulation of vital initiatives that engage with key contemporary issues crossing serious disciplinary divides; and to do so with current tools and emerging neuroscientific

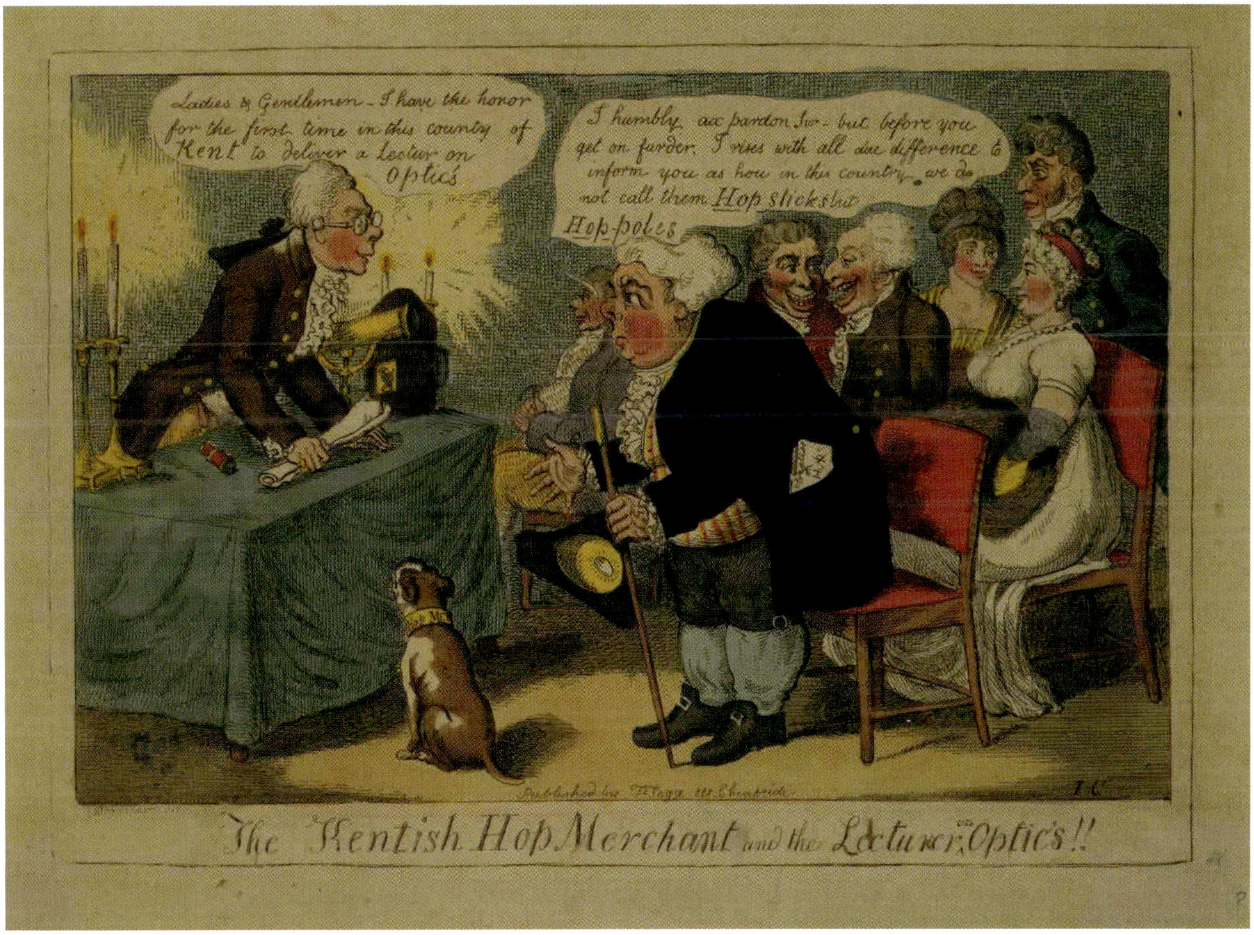

1 Woodward (artist) and Cruikshank (etcher), *A Kentish Hop Merchant and the Lecture on Optics*, c. 1809. Etching coloured by hand, 24.4 × 34.7 cm. London: Victoria and Albert Museum (E.1225–1990). Photo: Victoria and Albert Museum.

data in hand. I'm thinking of the sophisticated analysis of the variety, formats, and sensory effects not only of the high-order images of art but of the equally complex hybrid images produced by early modern optical technologies and beyond.

Specifically, as I've argued here, one such essential question should be: what do these diverse devices teach us about the hard-wired as well as the culturally modified operations of human attention – both in the past and in the present? What makes this ancient perceptual apparatus not obsolete but, in fact, cognitively informative and still eye-opening? In the *Devices of Wonder* exhibition (2001/02), we stated that, among other things, it might even become desirable to program their lost techniques and once-innovative technologies into today's speed-driven covert micro- and nano-machinery.

In an era fond of reducing all optical technologies to a single platform (the digital internet) and most compositional systems to illusionistic virtual-reality modalities, what could be more important than exhibiting how different experience-framing devices not only produce different kinds of information but make us think, feel, and desire differently. While more and more gadgets address our involuntary response mechanisms or secretly target our reward systems, I want to emphasize the value of those technologies that make us consciously aware that we are aware and fully attending. An additional conundrum is the as yet unresolved problem of how these two perceptual mechanisms interconnect. How do the varieties of art demonstrate this and why is it cognitively and aesthetically important that we hold both voluntary and involuntary (automatic/autopoietic) brain operations in some kind of balance?

There is something about the digital revolution that has not only made it harder to categorize cultural products, but has made it more difficult to categorize practitioners in the creative sector. Where once we would have spoken easily about artists, illustrators or designers we now have to work with a more expansive boundary-crossing vocabulary.

Where once we would have spoken about works of art, we now deal with the rise of 'creative industries'.[11] These networked crossings and mixtures of genres and devices are indicative that, for more than two decades, courses of media art and design are no longer locked into single media crafts such as drawing, painting, sculpture, film, graphic design, or web design. If such skill-mergers and intellectual hybridity are common in current art practice, why not also in conceptual art-historical practice?

On this point, it is noteworthy how many contemporary visual and sound artists are building on emerging neuroscience research to amplify their explorations of sensory perception, motion tracking, mirroring and simulation techniques, such as the new haptic and robotic devices that increasingly drive our neural circuits, inflecting our own experiences and how we resonate with or process the experiences of others. Tim Otto Roth's complex mergers of computing devices (acoustical, imaging, and neurological) in his 'Imachination' installations spring to mind. His recent *Sonapticon* project for ZKM's Sound Dome continues a quest to make otherwise invisible mechanisms – ranging from astronomical phenomena to micro-biological processes – visible as well as comprehensible because minutely embodied. In this case musicians and their audience, enveloped by an array of forty-three loud speakers registering the different frequencies of audio neurons, have their performance simultaneously interpreted as flashing colours revealing whether the impulse has

2 Detail of Tim Otto Roth, *Sonapticon*, November 2012, showing floor projection indicating the dynamics of the changing membrane potential of an audio neuron. Multimedia installation. Karlsruhe: Zentrum für Kunst und Medientechnologie. Photo: Tim Otto Roth.

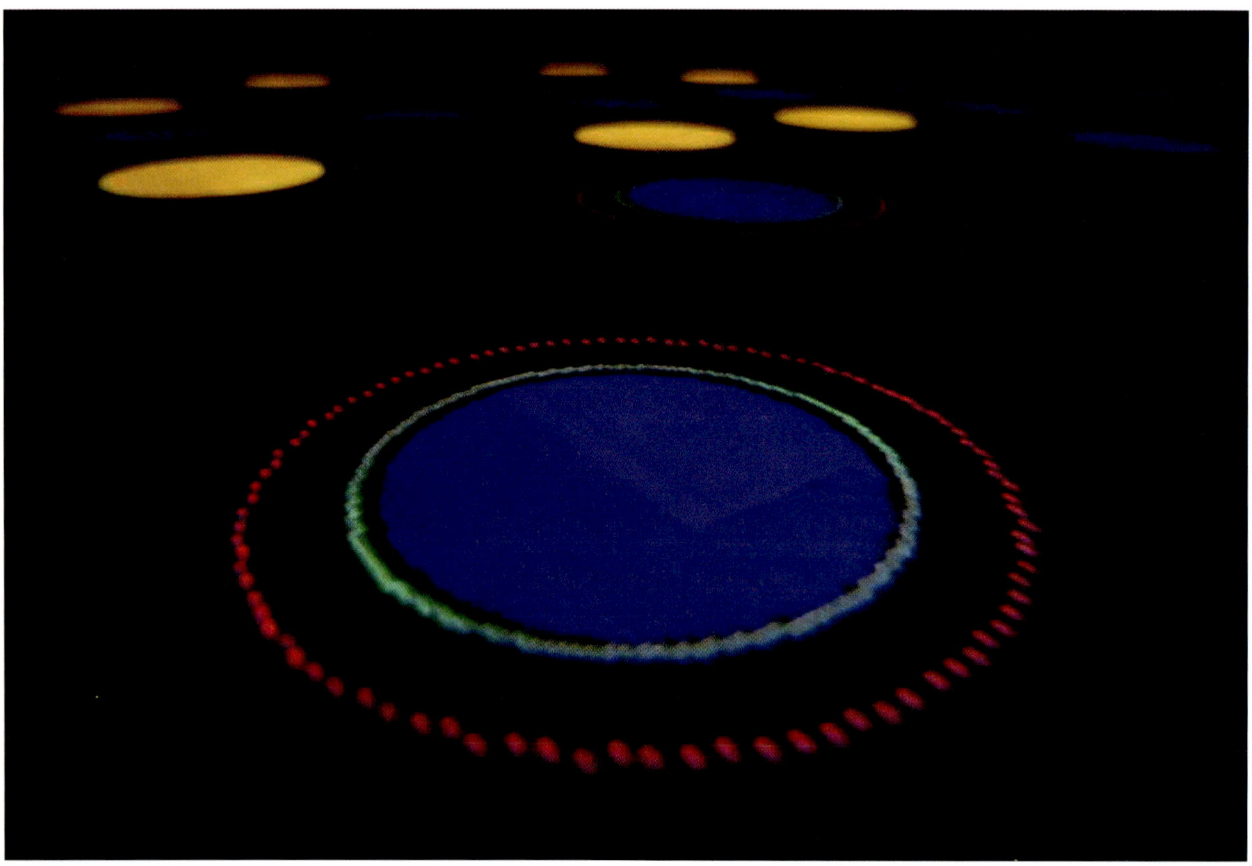

an inhibitory or excitatory effect (*plate* 2). Significantly, both internal space (i.e. the transitory mental activation driven by electric spiking) and external space (the shifting empirical environment of the Sound Dome) create a new sensual-surround experience. As Roth explains: 'You feel you stand in the middle of an organic process of self-organizing tones.'[12]

This transformation of the Sound Dome into a new instrument for innovative composition as well as for accessing the work of even single neurons radically redefines the meaning of art/science collaboration. It points to the need for at least dual educational degrees in seriously contrasting areas, for deeply interfusing historical studies with contemporary art and technology research initiatives, for making a serious start on defining higher-order image investigations that take into account the basis of all cognitive activity. In short, since data about neuronal and nervous system dynamics are constantly emerging, this means making the long durée of spatial/sensory studies forever alive to the present and thus newly relevant, even indispensable. This exciting neuroaesthetic process has begun but there are also many more singular and intersective devices to come – acoustic, affective, haptic, olfactory, optical, proprioceptive – all giving shape to our embodied brain and extended mind.

Notes

1 P. D. James, *Devices and Desires*, London and Boston, MA, 1989, 376.

2 See Barbara Maria Stafford and Frances Terpak, *Devices of Wonder: From the World in a Box to Images on a Screen*, exh. cat., Los Angeles, CA, 2001, and in particular my introductory essay: 'Revealing technologies/ magical domains', 1–142.

3 Barbara Maria Stafford, *Echo Objects: The Cognitive Work of Images*, Chicago, IL and London, 2007, 201–5.

4 Raffi Khatchadourian, 'We know how you feel', *The New Yorker*, 19 January 2015, 58.

5 A. Chatterjee, 'Cosmetic neurology: The controversy over enhancing movement, mentation, and mood', *Neurology*, 63, 2004, 968–74.

6 Khatchadourian, 'We know how you feel', 59. Perhaps this explosion of corporate interest in emotion is also tied to the explosion of affect studies launched by Antonio Damasio, *The Feeling of What Happens*, New York, 1999.

7 On the uses as well as the drawbacks of working memory (i.e. procedural memory), see Sian Beilock, *Choke: What the Secrets of the Brain Reveal About Getting It Right When You Have To*, New York, 2013, 66–7.

8 See Sian Beilock, 'Greening the brain: How the physical environment shapes thinking', in *How the Body Knows Its Mind*, New York, 2015, esp. 210–11; and M. G. Berman *et al.*, 'Interacting with nature improves cognition and affect in depressed individuals'. *Journal of Affective Disorders*, 140: 3, 2012, 300–5.

9 Chris Eliasmith, 'Building brains from bottom to top', Lecture: Center for Mind, Brain, Culture, Emory University, 25 March 2015.

10 For example, see the informative article on the fundamental problem of transforming 2D retinal images into 3D spatial representations: Ari Rosenberg and Dora E. Angelaki, 'Reliability-dependent contributions of visual orientation cues in parietal cortex', *Proceedings of the National Academy of Sciences, USA*, 111: 50, 2014, 18043–8. They demonstrate that when you look at a picture with perspective cues, you have a stronger sense of depth if you view it monocularly rather than binocularly – Leonardo, of course, was fascinated by this. Rosenberg and Angelaki's findings, however, demonstrate the first neural correlate of this, specifically in a subpopulation of neurons in the parietal cortex which combine binocular disparity and texture cues for visual orientation.

11 See David Garcia, http://new-tactical-research.co.uk/blog/notes-towards-reframing-creative-question/, accessed 7 September 2015.

12 See Tim Otto Roth, 'Sonapticon – Space as an acoustic neuronal network', http://www.mshparisnord.fr/JIM2013/actes/jim2013_19.pdf, accessed 29 September 2015.

Index

References to illustrations are in italics. References to notes are given as follows: 148n. Note that illustrations and notes are indexed separately only if there is no other information on that subject in the main text on that page.